Springer Texts in Business and Economics

More information about this series at http://www.springer.com/series/10099

Uwe Hassler

Stochastic Processes and Calculus

An Elementary Introduction with Applications

Springer

Uwe Hassler
Faculty of Economics and Business
Administration
Goethe University Frankfurt
Frankfurt, Germany

ISSN 2192-4333 ISSN 2192-4341 (electronic)
Springer Texts in Business and Economics
ISBN 978-3-319-23427-4 ISBN 978-3-319-23428-1 (eBook)
DOI 10.1007/978-3-319-23428-1

Library of Congress Control Number: 2015957196

Springer Cham Heidelberg New York Dordrecht London

Printed on acid-free paper

Springer International Publishing AG Switzerland is part of Springer Science+Business Media (www.springer.com)

I do not know what I may appear to the world, but to myself I seem to have been only like a boy playing on the sea-shore, and diverting myself in now and then finding a smoother pebble or a prettier shell than ordinary, whilst the great ocean of truth lay all undiscovered before me.

ISAAC NEWTON

Quoted from the novel Beyond Sleep *by Willem Frederik Hermans*

Preface

Over the past decades great importance has been placed on stochastic calculus and processes in mathematics, finance, and econometrics. This book addresses particularly readers from these fields, although students of other subjects as biology, engineering, or physics may find it useful, too.

Scope of the Book

By now there exist a number of books describing stochastic integrals and stochastic calculus in an accessible manner. Such introductory books, however, typically address an audience having previous knowledge about and interest in one of the following three fields exclusively: finance, econometrics, or mathematics. The textbook at hand attempts to provide an introduction into stochastic calculus and processes for students from each of these fields. Obviously, this can on no account be an exhaustive treatment. In the next chapter a survey of the topics covered is given. In particular, the book does neither deal with finance theory nor with statistical methods from the time series econometrician's toolkit; it rather provides a mathematical background for those readers interested in these fields.

The first part of this book is dedicated to discrete-time processes for modeling temporal dependence in time series. We begin with some basic principles of stochastics enabling us to define stochastic processes as families of random variables in general. We discuss models for short memory (so-called ARMA models), for long memory (fractional integration), and for conditional heteroscedasticity (so-called ARCH models) in respective chapters. One further chapter is concerned with the so-called frequency domain or spectral analysis that is often neglected in introductory books. Here, however, we propose an approach that is not technically too demanding. Throughout, we restrict ourselves to the consideration of stochastic properties and interpretation. The statistical issues of parameter estimation, testing, and model specification are not addressed due to space limitations; instead, we refer to, e.g., Mills and Markellos (2008), Kirchgässner, Wolters, and Hassler (2013), or Tsay (2005).

The second part contains an introduction to stochastic integration. We start with elaborations on the Wiener process $W(t)$ as we will define (almost) all integrals in terms of Wiener processes. In one chapter we consider Riemann integrals of the form $\int f(t)W(t)dt$, where f is a deterministic function. In another chapter Stieltjes integrals are constructed as $\int f(t)dW(t)$. More specifically, stochastic integrals as such result when a stochastic process is integrated with respect to the Wiener process, e.g., the Ito integral $\int W(t)dW(t)$. Solving stochastic differential equations is one task of stochastic integration for which we will need to use Ito's lemma. Our description aims at a similar compromise between concreteness and mathematical rigor as, e.g., Mikosch (1998). If the reader wants to address this matter more rigorously, we recommend Klebaner (2005) or Øksendal (2003).

The third part of the book applies previous results. The chapter on stochastic differential equations consists basically of applications of Ito's lemma. Concrete differential equations, as they are used, e.g., when modeling interest rate dynamics, will be covered in a separate chapter. The second area of application concerns certain limiting distributions of time series econometrics. A separate chapter on the asymptotics of integrated processes covers weak convergence to Wiener processes. The final two chapters contain applications for nonstationary processes without cointegration on the one hand and for the analysis of cointegrated processes on the other. Further details regarding econometric application can be found in the books by Banerjee, Dolado, Galbraith and Hendry (1993), Hamilton (1994), or Tanaka (1996).

The exposition in this book is elementary in the sense that knowledge of measure theory is neither assumed nor used. Consequently, mathematical foundations cannot be treated rigorously which is why, e.g., proofs of existence are omitted. Rather I had two goals in mind when writing this book. On the one hand, I wanted to give a basic and illustrative presentation of the relevant topics without many "troublesome" derivations. On the other hand, in many parts a technically advanced level has been aimed at: procedures are not only presented in form of recipes but are to be understood as far as possible which means they are to be proven. In order to meet both requirements jointly, this book is equipped with a lot of challenging problems at the end of each chapter as well as with the corresponding detailed solutions. Thus the virtual text – augmented with more than 60 basic examples and 45 illustrative figures – is rather easy to read while a part of the technical arguments is transferred to the exercise problems and their solutions. This is why there are at least two possible ways to work with the book. For those who are merely interested in applying the methods introduced, the reading of the text is sufficient. However, for an in-depth knowledge of the theory and its application, the reader necessarily needs to study the problems and their solution extensively.

Note to Students and Instructors

I have taught the material collected here to master students (and diploma students in the old days) of economics and finance or students of mathematics with a minor in those fields. From my personal experience I may say that the material presented here is too vast to be treated in a course comprising 45 contact hours. I used the textbook at hand for four slightly differing courses corresponding to four slightly differing routes through the parts of the book. Each of these routes consists of three stages: time series models, stochastic integration, and applications. After Part I on time series modeling, the different routes separate.

The finance route: When teaching an audience with an exclusive interest in finance, one may simply drop the final three chapters. The second stage of the course then consists of Chaps. 7, 8, 9, 10, and 11. This Part II on stochastic integration is finally applied to the solution of stochastic differential equations and interest rate modeling in Chaps. 12 and 13, respectively.

The mathematics route: There is a slight variant of the finance route for the mathematically inclined audience with an equal interest in finance or econometrics. One simply replaces Chap. 13 on interest rate modeling by Chap. 14 on weak convergence on function spaces, which is relevant for modern time series asymptotics.

The econometrics route: After Part I on time series modeling, the students from a class on time series econometrics should be exposed to Chaps. 7, 8, 9, and 10 on Wiener processes and stochastic integrals. The three chapters (Chaps. 11, 12, and 13) on Ito's lemma and its applications may be skipped to conclude the course with the last three chapters (Chaps. 14, 15, and 16) culminating in the topic of "cointegration."

The nontechnical route: Finally, the entire content of the textbook at hand can still be covered in one single semester; however, this comes with the cost of omitting technical aspects for the most part. Each chapter contains a rather technical section which in principle can be skipped without leading to a loss in understanding. When omitting these potentially difficult sections, it is possible to go through all the chapters in a single course. The following sections should be skipped for a less technical route:

$$\boxed{3.3} \,\&\, \boxed{4.3} \,\&\, \boxed{5.4} \,\&\, \boxed{6.4} \,\&\, \boxed{7.3} \,\&\, \boxed{8.4} \,\&\, \boxed{9.4}$$
$$\&\, \boxed{10.4} \,\&\, \boxed{11.4} \,\&\, \boxed{12.2} \,\&\, \boxed{13.4} \,\&\, \boxed{14.3} \,\&\, \boxed{15.4} \,\&\, \boxed{16.4}\ .$$

It has been mentioned that each chapter concludes with problems and solutions. Some of them are clearly too hard or lengthy to be dealt with in exams, while others are questions from former exams of my own or are representative of problems to be solved in my exams.

Frankfurt, Germany
July 2015

Uwe Hassler

References

Banerjee, A., Dolado, J. J., Galbraith, J. W., & Hendry, D. F. (1993). *Co-integration, error correction, and the econometric analysis of non-stationary data*. Oxford/New York: Oxford University Press.

Hamilton, J. (1994). *Time series analysis*. Princeton: Princeton University Press.

Kirchgässner, G., Wolters, J., & Hassler, U. (2013). *Introduction to modern time series analysis* (2nd ed.). Berlin/New York: Springer.

Klebaner, F. C. (2005). *Introduction to stochastic calculus with applications* (2nd ed.). London: Imperical College Press.

Mikosch, Th. (1998). *Elementary stochastic calculus with finance in view*. Singapore: World Scientific Publishing.

Mills, T. C., & Markellos, R. N. (2008). *The econometric modelling of financial time series* (3rd ed.). Cambridge/New York: Cambridge University Press.

Øksendal, B. (2003). *Stochastic differential equations: An introduction with applications* (6th ed.). Berlin/New York: Springer.

Tanaka, K. (1996). *Time series analysis: Nonstationary and noninvertible distribution theory*. New York: Wiley.

Tsay, R. S. (2005). *Analysis of financial time series* (2nd ed.). New York: Wiley.

Acknowledgments

This textbook grew out of lecture notes from which I taught over 15 years. Without my students' thirst for knowledge and their critique, I would not even have started the project. In particular, I thank Balázs Cserna, Matei Demetrescu, Eduard Dubin, Mehdi Hosseinkouchack, Vladimir Kuzin, Maya Olivares, Marc Pohle, Adina Tarcolea, and Mu-Chun Wang who corrected numerous errors in the manuscript. Originally, large parts of this text had been written in German, and I thank Verena Werkmann for her help when translating into English. Last but not least I am indebted to Goethe University Frankfurt for allowing me to take sabbatical leave. Without this support I would not have been able to complete this book at a time when academics are under pressure to publish in the first place primary research.

Contents

List of Figures

Introduction

1

1.1 Summary

Stochastic calculus is used in finance and econom(etr)ics for instance for solving stochastic differential equations and handling stochastic integrals. This requires stochastic processes. Although stemming from a rather recent area of mathematics, the methods of stochastic calculus have shortly come to be widely spread not only in finance and economics. Moreover, these techniques – along with methods of time series modeling – are central in the contemporary econometric tool box. In this introductory chapter some motivating questions are brought up being answered in the course of the book, thus providing a brief survey of the topics treated.

1.2 Finance

The names of two Nobel prize winners[1] dealing with finance are closely connected to one field of applications treated in the textbook at hand. The analysis and the modeling of stock prices and returns is central to this work.

Stock Prices

Let $S(t), t \geq 0$, be the continuous stock price of a stock with return $R(t) = S'(t)/S(t)$ expressed as growth rate. We assume constant returns,

$$R(t) = c \quad \Longleftrightarrow \quad S'(t) = c\,S(t) \quad \Longleftrightarrow \quad \frac{dS(t)}{dt} = cS(t).$$

[1] In 1997, R.C. Merton and M.S. Scholes were awarded the Nobel prize jointly, "for a new method to determine the value of derivatives" (according to the official statement of the Nobel Committee).

U. Hassler, *Stochastic Processes and Calculus*, Springer Texts in Business and Economics, DOI 10.1007/978-3-319-23428-1_1

This differential equation for the stock price is usually also written as follows:

$$dS(t) = c\, S(t)\, dt\,. \tag{1.1}$$

The corresponding solution is (see Problem 1.1)

$$S(t) = S(0)\, e^{ct}\,, \tag{1.2}$$

i.e. if $c > 0$ the exponential process is explosive. The assumption of a deterministic stock price movement is of course unrealistic which is why a stochastic differential equation consistent with (1.1) is often assumed since Black and Scholes (1973) and Merton (1973),

$$dS(t) = c\, S(t)\, dt + \sigma\, S(t)\, dW(t)\,, \tag{1.3}$$

where $dW(t)$ are the increments of a so-called Wiener process $W(t)$ (also referred to as Brownian motion, cf. Chap. 7). This is a stochastic process, i.e. a random process. Thus, for a fixed point in time t, $S(t)$ is a random variable. How does this random variable behave on average? How do the parameters c and σ affect the expected value and the variance as time passes by? We will find answers to these questions in Chap. 12 on stochastic differential equations.

Interest Rates

Next, $r(t)$ denotes an interest rate for $t \geq 0$. Assume it is given by the differential equation

$$dr(t) = c\,(r(t) - \mu)\, dt \tag{1.4}$$

with $c \in \mathbb{R}$ or equivalently by

$$r'(t) = \frac{dr(t)}{dt} = c\,(r(t) - \mu).$$

Expression (1.4) can alternatively be written as the following integral equation:

$$r(t) = r(0) + c \int_0^t (r(s) - \mu)\, ds. \tag{1.5}$$

The solution to this reads (see Problem 1.2)

$$r(t) = \mu + e^{ct}\,(r(0) - \mu)\,. \tag{1.6}$$

For $c < 0$ therefore it holds that the interest rate converges to μ as time goes by. Again, a deterministic movement is not realistic. This is why Vasicek (1977) specified a stochastic differential equation consistent with (1.4):

$$dr(t) = c\,(r(t) - \mu)\,dt + \sigma\,dW(t)\,. \tag{1.7}$$

As aforementioned, $dW(t)$ denotes the increments of a Wiener process. How is the interest rate movement (on average) affected by the parameter c? Which kind of stochastic process is described by (1.7)? The answers to these and similar questions will be obtained in Chap. 13 on interest rate models.

Empirical Returns

Looking at return time series one can observe that the variance (or volatility) fluctuates a lot as time passes by. Long quiet market phases characterized by only mild variation are followed by short periods characterized by extreme observations where extreme amplitudes again tend to entail extreme observations. Such a behavior is in conflict with the assumption of normally distributed data. It is an empirically well confirmed law ("stylized fact") that financial market data in general and returns in particular produce "outliers" with larger probability than it would be expected under normality.

It is crucial, however, that extreme observations occur in clusters (volatility clusters). Even though returns are not correlated over time in efficient markets, they are not independent as there exists a systematic time dependence of volatility. Engle (1982) suggested the so-called ARCH model (see Chap. 6) in order to capture the outlined effects. His work constituted an entire field of research known nowadays under the keyword "financial econometrics", and consequently he was awarded the Nobel prize in 2003.[2]

1.3 Econometrics

Clive Granger (1934–2009) was a British econometrician who created the concept of cointegration (Granger, 1981). He shared the Nobel prize "for methods of analyzing economic time series with common trends (cointegration)" (official statement of the Nobel Committee) with R.F. Engle. The leading example of trending time series he considered is the random walk.

[2]R.F. Engle shared the Nobel prize "for methods of analyzing economic time series with time-varying volatility (ARCH)" (official statement of the Nobel Committee) with C.W.J. Granger.

Random Walks

In econometrics, we are often concerned with time series not fluctuating with constant variance around a fixed level. A widely-used model for accounting for this nonstationarity are so-called integrated processes. They form the basis for the cointegration approach that has become an integral part of common econometric methodology since Engle and Granger (1987). Let's consider a special case – the random walk – as a preliminary model,

$$x_t = \sum_{j=1}^{t} \varepsilon_j\,, \quad t = 1, \ldots, n\,, \tag{1.8}$$

where $\{\varepsilon_t\}$ is a random process, i.e. ε_t and ε_s, $t \neq s$, are uncorrelated or even independent with zero expected value and constant variance σ^2. For a random walk with zero starting value $x_0 = 0$ it holds by definition that:

$$x_t = x_{t-1} + \varepsilon_t\,, \quad t = 1, \ldots, n\,, \qquad \text{with } \mathrm{Var}(x_t) = \sigma^2\, t\,. \tag{1.9}$$

The increments can also be written using the difference operator Δ,

$$\Delta x_t = x_t - x_{t-1} = \varepsilon_t\,.$$

Regressing two stochastically independent random walks on each other, a statistically significant relationship is identified which is a statistical artefact and therefore nonsense (see Chap. 15). Two random walks following a common trend, however, are called cointegrated. In this case the regression on each other does not only give the consistent estimation of the true relationship but the estimator is even "superconsistent" (cf. Chap. 16).

Dickey-Fuller Distribution

If one wants to test whether a given time series indeed follows a random walk, then equation (1.9) suggests to estimate the regression

$$x_t = \hat{a}\, x_{t-1} + \hat{\varepsilon}_t\,, \quad t = 1, \ldots, n\,.$$

From this, the (ordinary) least squares (LS) estimator under the null hypothesis (1.9), i.e. under $a = 1$, is obtained as

$$\hat{a} = \frac{\sum_{t=1}^{n} x_t\, x_{t-1}}{\sum_{t=1}^{n} x_{t-1}^2} = 1 + \frac{\sum_{t=1}^{n} x_{t-1}\, \varepsilon_t}{\sum_{t=1}^{n} x_{t-1}^2}\,.$$

This constitutes the basic ingredient for the test by Dickey and Fuller (1979). Under the null hypothesis of a random walk ($a = 1$) it holds asymptotically ($n \to \infty$)

$$n\,(\hat{a} - 1) \xrightarrow{d} \mathcal{DF}_a\,, \tag{1.10}$$

where "$\xrightarrow{d}$" stands for convergence in distribution and $\mathcal{DF}_a$ denotes the so-called Dickey-Fuller distribution. Corresponding modes of convergence will be explained in Chap. 14. Since Phillips (1987) an elegant way for expressing the Dickey-Fuller distribution by stochastic integrals is known (again, $W(t)$ denotes a Wiener process):

$$\mathcal{DF}_a = \frac{\int_0^1 W(t)\,dW(t)}{\int_0^1 W^2(t)\,dt}\,. \tag{1.11}$$

Note (and enjoy!) the formal correspondence of the sum of squares $\sum_{t=1}^{n} x_{t-1}^2$ in the denominator of $\hat{a} - 1$ and the integral over the squared Wiener process in the denominator of (1.11), $\int_0^1 W^2(t)\,dt$ (this is a Riemann integral, cf. Chap. 8). Just as well the sum $\sum_{t=1}^{n} x_{t-1}\,\varepsilon_t = \sum_{t=1}^{n} x_{t-1}\,\Delta x_t$ resembles the so-called Ito integral $\int_0^1 W(t)\,dW(t)$. But how are these integrals defined, what are they about? How is this distribution (and similar ones) attained? And why does there exist another equivalent representation,

$$\mathcal{DF}_a = \frac{W^2(1) - 1}{2\int_0^1 W^2(t)\,dt}\,, \tag{1.12}$$

of the Dickey-Fuller distribution? We concern ourselves with these questions in connection with Ito's lemma in Chap. 11.

Autocorrelation

The assumption of the increments $\Delta x_t = x_t - x_{t-1}$ of economic times series being free from serial (temporal) correlation – as it is true for the random walk – is too restrictive in practice. Thus, we have to learn how the Dickey-Fuller distribution is generalized with autocorrelated (i.e. serially correlated) increments. In practise, so-called ARMA models are used most frequently in order to model autocorrelation. This class of models will be discussed intuitively as well as rigorously in Chap. 3. The so-called spectral analysis translates autocorrelation patterns in oscillation patterns. In Chap. 4 we learn to determine which frequency's or period's oscillations add particularly intensely to a time series' variation. Often economists are refused access to spectral analysis because of the extensive use of complex numbers. Therefore, we suggest an approach that avoids complex numbers. Finally, Chap. 5 introduces a model where the temporal dependence is particularly persistent such that the autocorrelations die out more slowly than in the ARMA case. Such a feature has been called "long memory" and is observed with many economic and financial series.

1.4 Mathematics

Stochastic calculus, which will be applied here, is a rather recent area in mathematics. It was pioneered by Kiyoshi Ito[3] in a sequence of pathbreaking papers published in Japanese starting from the forties of the last century.[4] The Ito integral as a special case of stochastic integration is introduced in Chap. 10.

Ito Integrals

The aforementioned interest rate model by Vasicek (1977) leads to a stochastic process given by an integral constructed as $\int_0^t f(s)\, dW(s)$ where f is a deterministic function and again dW denotes the increments of a Wiener process. Such integrals – being in a sense classical integrals – will be defined as Stieltjes integrals in Chap. 9. Ito integrals are a generalization of these. At first glance, the deterministic function f is replaced by a stochastic process X, $\int_0^t X(s)\, dW(s)$. Mathematically, this results in a considerably more complicated object, the definition thereof being a problem on its own, cf. Chap. 10.

Ito's Lemma

At this point, the idea of Ito's lemma is briefly conveyed. For the moment, assume a deterministic (differentiable) function $f(t)$. Using the chain rule it holds for the derivative of the square f^2:

$$\frac{df^2(t)}{dt} = 2f(t)f'(t)$$

or rather

$$\frac{df^2(t)}{2} = f(t)f'(t)\, dt = f(t)\, df(t)\,. \tag{1.13}$$

[3]Alternative transcriptions of his name into the Latin alphabet, Itô or Itō, are frequently used in the literature and are equally accepted. In this textbook we follow the spelling of Ito's compatriot (Tanaka, 1996).

[4]In 2006, Ito received the inaugural Gauss Prize for Applied Mathematics by the International Mathematical Union, which is awarded every fourth year since then.

Thus, for the ordinary integral it follows

$$\int_0^t f(s)\, df(s) = \int_0^t f(s) f'(s)\, ds = \frac{1}{2} f^2(s)\Big|_0^t = \frac{1}{2}\left(f^2(t) - f^2(0)\right).$$

However, among other things, we will learn that the Wiener process is not a differentiable function with respect to time t. The ordinary chain rule does not apply and for the according Ito integral one obtains

$$\int_0^t W(s)\, dW(s) = \frac{1}{2}\left(W^2(s) - s\right)\Big|_0^t = \frac{1}{2}\left(W^2(t) - W^2(0) - t\right). \tag{1.14}$$

This result follows from the famous and fundamental lemma by Ito being a kind of "stochastified chain rule" for Wiener processes in its simplest case. Instead of (1.13) for Wiener processes it holds that

$$\frac{dW^2(t)}{2} = W(t)\, dW(t) + \frac{1}{2}\, dt. \tag{1.15}$$

Substantial generalizations and multivariate extensions will be discussed in Chap. 11. In particular, Ito's lemma will enable us to solve stochastic differential equations in Chap. 12, and it will turn out that $S(t)$ solving (1.3) is a so-called geometric Brownian motion. In Chap. 13 we will look in greater detail in models for interest rates as e.g. given by Eq. (1.7).

Starting point for all the considerations outlined is the Wiener process – often also called Brownian motion. Before turning to it and its properties, general stochastic processes need to be defined and classified beforehand. This is done – among other things – in the following chapter on basic concepts from probability theory.

1.5 Problems and Solutions

Problems

1.1 Solve the differential equation (1.1), i.e. obtain the solution (1.2).

1.2 Verify that $r(t)$ from (1.6) solves the differential equation (1.4).

1.3 Consider a simple regression model,

$$y_i = \alpha + \beta\, x_i + \varepsilon_i\,, \quad i = 1, \ldots, n\,,$$

with OLS estimator $\hat{\beta}$. Check that:

$$\hat{\beta} - \beta = \frac{\sum_{i=1}^{n}(x_i - \overline{x})\,\varepsilon_i}{\sum_{i=1}^{n}(x_i - \overline{x})^2},$$

with arithmetic mean $\overline{x}$.
Hint: $\sum_{i=1}^{n}(x_i - \overline{x}) = 0$.

1.4 Let f^n denote n-th power of a function f with derivative f', $n \in \mathbb{N}$. Show that:

$$f^n(t) = f^n(0) + n\int_0^t f^{n-1}(s)f'(s)\,ds\,.$$

Hint: Chain rule as in (1.13).

Solutions

1.1 Using equation (1.1) we get by integration

$$\int_0^t \frac{S'(r)}{S(r)}\,dr = \int_0^t c\,dr = ct.$$

Since[5]

$$\frac{d\log(S(t))}{dt} = \frac{S'(t)}{S(t)}$$

this implies

$$\log(S(t)) - \log(S(0)) = ct\,,$$

or

$$\begin{aligned} S(t) &= e^{\log(S(0))}\,e^{ct} \\ &= S(0)\,e^{ct}\,, \end{aligned}$$

which is the required solution.

[5] By "log" we denote the natural logarithm to the base e.

1.2 Taking the derivative of (1.6) yields:

$$\begin{aligned}\frac{dr(t)}{dt} &= c\, e^{ct}\,(r(0) - \mu)\\ &= c\,(r(t) - \mu)\,,\end{aligned}$$

where again the given form of $r(t)$ was used. By purely symbolically multiplying by dt the equation (1.4) is obtained. Hence, the problem is already solved.

1.3 It is well known that the OLS estimator is given by "covariance divided by variance of the regressor", i.e. it holds that:

$$\hat{\beta} = \frac{\frac{1}{n}\sum_{i=1}^{n}(x_i - \bar{x})(y_i - \bar{y})}{\frac{1}{n}\sum_{i=1}^{n}(x_i - \bar{x})^2}\,.$$

Because of $\sum_{i=1}^{n}(x_i - \bar{x}) = 0$ this simplifies to

$$\hat{\beta} = \frac{\frac{1}{n}\sum_{i=1}^{n}(x_i - \bar{x})y_i}{\frac{1}{n}\sum_{i=1}^{n}(x_i - \bar{x})^2}\,.$$

Assuming the model to be correct and substituting $y_i = \alpha + \beta\, x_i + \varepsilon_i$, one obtains

$$\hat{\beta} = \frac{\sum_{i=1}^{n}(x_i - \bar{x})(\alpha + \beta\, x_i + \varepsilon_i)}{\sum_{i=1}^{n}(x_i - \bar{x})^2}\,.$$

Again applying the argument $\sum_{i=1}^{n}(x_i - \bar{x}) = 0$ yields

$$\begin{aligned}\hat{\beta} &= \frac{\sum_{i=1}^{n}(x_i - \bar{x})(\beta(x_i - \bar{x}) + \varepsilon_i)}{\sum_{i=1}^{n}(x_i - \bar{x})^2}\\ &= \beta + \frac{\sum_{i=1}^{n}(x_i - \bar{x})\varepsilon_i}{\sum_{i=1}^{n}(x_i - \bar{x})^2}\,.\end{aligned}$$

This was exactly the claim.

1.4 We address the problem in a slightly more general way. Let g be a differentiable function with derivative g'. By the fundamental theorem of calculus it holds that[6]

$$\int_0^t g'(s)ds = g(t) - g(0),$$

[6]For an introduction to calculus we recommend Trench (2013); this book is available electronically for free as a textbook approved by the American Institute of Mathematics.

or

$$g(t) = g(0) + \int_0^t g'(s)ds.$$

If g describes a process over time, this last relation can be interpreted the following way: The value at time t is made up by the starting value $g(0)$ plus the sum or integral over all changes occurring between 0 and t. Now, choosing in particular $g(t) = f^n(t)$ with

$$g'(t) = nf^{n-1}(t)f'(t),$$

we obtain the required result.

References

Black, F., & Scholes, M. (1973). The pricing of options and corporate liabilities. *The Journal of Political Economy, 81*, 637–654.

Dickey, D. A., & Fuller, W. A. (1979). Distribution of the estimators for autoregressive time series with a unit root. *Journal of the American Statistical Association, 74*, 427–431.

Engle, R. F. (1982). Autoregressive conditional heteroskedasticity with estimates of the variance of U.K. inflation. *Econometrica, 50*, 987–1008.

Engle, R. F., & Granger, C. W. J. (1987). Co-integration and error correction: Representation, estimation, and testing. *Econometrica, 55*, 251–276.

Granger, C. W. J. (1981). Some properties of time series data and their use in econometric model specification. *Journal of Econometrics, 16*, 121–130.

Merton, R. C. (1973). Theory of rational option pricing. *The Bell Journal of Economics and Management Science, 4*, 141–183.

Phillips, P. C. B. (1987). Time series regression with a unit root. *Econometrica, 55*, 277–301.

Tanaka, K. (1996). *Time series analysis: Nonstationary and noninvertible distribution theory*. New York: Wiley.

Trench, W. F. (2013). *Introduction to real analysis*. Free Hyperlinked Edition 2.04 December 2013. Downloaded on 10th May 2014 from http://digitalcommons.trinity.edu/mono/7.

Vasicek, O. (1977). An equilibrium characterization of the term structure. *Journal of Financial Economics, 5*, 177–188.

Part I

Time Series Modeling

2 Basic Concepts from Probability Theory

2.1 Summary

This chapter reviews some basic material. We collect some elementary concepts and properties in connection with random variables, expected values, multivariate and conditional distributions. Then we define stochastic processes, both discrete and continuous in time, and discuss some fundamental properties. For a successful study of the remainder of this book, the reader is required to be familiar with all of these principles.

2.2 Random Variables

Stochastic processes are defined as families of random variables. This is why related concepts will be recapitulated to facilitate the definition of random variables. Measure theoretical aspects, however, will not be touched.[1]

Probability Space

We denote the possible **set of outcomes** of a random experiment by Ω. Subsets A, $A \subseteq \Omega$, are called **events**. These events are assigned probabilities to. The **probability** is a mapping

$$A \mapsto \mathrm{P}(A) \in [0, 1], \quad A \subseteq \Omega,$$

[1]Ross (2010) provides a nice introduction to probability, and so do Grimmett and Stirzaker (2001) with a focus on stochastic processes. For a short reference and refreshing e.g. the shorter appendix in Bickel and Doksum (2001) is recommended.

U. Hassler, *Stochastic Processes and Calculus*, Springer Texts in Business and Economics, DOI 10.1007/978-3-319-23428-1_2

which fulfills the axioms of probability,

- $\mathrm{P}(A) \geq 0$,
- $\mathrm{P}(\Omega) = 1$,
- $\mathrm{P}\left(\bigcup_i A_i\right) = \sum_i \mathrm{P}(A_i)$ for $A_i \cap A_j = \emptyset$ with $i \neq j$,

where $\{A_i\}$ may be a possibly infinite sequence of pairwise disjoint events. For a well-defined mapping, we do not consider every possible event but in particular only those being contained in σ-algebras. A **σ-algebra**[2] $\mathcal{F}$ of Ω is defined as a system of subsets containing

- the empty set $\emptyset$,
- the complement A^c of every subset $A \in \mathcal{F}$ (this is the set Ω without A, $A^c = \Omega \setminus A$),
- and the union $\bigcup_i A_i$ of a possibly infinite sequence of elements $A_i \in \mathcal{F}$.

Of course, a σ-algebra is not unique but can be constructed according to problems of interest. The interrelated triple of set of outcomes, σ-algebra and probability measure, $(\Omega, \mathcal{F}, \mathrm{P})$, is also called a **probability space**.

Example 2.1 (Game of Dice) Consider a fair hexagonal die with the set of outcomes

$$\Omega = \{1, 2, 3, 4, 5, 6\},$$

where each elementary event $\{\omega\} \subseteq \Omega$ is assigned the same probability to:

$$\mathrm{P}(\{1\}) = \ldots = \mathrm{P}(\{6\}) = \frac{1}{6}.$$

When $\#(A)$ denotes the number of elements of $A \subseteq \Omega$, it holds in the example of the die that

$$\mathrm{P}(A) = \frac{\#(A)}{\#(\Omega)} = \frac{\#(A)}{6}.$$

The probability for the occurrence of A hence equals the number of outcomes leading to A divided by the number of possible outcomes. If one is only interested in the event whether an even or an odd number occurs,

$$E = \{2, 4, 6\}, \quad E^c = \Omega \setminus E = \{1, 3, 5\},$$

[2]Sometimes also called a σ-field, which motivates the symbol $\mathcal{F}$.

then the σ-algebra obviously reads

$$\mathcal{F}_1 = \{\emptyset, E, E^c, \Omega\}.$$

If one is interested in all possible outcomes without any qualification, then the σ-algebra chosen will be the power set of Ω, $\mathcal{P}(\Omega)$. This is the set of all subsets of Ω:

$$\mathcal{F}_2 = \mathcal{P}(\Omega) = \{\emptyset, \{1\}, \ldots, \{6\}, \{1,2\}, \ldots, \{5,6\}, \{1,2,3\}, \ldots, \Omega\}.$$

Systematic counting shows that $\mathcal{P}(\Omega)$ contains exactly $2^{\#(\Omega)} = 2^6 = 64$ elements. With one and the same probability mapping one obtains for different σ-algebras different probability spaces:

$$(\Omega, \mathcal{F}_1, \mathrm{P}) \quad \text{and} \quad (\Omega, \mathcal{F}_2, \mathrm{P}) . \qquad \blacksquare$$

Random Variable

Often not the events themselves are of interest but some values associated with them, that is to say random variables. A real-valued one-dimensional **random variable** X maps the set of outcomes Ω of the space (Ω, $\mathcal{F}$, P) to the real numbers:

$$\begin{aligned} X: \quad \Omega &\to \mathbb{R} \\ \omega &\mapsto X(\omega) . \end{aligned}$$

Again, however, not all such possible mappings can be considered. In particular, a random variable is required to have the property of **measurability** (more precisely: $\mathcal{F}$-measurability). This implies the following: A subset $B \subseteq \mathbb{R}$ defines an event of Ω in such a way that:

$$X^{-1}(B) := \{\omega \in \Omega \mid X(\omega) \in B\} .$$

This so-called inverse image $X^{-1}(B) \subseteq \Omega$ of B contains exactly the very elements of Ω which are mapped by X to B. Let $\mathcal{B}$ be a family of sets consisting of subsets of $\mathbb{R}$. Then as measurability it is required from a random variable X that for all $B \in \mathcal{B}$ all inverse images are contained in the σ-algebra $\mathcal{F}$: $X^{-1}(B) \in \mathcal{F}$. Thereby the probability measure P on $\mathcal{F}$ is conveyed to $\mathcal{B}$, i.e. the probability function P_x assigning values to X is induced as follows:

$$\mathrm{P}_x(X \in B) = \mathrm{P}\left(X^{-1}(B)\right) , \quad B \in \mathcal{B} .$$

Thus, strictly speaking, X does not map from Ω to $\mathbb{R}$ but from one probability space to another:

$$X: \quad (\Omega, \mathcal{F}, \mathrm{P}) \rightarrow (\mathbb{R}, \mathcal{B}, \mathrm{P}_x) \,,$$

where $\mathcal{B}$ now denotes a σ-algebra named after Emile Borel. This Borel algebra $\mathcal{B}$ is the smallest σ-algebra over $\mathbb{R}$ containing all real intervals. In particular, for $x \in \mathbb{R}$ the event $X \leq x$ has an induced probability leading to the **distribution function** of X defined as follows:

$$F_x(x) := \mathrm{P}_x(X \leq x) = \mathrm{P}_x\left(X \in (-\infty, x]\right) = \mathrm{P}\left(X^{-1}\left((-\infty, x]\right)\right) , \quad x \in \mathbb{R} .$$

Example 2.2 (Game of Dice) Let us continue the example of dice and let us define a random variable X assigning a gain of 50 monetary units to an even number and assigning a loss of 50 monetary units to an odd number,

$$X: \begin{array}{l} 1 \mapsto -50 \\ 2 \mapsto +50 \\ 3 \mapsto -50 \\ 4 \mapsto +50 \\ 5 \mapsto -50 \\ 6 \mapsto +50 \end{array}$$

The random variable X operates on the probability space $(\Omega, \mathcal{F}_1, \mathrm{P})$ known from Example 2.1. For arbitrary real intervals probabilities P_x with $\mathcal{F}_1 = \{\emptyset, E, E^c, \Omega\}$ are induced, e.g.:

$$\mathrm{P}_x\left(X \in [-100, -50]\right) = \mathrm{P}\left(X^{-1}\left([-100, -50]\right)\right) = \mathrm{P}(E^c) = \frac{1}{2},$$

$$F_x(60) = \mathrm{P}_x\left(X \in (-\infty, 60]\right) = \mathrm{P}\left(X^{-1}\left((-\infty, 60]\right)\right) = \mathrm{P}(\Omega) = 1.$$

Let a second random variable Y model the following gain or loss function:

$$Y: \begin{array}{l} 1 \mapsto -10 \\ 2 \mapsto -20 \\ 3 \mapsto -30 \\ 4 \mapsto -40 \\ 5 \mapsto \quad 0 \\ 6 \mapsto \ 100 \end{array}$$

As in this case each outcome leads to another value of the random variable, the probability space chosen is $(\Omega, \mathcal{F}_2, \mathrm{P})$ with the power set $\mathcal{F}_2 = \mathcal{P}(\Omega)$ being the

σ-algebra. Then we obtain for Y for instance the following probabilities:

$$F_y(0) = \mathrm{P}_y(Y \leq 0) = \mathrm{P}\left(Y^{-1}(-\infty, 0]\right) = \mathrm{P}(\{1,2,3,4,5\}) = \frac{5}{6},$$

$$\mathrm{P}_y(Y \in [-20, 20)) = \mathrm{P}\left(Y^{-1}[-20, 20)\right) = \mathrm{P}(\{1,2,5\}) = \frac{1}{2}.$$

For another probability space the mapping Y is possibly not measurable and therefore it cannot be a random variable. E.g. Y is not $\mathcal{F}_1$-measurable. This is due to the fact that the image $Y = 0$ has the inverse image $Y^{-1}(0) = \{5\} \subseteq \Omega$ which is not contained in $\mathcal{F}_1$ as an elementary event: $\{5\} \notin \mathcal{F}_1$. ■

Continuous Random Variables

For most of all problems in practice we do not explicitly construct a random experiment with probability P in order to derive probabilities P_x of a random variable X. Typically we start directly with the quantity of interest X modeling a probability distribution without inducing it. In particular, this is the case for so-called continuous variables. For a continuous random variable every value taken from a real interval is a possible realization. As a continuous random variable can therefore take uncountably many values it is not possible to calculate a probability $\mathrm{P}(x_1 < X \leq x_2)$ by summing up the individual probabilities. Instead, probabilities are calculated by integrating a probability density. We assume the function $f(x)$ to be continuous (or at least Riemann-integrable) and to be nonnegative for all $x \in \mathbb{R}$. Then f is called (probability) density (or **density function**) of X if it holds for arbitrary numbers $x_1 < x_2$ that

$$\mathrm{P}(x_1 < X \leq x_2) = \int_{x_1}^{x_2} f(x)\,dx.$$

The area beneath the density function therefore measures the probability with which the continuous random variable takes on values of the interval considered. In general, a density is defined by two properties:

1. $f(x) \geq 0$,
2. $\int_{-\infty}^{+\infty} f(x)\,dx = 1$.

Thus, the distribution function $F(x) = \mathrm{P}(X \leq x)$ of a continuous random variable X is calculated as follows:

$$F(x) = \int_{-\infty}^{x} f(t)\,dt.$$

If there is the danger of a confusion, we sometimes subscript the distribution function, e.g. $F_x(0) = \mathrm{P}(X \leq 0)$.

Expected Value and Higher Moments

As is well known, the **expected value** $\mathrm{E}(X)$ (also called **expectation**) of a continuous random variable X with continuous density f is defined as follows:

$$\mathrm{E}(X) = \int_{-\infty}^{\infty} xf(x)\,dx.$$

For (measurable) mappings g, transformations $g(X)$ are again random variables, and the expected value is given by:

$$\mathrm{E}\left[g(X)\right] = \int_{-\infty}^{\infty} g(x)f(x)\,dx.$$

In particular, for each power of X so-called **moments** are defined for $k = 1, 2, \ldots$:

$$\mu_k = \mathrm{E}\left[X^k\right].$$

Note that this term represents integrals which are not necessarily finite (then one says: the respective moments do not exist). There are even random variables whose density f allows for very large observations in absolute value with such a high probability that even the expected value μ_1 is not finite.[3] If nothing else is suggested, we will always assume random variables with finite moments without pointing out explicitly.

Often we consider so-called centered moments where $g(X)$ is chosen as $(X - \mathrm{E}(X))^k$. For $k = 2$ the **variance** is obtained (often denoted by σ^2)[4]:

$$\sigma^2 = \mathrm{Var}(X) = \int_{-\infty}^{\infty} (x - \mathrm{E}(X))^2 f(x)\,dx.$$

Elementarily, the following additive decomposition is shown:

$$\mathrm{Var}(X) = \mathrm{E}(X^2) - (\mathrm{E}(X))^2 = \mu_2 - \mu_1^2. \tag{2.1}$$

[3] An example for this is the Cauchy distribution, i.e. the t-distribution with one degree of freedom. For the Pareto distribution, as well, the existence of moments is dependent on the parameter value; this is shown in Problem 2.2.

[4] Then σ describes the square root of $\mathrm{Var}(X)$ with positive sign.

Since $\text{Var}(X) \geq 0$ by construction, this gives rise to the following inequality:

$$(\text{E}(X))^2 \leq \text{E}\left(X^2\right) . \tag{2.2}$$

In addition to centering, for higher moments a standardization is typically considered. The following measures of **skewness** and **kurtosis** with $k = 3$ and $k = 4$, respectively, are widely used:

$$\gamma_1 = \frac{\text{E}[(X-\mu_1)^3]}{\sigma^3}, \quad \gamma_2 = \frac{\text{E}[(X-\mu_1)^4]}{\sigma^4}.$$

The skewness coefficient is used to measure deviations from symmetry. If X exhibits a density f which is symmetric around the expected value, it obviously follows that $\gamma_1 = 0$. The interpretation of the kurtosis coefficient is more difficult. Generally, γ_2 is taken as a measure for a distribution's "peakedness", or alternatively, for how probable extreme observations ("outliers") are. Frequently, the normal distribution is taken as a reference. For every normal distribution (also called Gaussian distribution, see Example 2.4) it holds that the kurtosis takes the value 3. Furthermore, it can be shown that it holds always true that

$$\gamma_2 \geq 1 ,$$

which is verified in Problem 2.1.

Example 2.3 (Kurtosis of a Continuous Uniform Distribution) The random variable X is assumed to be uniformly distributed on $[0, b]$ with density

$$f(x) = \begin{cases} \frac{1}{b}, & x \in [0, b] \\ 0, & \text{else} \end{cases} .$$

As is well known, it then holds that

$$\mu_1 = \text{E}(X) = \frac{b}{2}, \qquad \sigma^2 = \text{Var}(X) = \frac{b^2}{12}.$$

In order to calculate the kurtosis γ_2 we are interested in the fourth centered moment:

$$\text{E}\left[(X-\mu_1)^4\right] = \int_0^b \left(x - \frac{b}{2}\right)^4 \frac{1}{b}\, dx.$$

For this we determine (binomial theorem):

$$\left(x-\frac{b}{2}\right)^4 = x^4 + 4x^3\left(-\frac{b}{2}\right) + 6x^2\frac{b^2}{4} + 4x\left(-\frac{b}{2}\right)^3 + \left(\frac{b}{2}\right)^4$$
$$= x^4 - 2x^3b + \frac{3}{2}x^2b^2 - \frac{1}{2}xb^3 + \frac{b^4}{16}.$$

From this it is obtained that

$$\int_0^b \left(x-\frac{b}{2}\right)^4 dx = \frac{b^5}{5} - \frac{b^5}{2} + \frac{b^5}{2} - \frac{b^5}{4} + \frac{b^5}{16}$$
$$= \frac{b^5}{80},$$

and hence

$$\mathrm{E}\left[(X-\mu_1)^4\right] = \frac{b^4}{80}.$$

The kurtosis coefficient is therefore determined as

$$\gamma_2 = \frac{\mathrm{E}\left[(X-\mu_1)^4\right]}{\sigma^4} = \frac{b^4}{80}\left(\frac{12}{b^2}\right)^2 = 1.8.$$

It is obvious that the kurtosis is independent of b. The value 1.8 is clearly smaller than 3 indicating that the uniform distribution's curve exhibits a flatter behavior than that of the normal distribution. ■

Markov's and Chebyshev's Inequality

Consider again the random variable X with variance $\sigma^2 = \mathrm{Var}(X)$. Depending on σ^2, Chebyshev's inequality allows to bound the probability with which the random variable is distributed around its expected value. In fact, this result is a special case of the more general Markov's inequality, see (2.3), which is established e.g. in Ross (2010, Sect. 8.2). A proof of Chebyshev's result given in (2.4) will be provided in Problem 2.3.

Lemma 2.1 (Markov's and Chebyshev's Inequality) *Let X be a random variable.*

(a) If X takes only nonnegative values, then it holds for any real constant $a > 0$:

$$P(X \geq a) \leq \frac{E(X)}{a}; \tag{2.3}$$

(b) with $\sigma^2 = Var(X) < \infty$ it holds that

$$P(|X - E(X)| \geq \varepsilon) \leq \frac{\sigma^2}{\varepsilon^2}, \tag{2.4}$$

where $\varepsilon > 0$ is an arbitrary real constant.

Example 2.4 (Normal Distribution) The density of a random variable X with normal or Gaussian distribution with parameters μ and $\sigma > 0$ goes back to Gauss[5] and is, as is well known,

$$f(x) = \frac{1}{\sqrt{2\pi}\sigma} \exp\left(-\frac{1}{2}\left(\frac{x-\mu}{\sigma}\right)^2\right), \quad x \in \mathbb{R},$$

with

$$\mathrm{E}(X) = \mu \text{ and } \mathrm{Var}(X) = \sigma^2.$$

In symbols we also write $X \sim \mathcal{N}(\mu, \sigma^2)$. As the density function is symmetric around μ it follows that $\gamma_1 = 0$. The kurtosis we adopt from the literature without calculation as $\gamma_2 = 3$. Sometimes we use this result for determining the fourth centered moment. Under normality it holds that:

$$\mathrm{E}[(X - \mu_1)^4] = 3\,(\mathrm{Var}(X))^2\,.$$

We want to use this example to show that Chebyshev's inequality may be not very sharp. For example,

$$\mathrm{P}(|X - \mu| \geq 2\,\sigma) \leq \frac{\sigma^2}{4\,\sigma^2} = 0.25.$$

[5]The traditional German spelling is Gauß. Carl Friedrich Gauß lived from 1777 to 1855 and was a professor in Göttingen. His name is connected to many discoveries and inventions in theoretical and applied mathematics. His portrait and a graph of the density of the normal distribution decorated the 10-DM-bill in Germany prior to the Euro.

When using the standard normal distribution, however, one obtains a much smaller probability than the bound due to (2.4):

$$\mathrm{P}\left(|X-\mu| \geq 2\sigma\right) = \mathrm{P}\left(\frac{|X-\mu|}{\sigma} \geq 2\right) = 2\,\mathrm{P}\left(\frac{X-\mu}{\sigma} \leq -2\right) \approx 0.044. \quad \blacksquare$$

2.3 Joint and Conditional Distributions

In this section we first recapitulate some widely known results. At the end we introduce the more involved theory of conditional expectation.

Joint Distribution and Independence

In order to restrict the notational burden, we only consider the three-dimensional case of continuous random variables X, Y and Z with the joint density function $f_{x,y,z}$ mapping from $\mathbb{R}^3$ to $\mathbb{R}$. For arbitrary real numbers a, b and c, probabilities are defined as multiple (or iterated) integrals:

$$\mathrm{P}(X \leq a, Y \leq b, Z \leq c) = \int_{-\infty}^{c}\int_{-\infty}^{b}\int_{-\infty}^{a} f_{x,y,z}(x,y,z)dxdydz.$$

As long as f is a continuous function, the order of integration does not matter, i.e. one obtains e.g.

$$\begin{aligned}\mathrm{P}(X \leq a, Y \leq b, Z \leq c) &= \int_{-\infty}^{a}\int_{-\infty}^{b}\int_{-\infty}^{c} f_{x,y,z}(x,y,z)dzdydx \\ &= \int_{-\infty}^{b}\int_{-\infty}^{a}\int_{-\infty}^{c} f_{x,y,z}(x,y,z)dzdxdy\,.\end{aligned}$$

This reversibility is sometimes called **Fubini's theorem**.[6]

Univariate and bivariate marginal distributions arise from integrating the respective variable:

$$\begin{aligned}f_x(x) &= \int_{-\infty}^{\infty}\int_{-\infty}^{\infty} f_{x,y,z}(x,y,z)dydz, \\ f_{x,y}(x,y) &= \int_{-\infty}^{\infty} f_{x,y,z}(x,y,z)dz.\end{aligned}$$

[6] Cf. Sydsæter, Strøm, and Berck (1999, p. 53). A proof is contained e.g. in the classical textbook by Rudin (1976, Thm. 10.2), or in Trench (2013, Coro. 7.2.2); the latter book may be recommended since it is downloadable free of charge.

The variables are called **stochastically independent** if, for arbitrary arguments, the joint distribution is given as the product of the marginal densities:

$$f_{x,y,z}(x,y,z) = f_x(x)f_y(y)f_z(z)\,,$$

which implies pairwise independence:

$$f_{x,y}(x,y) = f_x(x)f_y(y)\,.$$

The joint probability

$$\mathrm{P}(X \le a, Y \le b, Z \le c) = \int_{-\infty}^{c}\int_{-\infty}^{b}\int_{-\infty}^{a} f_x(x)f_y(y)f_z(z)dxdydz$$

is, under independence, factorized to

$$\begin{aligned}
\mathrm{P}(X \le a, Y \le b, Z \le c) &= \int_{-\infty}^{c}\int_{-\infty}^{b} f_y(y)f_z(z)\left[\int_{-\infty}^{a} f_x(x)\,dx\right]dydz\\
&= \int_{-\infty}^{c} f_z(z)\left\{\int_{-\infty}^{b} f_y(y)\,dy\right\}\left[\int_{-\infty}^{a} f_x(x)\,dx\right]dz\\
&= \int_{-\infty}^{a} f_x(x)dx \int_{-\infty}^{b} f_y(y)dy \int_{-\infty}^{c} f_z(z)dz\\
&= \mathrm{P}(X \le a)\,\mathrm{P}(Y \le b)\,\mathrm{P}(Z \le c).
\end{aligned}$$

Covariance

In particular for only two variables a generalization of the expectation operator is considered. Let h be a real-valued function of two variables, h: $\mathbb{R}^2 \to \mathbb{R}$, then we define as a double integral:

$$\mathrm{E}[h(X,Y)] = \int_{-\infty}^{\infty}\int_{-\infty}^{\infty} h(x,y)f_{x,y}(x,y)dxdy.$$

Hence, the **covariance** between X and Y can be defined as follows:

$$\begin{aligned}
\mathrm{Cov}(X,Y) &:= \mathrm{E}[(X-\mathrm{E}(X))(Y-\mathrm{E}(Y))]\\
&= \mathrm{E}(XY) - \mathrm{E}(X)\mathrm{E}(Y)\,,
\end{aligned}$$

where the finiteness of these integrals is again assumed tacitly. It can be easily shown that the independence of two variables implies their uncorrelatedness, i.e. $\mathrm{Cov}(X,Y) = 0$, whereas the reverse does not generally hold true. In particular, the covariance only measures the linear relation between two variables. In order to have the measure independent of the units, it is usually standardized as follows:

$$\rho_{xy} = \frac{\mathrm{Cov}(X,Y)}{\sqrt{\mathrm{Var}(X)}\sqrt{\mathrm{Var}(Y)}}.$$

The **correlation coefficient** ρ_{xy} is smaller than or equal to one in absolute value, see Problem 2.7.

Example 2.5 (Bivariate Normal Distribution) Let X and Y be two Gaussian random variables,

$$X \sim \mathcal{N}(\mu_x, \sigma_x^2), \quad Y \sim \mathcal{N}(\mu_y, \sigma_y^2),$$

with correlation coefficient ρ. We talk about a bivariate normal distribution if the joint density takes the following form:

$$f_{x,y}(x,y) = \frac{1}{2\pi\sigma_x\sigma_y\sqrt{1-\rho^2}}\,\varphi_{x,y}(x,y)$$

with $\varphi_{x,y}(x,y)$ equal to

$$\exp\left\{-\frac{1}{2(1-\rho^2)}\left[\left(\frac{x-\mu_x}{\sigma_x}\right)^2 - 2\rho\left(\frac{x-\mu_x}{\sigma_x}\right)\left(\frac{y-\mu_y}{\sigma_y}\right) + \left(\frac{y-\mu_y}{\sigma_y}\right)^2\right]\right\}.$$

Symbolically, we denote the vector as

$$\begin{pmatrix} X \\ Y \end{pmatrix} \sim \mathcal{N}_2(\mu, \Sigma),$$

where μ is a vector and Σ stands for a symmetric matrix:

$$\mu = \begin{pmatrix} \mu_x \\ \mu_y \end{pmatrix}, \quad \Sigma = \begin{pmatrix} \sigma_x^2 & \mathrm{Cov}(X,Y) \\ \mathrm{Cov}(X,Y) & \sigma_y^2 \end{pmatrix}.$$

In general, the covariance matrix is defined as follows:

$$\Sigma = \mathrm{E}\left[\begin{pmatrix} X - \mathrm{E}(X) \\ Y - \mathrm{E}(Y) \end{pmatrix}(X - \mathrm{E}(X),\ Y - \mathrm{E}(Y))\right].$$

Note that in the case of uncorrelatedness ($\rho = 0$) it holds that

$$\begin{aligned} f_{x,y}(x,y) &= \frac{1}{\sqrt{2\pi}\sigma_x} \exp\left\{-\frac{(x-\mu_x)^2}{2\sigma_x^2}\right\} \frac{1}{\sqrt{2\pi}\sigma_y} \exp\left\{-\frac{(y-\mu_y)^2}{2\sigma_y^2}\right\} \\ &= f_x(x) f_y(y). \end{aligned}$$

The joint density function is then determined as the product of the individual densities. Consequently, the random variables X and Y are independent. Therefore it follows, in particular for the normal distribution, that uncorrelatedness is equivalent to stochastic independence. Furthermore, bivariate Gaussian random variables have the property that each linear combination is univariate normally distributed. More precisely, it holds for $\lambda \in \mathbb{R}^2$ with[7] $\lambda' = (\lambda_1, \lambda_2)$ that:

$$\lambda' \begin{pmatrix} X \\ Y \end{pmatrix} = \lambda_1 X + \lambda_2 Y \sim \mathcal{N}(\lambda'\mu, \lambda'\Sigma\lambda).$$

Interesting special cases are obtained with $\lambda' = (1, 1)$ and $\lambda' = (1, -1)$ for sums and differences. Note that furthermore for multivariate normal distributions necessarily all marginal distributions are normal (with $\lambda' = (1, 0)$ and $\lambda' = (0, 1)$). The reverse does not hold. A bivariate example for Gaussian marginal distributions *without* joint normal distributions is given by Bickel and Doksum (2001, p. 533). ■

Cauchy-Schwarz Inequality

The inequality by Cauchy and Schwarz is the reason why $|\rho_{xy}| \leq 1$ applies. The following statement is verified in Problem 2.6.

Lemma 2.2 (Cauchy-Schwarz Inequality) *For arbitrary random variables Y and Z it holds that*

$$|E(YZ)| \leq \sqrt{E(Y^2)}\sqrt{E(Z^2)}, \tag{2.5}$$

where finite moments are assumed.

We want to supplement the Cauchy-Schwarz inequality by an intermediate inequality, see (2.8). For this purpose we remember the so-called **triangle inequality** for

[7] Up to this point a superscript prime at a function has denoted its derivative. In the rare cases in which we are concerned with matrices or vectors, the symbol will also be used to indicate transposition. Bearing in mind the respective context, there should not occur any ambiguity.

two real numbers:

$$|a_1 + a_2| \leq |a_1| + |a_2|.$$

Obviously, this can be generalized to:

$$\left|\sum_{i=1}^{n} a_i\right| \leq \sum_{i=1}^{n} |a_i|.$$

If the sequence is absolutely summable, it is allowed to set $n = \infty$. This suggests that an analogous inequality also applies for integrals. If the function g is continuous, this implies continuity of $|g|$ and one obtains:

$$\left|\int g(x)dx\right| \leq \int |g(x)|dx.$$

This implies for the expected value of a random variable X:

$$|\mathrm{E}(X)| \leq \mathrm{E}(|X|). \tag{2.6}$$

This relation resembles (2.2); in fact, both relations are special cases of **Jensen's inequality**.[8] A random variable is called **integrable** if $\mathrm{E}(|X|) < \infty$. Of course this implies a finite expected value. For integrability a finite second moment is sufficient, which follows from (2.5) with $Y = |X|$ and $Z = 1$:

$$\mathrm{E}(|X|) \leq \sqrt{\mathrm{E}|X|^2}\sqrt{1^2} = \sqrt{\mathrm{E}(X)^2}.$$

Now, if setting $X = YZ$ in (2.6), it follows that: $|\mathrm{E}(YZ)| \leq \mathrm{E}(|Y||Z|)$. This is the bound added to (2.5):

$$|\mathrm{E}(YZ)| \leq \mathrm{E}(|Y||Z|) \leq \sqrt{\mathrm{E}(Y^2)}\sqrt{\mathrm{E}(Z^2)}. \tag{2.8}$$

The first inequality follows from (2.6). The second one will be verified in the problem section.

[8]The general statement is: for a convex function g it holds

$$g(\mathrm{E}(X)) \leq \mathrm{E}(g(X)) ; \tag{2.7}$$

see e.g. Sydsæter et al. (1999, p. 181), while a proof is given e.g. in Davidson (1994, Ch. 9) or Ross (2010, p. 409).

Conditional Distributions

Conditional distributions and densities, respectively, are defined as the ratio of the joint density and the "conditioning density", i.e. they are defined by the following density functions (where positive denominators are assumed):

$$f_{x|y}(x) = \frac{f_{x,y}(x,y)}{f_y(y)},$$

$$f_{x|y,z}(x) = \frac{f_{x,y,z}(x,y,z)}{f_{y,z}(y,z)},$$

$$f_{x,y|z}(x,y) = \frac{f_{x,y,z}(x,y,z)}{f_z(z)}.$$

It should be clear that these conditional densities are in fact density functions. In case of independence it holds by definition that the conditional and the unconditional densities are equal, e.g.

$$f_{x|y}(x) = f_x(x).$$

This is very intuitive: In case of two independent random variables, one does not have any influence on the probability with which the other takes on values.

Conditional Expectation

If the random variables X and Y are not independent and if the realization of Y is known, $Y = y$, then the expectation of X will be affected:

$$\mathrm{E}(X|Y=y) = \int_{-\infty}^{\infty} x f_{x|y}(x) dx.$$

Analogously, we define the conditional expectation of a random variable Z, $Z = h(X,Y)$, $h : \mathbb{R}^2 \to \mathbb{R}$, given $Y = y$ as:

$$\begin{aligned} \mathrm{E}(Z|Y=y) &= \mathrm{E}\left(h(X,Y) \mid Y=y\right) \\ &= \int_{-\infty}^{\infty} h(x,y) f_{x|y}(x)\, dx. \end{aligned}$$

In particular, for $h(X,Y) = X\,g(Y)$ with $g : \mathbb{R} \to \mathbb{R}$ one therefore obtains

$$\begin{aligned} \mathrm{E}(X\,g(Y) \mid Y=y) &= g(y) \int_{-\infty}^{\infty} x f_{x|y}(x)\, dx \\ &= g(y)\,\mathrm{E}(X|Y=y). \end{aligned}$$

Here, the marginal density of X is replaced by the conditional density conditioned on the value $Y = y$.

Technically, we can calculate the density conditioned on the random variable Y instead of conditioned on a value[9] $Y = y$:

$$f_{x|Y}(x) = \frac{f_{x,y}(x, Y)}{f_y(Y)}.$$

By $f_{x|Y}(x)$ a transformation of the random variable Y and consequently a new random variable is obtained. This is also true for the related conditional expectations:

$$\begin{aligned} \mathrm{E}(X|Y) &= \int_{-\infty}^{\infty} x f_{x|Y}(x)dx\,, \\ \mathrm{E}(h(X, Y)|Y) &= \int_{-\infty}^{\infty} h(x, Y) f_{x|Y}(x)dx. \end{aligned}$$

As this is about random variables, it is absolutely reasonable to determine the expected value over the conditional expectation. This calculation can be carried out applying a rule called the "law of iterated expectations (LIE)" in the literature; it is given in Proposition 2.1. In order to prevent confusion whether X or Y is integrated, it is advisable to subscript the expectation operator accordingly:

$$\mathrm{E}_y[\mathrm{E}_x(X|Y)] = \int_{-\infty}^{\infty} [\mathrm{E}_x(X|y)]\, f_y(y)dy = \int_{-\infty}^{\infty} \left[\int_{-\infty}^{\infty} x f_{x|y}(x)dx\right] f_y(y)dy.$$

Although Y and $g(Y)$ are random variables, after conditioning on Y they can be treated as constants and in case of a multiplicative composition, they can be put in front of the expected value when integration is with respect to X. This is the second statement in the following proposition, also cf. Davidson (1994, Theorem 10.10). The first statement will be derived in Problem 2.9.

Proposition 2.1 (Conditional Expectation) *With the notation introduced above, it holds that:*

(a) $E_y[E_x(X|Y)] = E_x(X)$,
(b) $E_h(g(Y)X|Y) = g(Y)E_x(X|Y)$ *for* $h(X, Y) = X\,g(Y)$.

[9]This is not a really rigorous way of introducing expectations conditioned on random variables. A mathematically correct exposition, however, requires measure theoretical arguments not being available at this point; cf. for example Davidson (1994, Ch. 10), or Klebaner (2005, Ch. 2). More generally, one may define expectations conditioned on a σ-algebra, $\mathrm{E}(X|\mathcal{G})$, where $\mathcal{G}$ could be the σ-algebra generated by Y: $\mathcal{G} = \sigma(Y)$.

Frequently, we formulate these statements in a shorter way,

$$\begin{aligned}
\mathrm{E}[\mathrm{E}(X|Y)] &= \mathrm{E}(X), \\
\mathrm{E}\,(g(Y)X|Y) &= g(Y)\,\mathrm{E}(X|Y),
\end{aligned}$$

if there is no risk of misunderstanding.

2.4 Stochastic Processes (SP)

In this section stochastic processes are defined and classified. In the following chapters we will be confronted with concrete types of stochastic processes.

Definition

A univariate **stochastic process** (SP) is a family of (real-valued) random variables, $\{X(t;\omega)\}_{t\in\mathbb{T}}$, for a given **index set** $\mathbb{T}$:

$$\begin{aligned}
X: \quad \mathbb{T}\times\Omega &\to \mathbb{R} \\
(t;\,\omega) &\mapsto X(t;\omega)\,.
\end{aligned}$$

The subscript $t \in \mathbb{T}$ is always to be interpreted as "time". At a fixed point in time t_0 the stochastic process is therefore simply a random variable,

$$\begin{aligned}
X: \quad \Omega &\to \mathbb{R} \\
\omega &\mapsto X(t_0;\omega)\,.
\end{aligned}$$

A fixed ω_0, however, results in a path, a trajectory or a realization of a process which is also often referred to as **time series**,

$$\begin{aligned}
X: \quad \mathbb{T} &\to \mathbb{R} \\
t &\mapsto X(t;\omega_0)\,.
\end{aligned}$$

In fact, a stochastic process is a rather complex object. In order to characterize it mathematically, random vectors of arbitrary, finite length n at arbitrary points in time $t_1 < \cdots < t_n$ have to be considered:

$$X_n(t_i) := (X(t_1;\omega),\,\ldots,\,X(t_n;\omega))'\,,\quad t_1 < \cdots < t_n\,.$$

The multivariate distribution of such an arbitrary random vector characterizes a stochastic process. In particular, certain minimal requirements for the finite-dimensional distribution of $X_n(t_i)$ guarantee that a stochastic process exists at all.[10]

Depending on the countability or non-countability of the index set $\mathbb{T}$, discrete-time and continuous-time SPs are distinguished. In the case of sequences of random variables, we talk about **discrete-time processes**, where the index set consists of integers , $\mathbb{T} \subseteq \mathbb{N}$ or $\mathbb{T} \subseteq \mathbb{Z}$. For discrete-time processes we agree upon lower case letters as an abbreviation without explicitly denoting the dependence on ω,

$$\{x_t\}\,,\ t \in \mathbb{T} \quad \text{for } \{X(t;\omega)\}_{t\in\mathbb{T}}\,.$$

For so-called **continuous-time processes** the index set $\mathbb{T}$ is a real interval, $\mathbb{T} = [a, b] \subseteq \mathbb{R}$, frequently $\mathbb{T} = [0, T]$ or $\mathbb{T} = [0, 1]$, however, open intervals are also admitted. For continuous-time processes we also suppress the dependence on ω notationally and write in a shorter way[11]

$$X(t)\,,\ t \in \mathbb{T} \quad \text{for } \{X(t;\omega)\}_{t\in\mathbb{T}}\,.$$

Stationary and Gaussian Processes

Consider again generally an arbitrary vector of the length n,

$$X_n(t_i) = (X(t_1;\omega),\ \ldots,\ X(t_n;\omega))'\,.$$

If $X_n(t_i)$ is jointly normally distributed for all n and t_i, then $X(t;\omega)$ is called a **normal process** (also: **Gaussian process**). Furthermore, we talk about a **strictly stationary process** if the distribution is invariant over time. More precisely, $X_n(t_i)$ follows the same distribution as a vector which is shifted by s units on the time axis.

$$X_n'(t_i + s) = (X(t_1 + s;\omega),\ \ldots,\ X(t_n + s;\omega))\,.$$

The distributional properties of a strictly stationary process do not depend on the location on the time axis but only on how far the individual components $X(t_i;\omega)$ are apart from each other temporally. Strict stationarity therefore implies

[10] These "consistency" requirements due to Kolmogorov are found e.g. in Brockwell and Davis (1991, p. 11) or Grimmett and Stirzaker (2001, p. 372). A proof of Kolmogorov's existence theorem can be found e.g. in Billingsley (1986, Sect. 36).

[11] The convention of using upper case letters for continuous-time process is not universal.

the expected value and the variance to be constant (assuming they are finite) and the autocovariance for two points in time to depend only on the temporal interval:

1. $\mathrm{E}(X(t;\omega)) = \mu_x$ for $t \in \mathbb{T}$,
2. $\mathrm{Cov}(X(t;\omega), X(t+h;\omega)) = \gamma_x(h)$ for all $t, t+h \in \mathbb{T}$,

and therefore in particular

$$\mathrm{Var}(X(t;\omega)) = \gamma_x(0) \quad \text{for all } t \in \mathbb{T} .$$

A process (with finite second moments $\mathrm{E}[(X(t;\omega))^2]$) fulfilling these two conditions (without necessarily being strictly stationary) is also called **weakly stationary** (or: second-order stationary). Under stationarity, we define as **autocorrelation** coefficient also independent of t:

$$\rho_x(h) = \frac{\gamma_x(h)}{\gamma_x(0)} .$$

Synonymously to autocorrelation we also speak of serial or temporal correlation. For weak stationarity not necessarily the whole distribution is invariant over time, however, at least the expected value and the autocorrelation structure are constant.

In the following, the term "stationarity" always refers to the weak form unless stated otherwise.

Example 2.6 (White Noise Process) In the following chapters, $\{\varepsilon_t\}$ often denotes a discrete-time process $\{\varepsilon(t;\omega)\}$ free of serial correlation. In addition we assume a mean of zero and a constant variance $\sigma^2 > 0$, i.e.

$$\mathrm{E}(\varepsilon_t) = 0 \quad \text{and} \quad \mathrm{E}(\varepsilon_t \varepsilon_s) = \begin{cases} \sigma^2 , & t = s \\ 0 , & t \neq s \end{cases} .$$

By definition such a process is weakly stationary. We typically denote it as

$$\{\varepsilon_t\} \sim \mathrm{WN}(0, \sigma^2).$$

The reason why such a process is called **white noise** will be provided in Chap. 4. ■

Example 2.7 (Pure Random Process) Sometimes $\{\varepsilon_t\}$ from Example 2.6 will meet the stronger requirements of being identically and independently distributed. Identically distributed implies that the marginal distribution

$$F_i(\varepsilon) = \mathrm{P}(\varepsilon_{t_i} \leq \varepsilon) = F(\varepsilon) , \quad i = 1, \ldots, n ,$$

does not vary over time. Independence means that the joint distribution of the vector

$$\varepsilon'_{n,t_i} = (\varepsilon_{t_1}, \ldots, \varepsilon_{t_n})$$

equals the product of the marginal distributions. As the marginal distributions are invariant with respect to time, this also holds for their product. Thus, $\{\varepsilon_t\}$ is strictly stationary. In the following it is furthermore assumed that ε_t has zero expectation and the finite variance σ^2. Symbolically, we also write[12]:

$$\{\varepsilon_t\} \sim \text{iid}(0, \sigma^2).$$

A stochastic process with these properties is frequently called a **pure random process**. Clearly, an iid (or pure random) process is white noise. ■

Markov Processes and Martingales

A SP is called a Markov process if all information of the past about its future behavior is entirely concentrated in the present. In order to capture this concept more rigorosly, the set of information about the past of the process available up to time t is denoted by $\mathcal{I}_t$. Frequently, the **information set** is also referred to as

$$\mathcal{I}_t = \sigma\left(X(r;\omega),\ r \le t\right) ,$$

because it is the smallest σ-algebra generated by the past and presence of the process $X(r;\omega)$ up to time t.[13] The entire information about the process up to time t is contained in $\mathcal{I}_t$. A **Markov process**, so to speak, does not remember how it arrived at the present state: The probability that the process takes on a certain value at time $t+s$ depends only on the value at time t ("present") and does not depend on the past behavior. In terms of conditional probabilities, for $s > 0$ the corresponding property reads:

$$\mathrm{P}\left(X(t+s;\omega) \le x \mid \mathcal{I}_t\right) = \mathrm{P}\left(X(t+s;\omega) \le x \mid X(t;\omega)\right) . \tag{2.9}$$

A process is called a martingale if the present value is the best prediction for the future. A **martingale** technically fulfills two properties. In the first place, it has to be (absolutely) integrable, i.e. it is required that (2.10) holds. Secondly, given all

[12]The acronym stands for "independently identically distributed".

[13]By assumption, the information at an earlier point in time is contained in the information set at a subsequent point in time: $\mathcal{I}_t \subseteq \mathcal{I}_{t+s}$ for $s \ge 0$. A family of such nested σ-algebras is also called "filtration".

information $\mathcal{I}_t$, the conditional expectation only uses the information at time t. More precisely, the expected value for the future is equal to today's value. Technically, this amounts to:

$$\mathrm{E}\left(|X(t;\omega)|\right) < \infty\,, \tag{2.10}$$

$$\mathrm{E}\left(X(t+s;\omega)\,|\,\mathcal{I}_t\right) = X(t;\omega)\,,\ s \geq 0. \tag{2.11}$$

Note that the conditional expectation is a random variable. Therefore, strictly speaking, equation (2.11) only holds with probability one.

Martingale Differences

Now, let us focus on the discrete-time case. A discrete-time martingale is defined by the expectation at time t for $t+1$ being given by the value at time t. This is equivalent to expecting a zero increment from t to $t+1$. Therefore, this concept is frequently expressed in form of differences. We then talk about martingale differences. As we will see, in a sense, such a property is settled between uncorrelatedness and independence and is interesting from both an economic and a statistical point of view.

We again assume an integrable process, i.e. $\{x_t\}$ fulfills (2.10). It is called a **martingale difference** (or martingale difference sequences) if the conditional expectation (given its own past) is zero:

$$\mathrm{E}(x_{t+1}|\sigma(x_t, x_{t-1}, \ldots)) = 0.$$

This condition states concretely that the past does not have any influence on predictions (conditional expectation); i.e. knowing the past does not lead to an improvement of the prediction, the forecast is always zero. Not surprisingly, this also applies if only one single past observation is known (see Proposition 2.2(a)). Two further conclusions for unconditional moments contained in the proposition can be verified,[14] see Problem 2.10: martingale differences are zero on average and free of serial correlation. In spite of serial uncorrelatedness, martingale differences in general are on no account independent over time. What is more, they do not even have to be stationary as it is not ruled out that their variance function depends on t.

[14]We cannot prove the first statement rigorously, which would require a generalization of Proposition 2.1(a). The more general statement taken e.g. from Breiman (1992, Prop. 4.20) or Davidson (1994, Thm. 10.26) reads in our setting as

$$\mathrm{E}\left[\mathrm{E}(x_t|\mathcal{I}_{t-1})|x_{t-h}\right] = \mathrm{E}(x_t|x_{t-h})\,.$$

Proposition 2.2 (Martingale Differences) *For a martingale difference sequence* $\{x_t\}$ *with* $\mathcal{I}_t = \sigma(x_s\,,\ s \le t)$ *it holds that*

(a) $E(x_t|x_{t-h}) = 0$ *for* $h > 0$,
(b) $E(x_t) = 0$,
(c) $Cov(x_t, x_{t+h}) = E(x_t x_{t+h}) = 0$ *for* $h \neq 0$

for all $t \in \mathbb{T}$.

Note that a stationary martingale difference sequence has a constant variance and is thus white noise by Proposition 2.2. The concept should be further clarified by means of an example.

Example 2.8 (Martingale Difference) Consider the process given by

$$x_t = x_{t-1}\,\frac{\varepsilon_t}{\varepsilon_{t-2}}\,, \qquad t \in \{2,\ldots,n\}\,, \qquad \{\varepsilon_t\} \sim \text{iid}(0,\sigma^2)\,,$$

with $x_1 = \varepsilon_1$ and $\varepsilon_0 = 1$. From this it follows that $x_2 = x_1\,\frac{\varepsilon_2}{\varepsilon_0} = \varepsilon_1\varepsilon_2$ and by continued substitution:

$$x_t = \varepsilon_{t-1}\varepsilon_t\,, \qquad t = 2,\ldots,n.$$

We want to show that this is a martingale difference sequence. Therefore, we note that the past of the pure random process can be reconstructed from the past of x_t:

$$\varepsilon_2 = \frac{x_2}{\varepsilon_1} = \frac{x_2}{x_1}\,,\ \varepsilon_3 = \frac{x_3}{\varepsilon_2}\,,\ \ldots\ ,\ \varepsilon_t = \frac{x_t}{\varepsilon_{t-1}}\,.$$

Therefore, the information set $\mathcal{I}_t$ constructed from $\{x_t,\ldots,x_1\}$ contains not only the past values of x_{t+1}, but also the ones of the iid process up to time t. Thus, it holds that

$$\begin{aligned}
\mathrm{E}(x_{t+1}|\mathcal{I}_t) &= \mathrm{E}(\varepsilon_t\varepsilon_{t+1}|\mathcal{I}_t)\\
&= \varepsilon_t \mathrm{E}(\varepsilon_{t+1}|\mathcal{I}_t)\\
&= \varepsilon_t \mathrm{E}(\varepsilon_{t+1})\\
&= 0\,.
\end{aligned}$$

The first equality follows from the definition of the process. The second equality is accounted for by Proposition 2.1(b). The third step is due to the independence of ε_{t+1} of the past up to t, that is why conditional and unconditional expectation coincide. Finally, by assumption, ε_{t+1} is zero on average. All in all, by this the property of martingale differences is established. Therefore, $\{x_t\}$ is free of serial correlation, however, it is serially (i.e. temporally) dependent, which is obvious from the recursive definition. ■

A prominent class of martingale differences are the ARCH processes treated in Chap. 6.

2.5 Problems and Solutions

Problems

2.1 Prove for the kurtosis coefficient: $\gamma_2 \geq 1$.

2.2 Let X follow a Pareto distribution with

$$f(x) = \theta\, x^{-\theta-1}, \quad x \geq 1, \quad \theta > 0.$$

Prove that X has finite k-th moments if and only if $\theta > k$.

2.3 Prove Chebyshev's inequality (2.4).

2.4 Consider a bivariate distribution with:

$$f_{x,y}(x,y) = \begin{cases} \frac{1}{ab}, & (x,y) \in [0,a] \times [0,b] \\ 0, & \text{else} \end{cases}.$$

Prove that X and Y are stochastically independent.

2.5 Calculate the expected values, variances and the correlation of X and Y from Example 2.2.

2.6 Prove the second inequality from (2.8).

2.7 Prove for the correlation coefficient that $|\rho_{xy}| \leq 1$.

2.8 Consider a bivariate logistic distribution function for X and Y:

$$F_{x,y}(x,y) = (1 + e^{-x} + e^{-y})^{-1},$$

where x and y from $\mathbb{R}$ are arbitrary. What does the conditional density function of X given $Y = y$ look like?

2.9 Prove statement (a) from Proposition 2.1.

2.10 Derive the properties (b) and (c) from Proposition 2.2.
Hint: Use statement (a).

Solutions

2.1 Assuming finite fourth moments we define for a random variable X with $\mu_1 = \mathrm{E}(X)$:

$$\gamma_2 = \frac{\mathrm{E}(X-\mu_1)^4}{\sigma^4}.$$

Consider the standardized random variable Z with expectation 0 and variance 1:

$$Z = \frac{X-\mu_1}{\sigma} \quad \text{with} \quad \mathrm{E}(Z^2) = 1.$$

For this random variable, it holds that $\gamma_2 = \mathrm{E}(Z^4)$. Replacing X by Z^2 in (2.2), it follows

$$1 = \left(\mathrm{E}(Z^2)\right)^2 \leq \mathrm{E}\left(Z^4\right) = \gamma_2\,,$$

which proves the claim.

2.2 For the k-th moment it holds:

$$\mathrm{E}(X^k) = \int_{-\infty}^{\infty} x^k f(x) dx = \int_1^{\infty} \theta\, x^{k-\theta-1} dx.$$

1. case: If $\theta \neq k$, then the antiderivative results in

$$\int \theta\, x^{k-\theta-1}\, dx = \frac{\theta}{k-\theta}\, x^{k-\theta}.$$

The corresponding improper integral is defined as limit:

$$\int_1^{\infty} \theta\, x^{k-\theta-1} dx = \lim_{M\to\infty} \frac{\theta}{k-\theta} \left[x^{k-\theta}\right]_1^M.$$

For $\theta > k$ it follows that

$$\int_1^{\infty} \theta\, x^{k-\theta-1} dx = 0 - \frac{\theta}{k-\theta} = \frac{\theta}{\theta-k} < \infty.$$

For $\theta < k$, however, no finite value is obtained as $M^{k-\theta}$ goes off to infinity.

2. case: For $\theta = k$ the antiderivative takes on another form:

$$\int \theta\, x^{k-\theta-1} dx = \int \theta\, x^{-1}\, dx = \theta\, \log(x).$$

As the logarithm is unbounded, or $\log(M) \to \infty$ for $M \to \infty$, one cannot obtain a finite expectation either, as the upper bound of integration is ∞.

Both cases jointly prove the claim.

2.3 We provide two proofs. The first one builds on the fact that (2.4) is a special case of (2.3). The second one is less abstract and more elementary, and hence instructive, too.

1. Note that $(X-\mu)^2$ is a nonnegative random variable. Therefore, (2.3) applies with $a = \varepsilon^2$:

$$P((X-\mu)^2 \geq \varepsilon^2) \;\leq\; \frac{E((X-\mu)^2)}{\varepsilon^2}.$$

   The event $(X-\mu)^2 \geq \varepsilon^2$, however, is equivalent to $|X-\mu| \geq \varepsilon$, which establishes (2.4).
2. Elementarily, we prove the claim for the case that X is a continuous random variable with density function f; the discrete case can be accomplished analogously. Note the following sequence of inequalities:

$$\begin{aligned}
\operatorname{Var}(X) &= \int_{-\infty}^{\infty} (x-\mu)^2 f(x)dx \\
&\geq \int_{-\infty}^{\mu-\varepsilon} (x-\mu)^2 f(x)dx + \int_{\mu+\varepsilon}^{\infty} (x-\mu)^2 f(x)dx \\
&\geq \int_{-\infty}^{\mu-\varepsilon} \varepsilon^2 f(x)dx + \int_{\mu+\varepsilon}^{\infty} \varepsilon^2 f(x)dx.
\end{aligned}$$

   The first inequality is of course due to the omittance of

$$\int_{\mu-\varepsilon}^{\mu+\varepsilon} (x-\mu)^2 f(x)dx \geq 0.$$

   The second one is accounted for by the fact that for the integrands of the respective integrals it holds that:

$$x-\mu < -\varepsilon \quad \text{for} \quad x < \mu-\varepsilon$$

and

$$x - \mu > \varepsilon \quad \text{for} \quad x > \mu + \varepsilon.$$

Up to this point, it is therefore shown that:

$$\begin{aligned}\mathrm{Var}(X) &\geq \varepsilon^2 \mathrm{P}(X \leq \mu - \varepsilon) + \varepsilon^2 \mathrm{P}(X \geq \mu + \varepsilon)\\ &= \varepsilon^2 \mathrm{P}(|X - \mu| \geq \varepsilon).\end{aligned}$$

This is equivalent to the claim.

2.4 The marginal density is obtained as follows:

$$\begin{aligned}f_x(x) &= \int_{-\infty}^{\infty} f_{x,y}(x,y)dy\\ &= \int_0^b \frac{1}{ab}\,dy\\ &= \frac{b-0}{ab}\\ &= \frac{1}{a} \quad \text{for } x \in [0,a]\,,\end{aligned}$$

and $f_x(x) = 0$ for $x \notin [0, a]$. It also holds that

$$f_y(y) = \begin{cases} \frac{1}{b}\,, & y \in [0,b] \\ 0\,, & \text{else} \end{cases} \quad .$$

Hence, one immediately obtains for all x and y:

$$f_{x,y}(x,y) = f_x(x) f_y(y),$$

which was to be proved.

2.5 Obviously, the expected value of X is zero,

$$\mathrm{E}(X) = 50 \cdot \mathrm{P}_x(X = 50) - 50 \cdot \mathrm{P}_x(X = -50) = 0.$$

Therefore, it holds for the variance that:

$$\begin{aligned}\mathrm{Var}(X) &= \mathrm{E}[(X - \mathrm{E}(X))^2] = \mathrm{E}(X^2)\\ &= 50^2 \cdot \mathrm{P}_x(X = 50) + (-50)^2 \cdot \mathrm{P}_x(X = -50)\\ &= \frac{2500}{2} + \frac{2500}{2} = 2500.\end{aligned}$$

Also Y is zero on average:

$$\begin{aligned} \mathrm{E}(Y) &= -10 \cdot \mathrm{P}_y(Y = -10) - 20 \cdot \mathrm{P}_y(Y = -20) - 30 \cdot \mathrm{P}_y(Y = -30) \\ &\quad - 40 \cdot \mathrm{P}_y(Y = -40) + 0 \cdot \mathrm{P}_y(Y = 0) + 100 \cdot \mathrm{P}_y(Y = 100) \\ &= \frac{1}{6}\,(-10 - 20 - 30 - 40 + 100) = 0\,. \end{aligned}$$

Hence, the variance reads

$$\mathrm{Var}(Y) = \frac{1}{6}\left((-10)^2 + (-20)^2 + (-30)^2 + (-40)^2 + 0^2 + 100^2\right) = 2166.67.$$

For the covariance we obtain

$$\begin{aligned} \mathrm{Cov}(X, Y) &= \mathrm{E}\left[(X - \mathrm{E}(X))\,(Y - \mathrm{E}(Y))\right] \\ &= \mathrm{E}(X\,Y) \\ &= \sum_{i=1}^{2}\sum_{j=1}^{6} x_i y_j \mathrm{P}_{x,y}(X = x_i, Y = y_j). \end{aligned}$$

In order to compute it, the entire joint probability distribution is to be established:

$$\mathrm{P}_{x,y}(X = -50, Y = -40) = \mathrm{P}(\{1,3,5\} \cap \{4\}) = \mathrm{P}(\emptyset) = 0,$$

$$\mathrm{P}_{x,y}(X = 50, Y = -40) = \mathrm{P}(\{2,4,6\} \cap \{4\}) = \mathrm{P}(\{4\}) = \frac{1}{6},$$

$$\mathrm{P}_{x,y}(X = -50, Y = -30) = \mathrm{P}(E^c \cap \{3\}) = \mathrm{P}(\{3\}) = \frac{1}{6},$$

$$\mathrm{P}_{x,y}(X = 50, Y = -30) = \mathrm{P}(E \cap \{3\}) = \mathrm{P}(\emptyset) = 0.$$

We may collect those numbers in a table:

$Y =$	-40	-30	-20	-10	0	100
$X = -50$	0	$\frac{1}{6}$	0	$\frac{1}{6}$	$\frac{1}{6}$	0
$X = 50$	$\frac{1}{6}$	0	$\frac{1}{6}$	0	0	$\frac{1}{6}$

Plugging in yields

$$\mathrm{E}(X\,Y) = \frac{1}{6}\,[-50 \cdot 40 + 50 \cdot 30 - 50 \cdot 20 + 50 \cdot 10 + 50 \cdot 0 + 50 \cdot 100] = 666.67.$$

Therefore one obtains for the correlation coefficient apart from rounding errors $\rho_{xy} = 0.286$.

2.6 It only remains to be shown that:

$$\mathrm{E}(|Y||Z|) \leq \sqrt{\mathrm{E}(Y^2)}\sqrt{\mathrm{E}(Z^2)}.$$

In order to see that we use the binomial formula and obtain

$$\frac{Y^2}{\mathrm{E}(Y^2)} - \frac{2|Y||Z|}{\sqrt{\mathrm{E}(Y^2)}\sqrt{\mathrm{E}(Z^2)}} + \frac{Z^2}{\mathrm{E}(Z^2)} = \left(\frac{|Y|}{\sqrt{\mathrm{E}(Y^2)}} - \frac{|Z|}{\sqrt{\mathrm{E}(Z^2)}}\right)^2 \geq 0.$$

Therefore, the expectation of the left hand side cannot become negative, which yields:

$$1 - \frac{2\mathrm{E}(|Y||Z|)}{\sqrt{\mathrm{E}(Y^2)}\sqrt{\mathrm{E}(Z^2)}} + 1 = 2\left(1 - \frac{\mathrm{E}(|Y||Z|)}{\sqrt{\mathrm{E}(Y^2)}\sqrt{\mathrm{E}(Z^2)}}\right) \geq 0.$$

In particular, it can be observed that the expression is always positive except for the case $Y = Z$. Rearranging terms verifies the second inequality from (2.8).

2.7 Plugging in $X - \mathrm{E}(X)$ and $Y - \mathrm{E}(Y)$ instead of Y and Z in (2.5) by Cauchy-Schwarz it follows that

$$|\mathrm{E}[(X - \mathrm{E}(X))(Y - \mathrm{E}(Y))]| \leq \sqrt{\mathrm{E}[(X - \mathrm{E}(X))^2]}\sqrt{\mathrm{E}[(Y - \mathrm{E}(Y))^2]},$$

which is the same as:

$$|\mathrm{Cov}(X, Y)| \leq \sqrt{\mathrm{Var}(X)}\sqrt{\mathrm{Var}(Y)}.$$

This verifies the claim.

2.8 Due to

$$F_{x,y}(x, y) = \int_{-\infty}^{y}\int_{-\infty}^{x} f_{x,y}(r, s)dr\,ds,$$

$f_{x,y}$ is determined by taking the partial derivative of $F_{x,y}$ with respect to both arguments:

$$\begin{aligned}\frac{\partial^2 F_{x,y}(x, y)}{\partial x \partial y} &= \frac{\partial(1 + e^{-x} + e^{-y})^{-2}e^{-x}}{\partial y} \\ &= \frac{2e^{-x}e^{-y}}{(1 + e^{-x} + e^{-y})^3} \\ &= f_{x,y}(x, y).\end{aligned}$$

The marginal distribution of Y is determined by

$$\begin{aligned} F_y(y) &= \int_{-\infty}^{y}\int_{-\infty}^{\infty} f_{x,y}(x,s)dxds \\ &= \lim_{x\to\infty} F_{x,y}(x,y) = (1+e^{-y})^{-1}. \end{aligned}$$

The marginal density therefore reads

$$f_y(y) = \frac{e^{-y}}{(1+e^{-y})^2}.$$

Division yields the conditional density:

$$\begin{aligned} f_{x|y}(x) &= \frac{f_{x,y}(x,y)}{f_y(y)} \\ &= \frac{2e^{-x}(1+e^{-y})^2}{(1+e^{-x}+e^{-y})^3}. \end{aligned}$$

2.9 The following sequence of equalities holds and will be justified in detail. The first two equations define exactly the corresponding (conditional) expectations. For the third equality, the order of integration is reversed; this is due to Fubini's theorem. The fourth equation is again by definition (conditional density), whereas in the fifth equation only the density of Y is cancelled out. In the sixth equation, the influence of Y on the joint density is integrated out such that the marginal density of X remains. This again yields the expectation of X by definition. Therefore, it holds that

$$\begin{aligned} \mathrm{E}_y\left(\mathrm{E}_x\left(X|Y\right)\right) &= \int_{-\infty}^{\infty} \mathrm{E}_x\left(X|y\right) f_y\left(y\right) dy \\ &= \int_{-\infty}^{\infty}\left[\int_{-\infty}^{\infty} x f_{x|y}\left(x\right) dx\right] f_y\left(y\right) dy \\ &= \int_{-\infty}^{\infty} x\left[\int_{-\infty}^{\infty} f_{x|y}\left(x\right) f_y\left(y\right) dy\right] dx \\ &= \int_{-\infty}^{\infty} x\left[\int_{-\infty}^{\infty} \frac{f_{x,y}\left(x,y\right)}{f_y(y)} f_y\left(y\right) dy\right] dx \end{aligned}$$

$$\begin{aligned}
&= \int_{-\infty}^{\infty} x \left[\int_{-\infty}^{\infty} f_{x,y}(x,y)\, dy \right] dx \\
&= \int_{-\infty}^{\infty} x f_x(x)\, dx \\
&= \mathrm{E}_x(X)\ ,
\end{aligned}$$

which was to be verified.

2.10 We use statement (a), $\mathrm{E}(x_t|x_{t-h}) = 0$ for $h > 0$, connected with the law of iterated expectations:

$$\mathrm{E}(x_t) = \mathrm{E}[\mathrm{E}(x_t|x_{t-h})] = \mathrm{E}(0) = 0.$$

This proves (b), that martingale differences are also unconditionally zero on average. By applying both results of Proposition 2.1 for $h > 0$ again with (a), one arrives at:

$$\begin{aligned}
\mathrm{E}(x_t\, x_{t+h}) &= \mathrm{E}[\,\mathrm{E}(x_t\, x_{t+h}|x_t)] \\
&= \mathrm{E}[x_t \mathrm{E}(x_{t+h}|x_t)] \\
&= \mathrm{E}[x_t \cdot 0] \\
&= 0.
\end{aligned}$$

Therefore, $\mathrm{Cov}(x_t,\ x_{t+h}) = 0$ for $h > 0$. However, as the covariance function is symmetric in h, the result holds for arbitrary $h \neq 0$ which was to be verified to show (c).

References

Bickel, P. J., & Doksum, K. A. (2001). *Mathematical statistics: Basic ideas and selected topics, volume 1* (2nd ed.). Upper Saddle River: Prentice-Hall.

Billingsley, P. (1986). *Probability and measure* (2nd ed.). New York: Wiley.

Breiman, L. (1992). *Probability* (2nd ed.). Philadelphia: Society for Industrial and Applied Mathematics.

Brockwell, P. J., & Davis, R. A. (1991). *Time series: Theory and methods* (2nd ed.). New York: Springer.

Davidson, J. (1994). *Stochastic limit theory: An introduction for econometricians*. Oxford/New York: Oxford University Press.

Grimmett, G. R., & Stirzaker, D. R. (2001). *Probability and random processes* (3rd ed.). Oxford: Oxford University Press.

Klebaner, F. C. (2005). *Introduction to stochastic calculus with applications* (2nd ed.). London: Imperical College Press.

Ross, S. (2010). *A first course in probability* (8th ed.). Upper Saddle River: Prentice-Hall.

Rudin, W. (1976). *Principles of mathematical analyis* (3rd ed.). New York: McGraw-Hill.

Sydsæter, K., Strøm, A., & Berck, P. (1999). *Economists' mathematical manual* (3rd ed.). Berlin/New York: Springer.

Trench, W. F. (2013). *Introduction to real analysis*. Free Hyperlinked Edition 2.04 December 2013. Downloaded on 10th May 2014 from http://digitalcommons.trinity.edu/mono/7.

Autoregressive Moving Average Processes (ARMA) 3

3.1 Summary

This chapter is concerned with the modelling of serial correlation (or autocorrelation) that is characteristic of many time series dynamics. To that end we cover a class of stochastic processes widely used in practice. They are discrete-time processes: $\{x_t\}_{t\in\mathbb{T}}$ with $\mathbb{T} \subseteq \mathbb{Z}$. Throughout this chapter, the innovations or shocks $\{\varepsilon_t\}$ behind $\{x_t\}$ are assumed to form a white noise sequence as defined in Example 2.6. The next section treats the rather simple moving average structure. The third section addresses the inversion of lag polynomials at a general level. The fourth section breaks down the technical aspects to the application with ARMA processes.

3.2 Moving Average Processes

We define the **moving average process** of order q (MA(q)) as

$$x_t = \mu + b_0\,\varepsilon_t + b_1\,\varepsilon_{t-1} + \cdots + b_q\,\varepsilon_{t-q}\,, \quad b_0 = 1\,,\, t \in \mathbb{T}\,. \tag{3.1}$$

In the following, we assume a white noise process for the so-called innovation $\{\varepsilon_t\}$ governing the processes considered. In general, we assume $b_q \neq 0$. Let us consider a special case before we enter the general discussion of the model.

MA(1)

We set μ to zero and q to one in (3.1) and thereby obtain

$$x_t = \varepsilon_t + b\,\varepsilon_{t-1}\,.$$

U. Hassler, *Stochastic Processes and Calculus*, Springer Texts in Business and Economics, DOI 10.1007/978-3-319-23428-1_3

As the innovations are $\mathrm{WN}(0, \sigma^2)$, the moments are independent of time: Firstly, it holds that $\mathrm{E}(x_t) = \mu = 0$. Secondly, one obtains from

$$\gamma(h) = \mathrm{Cov}(x_t, x_{t+h}) = \mathrm{E}\left((\varepsilon_t + b\,\varepsilon_{t-1})\,(\varepsilon_{t+h} + b\,\varepsilon_{t+h-1})\right)$$

immediately

$$\gamma(0) = \sigma^2(1 + b^2)\,, \quad \gamma(1) = \sigma^2 b\,,$$

and

$$\gamma(h) = 0 \quad \text{for} \quad h > 1\,.$$

For the MA(1) process considered the autocorrelation function $\rho(h) = \frac{\gamma(h)}{\gamma(0)}$ therefore is:

$$\rho(1) = \frac{b}{1 + b^2}, \quad \rho(h) = 0,\ h > 1.$$

This is why the process is always (weakly) stationary without any assumptions with respect to b or to the index set $\mathbb{T}$. Elementary curve sketching shows that $\rho(1)$ considered with respect to b,

$$\rho(1; b) = \frac{b}{1 + b^2},$$

becomes extremal for $b = \pm 1$. For $b = -1$, $\rho(1)$ takes the minimum $-\frac{1}{2}$, and for $b = 1$ it takes the maximum $\frac{1}{2}$ (see Problem 3.1).

In Fig. 3.1 realizations were simulated by means of so-called pseudo random number with each 50 observations of moving average processes. All three graphs were generated by the same realizations of $\{\varepsilon_t\}$. The first graph with $b_1 = b = 0$ shows the simulation of a pure random process $\{\varepsilon_t\}$: The times series oscillates arbitrarily around the expected value zero. The third graph with $b_1 = b = 0.9$ depicts the case of (strong) positive autocorrelation of first order: Positive values are followed by positive values whereas negative values tend to entail negative values, i.e. the zero line is less often crossed than in the first case. Finally, the graph in the middle lies in between both extreme cases as weak positive autocorrelation ($b_1 = b = 0.3$) is present.

MA(q)

The results obtained for the MA(1) process can be generalized. Every MA process is – for all parameter values independent of starting value conditions – always stationary. For the MA(q) process, it holds that its autocorrelation sequence vanishes from the order q on (i.e. it becomes zero). The proof is elementary and is therefore omitted.

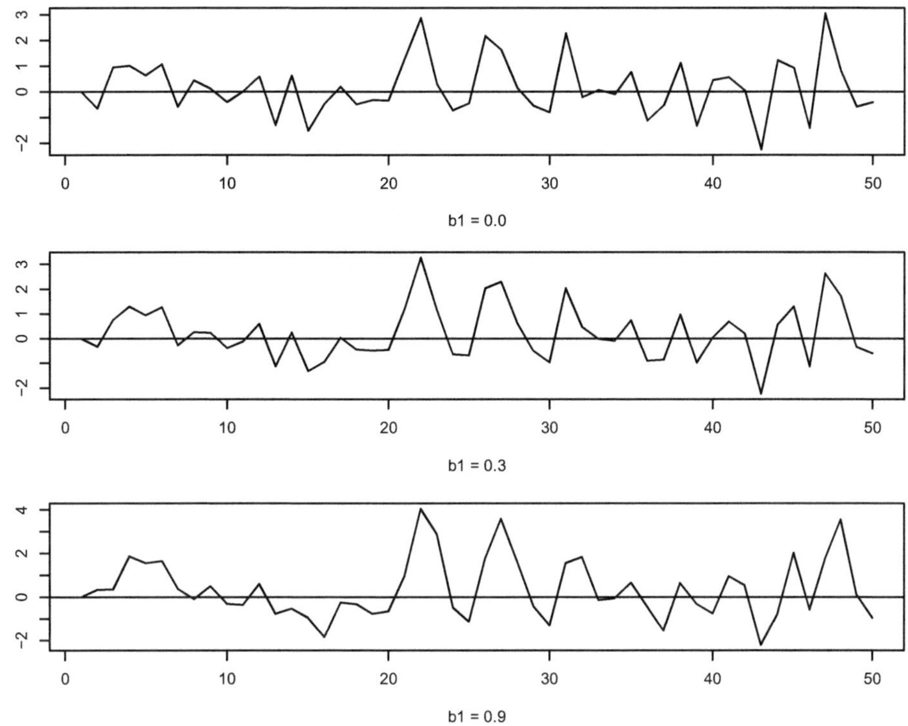

Fig. 3.1 Simulated MA(1) processes with $\sigma^2 = 1$ and $\mu = 0$

Proposition 3.1 (MA(q)) *Assume an MA(q) process from (3.1).*

(a) The process is stationary with expectation μ.
(b) For the autocovariances it holds that,

$$\gamma(h) = \sigma^2(b_h + b_{h+1}b_1 + \ldots + b_q b_{q-h}), \quad h = 0, 1, \ldots q,$$

and $\gamma(h) = 0$ for $h > q$.

Example 3.1 (Seasonal MA) Let S denote the seasonality, e.g. $S = 4$ or $S = 12$ for quarterly or monthly data, or $S = 5$ for (work) daily observations. Hence, we define as a special MA process[1]

$$x_t = \varepsilon_t + b\,\varepsilon_{t-S}.$$

[1]For $S = 1$ we obtain the MA(1) case, however, without a seasonal interpretation.

In this context, $q = S$ and $b_q = b$ hold and

$$b_1 = b_2 = \ldots = b_{q-1} = 0 .$$

Proposition 3.1 (b) yields:

$$\gamma(0) = \sigma^2(b_0 + b_1^2 + \cdots + b_q^2) = \sigma^2(1 + b^2) ,$$
$$\gamma(1) = \sigma^2(b_1 + b_2\, b_1 + \cdots + b_q\, b_{q-1}) = 0 ,$$

and also

$$\gamma(h) = 0 \quad \text{for} \quad h = 1, 2, \ldots, S - 1 .$$

The case $h = S$, however, yields

$$\gamma(S) = \sigma^2 b_S b_0 = \sigma^2 b .$$

For $h > S$, according to Proposition 3.1 it holds that $\gamma(h) = 0$. Hence, the process at hand is exclusively autocorrelated at lag S, which is why it is also called a seasonal MA process. ■

MA(∞) Processes

We now let q go off to infinity. For reasons that become obvious in the fourth section, the MA coefficients, however, are not denoted by b_j anymore. Instead, consider the infinite real sequence $\{c_j\}, j \in \mathbb{N}$, to define:

$$x_t = \mu + \sum_{j=0}^{\infty} c_j\, \varepsilon_{t-j} , \quad \sum_{j=0}^{\infty} |c_j| < \infty , \; c_0 = 1 . \tag{3.2}$$

Sometimes the process from (3.2) is called "causal" as there are only past or contemporaneous random variables $\varepsilon_{t-j}, j \geq 0$, entering the process at time t. The condition on the coefficients $\{c_j\}$ of being absolutely summable guarantees that $\sum_{j=0}^{\infty} c_j\, \varepsilon_{t-j}$ in (3.2) is a well-defined random variable, which then can be called x_t, see e.g. Fuller (1996, Theorem 2.2.1). Absolute summability naturally implies square summability,

$$\sum_{j=0}^{\infty} c_j^2 < \infty ,$$

and, indeed, sometimes we define the MA(∞) process upon this weaker assumption. Square summability of $\{c_j\}$ is sufficient for stationarity with $\mathrm{E}(x_t) = \mu$ and

$$\gamma(h) = \sigma^2 \sum_{j=0}^{\infty} c_j \, c_{j+h} \, ,$$

see Fuller (1996, Theorem 2.2.3), which is the first result of the follow proposition. The second one is established in Problem 3.3.

Proposition 3.2 (Infinite MA) *Assume an MA(∞) process,*

$$x_t = \mu + \sum_{j=0}^{\infty} c_j \, \varepsilon_{t-j} \, , \quad \{\varepsilon_t\} \sim WN(0, \sigma^2) \, , \; c_0 = 1 \, ,$$

with $\sum_{j=0}^{\infty} c_j^2 < \infty$.

(a) The process is stationary with expected value μ, *and for the autocovariances it holds that*

$$\gamma(h) = \sigma^2 \sum_{j=0}^{\infty} c_j \, c_{j+h} \, , \quad h = 0, 1, \ldots \, .$$

(b) Under absolute summability, $\sum_{j=0}^{\infty} |c_j| < \infty$, *the sequence of autocovariances is absolutely summable:*

$$\sum_{h=0}^{\infty} |\gamma(h)| < \infty \, .$$

The fact that the sequence of autocovariances is absolutely summable under (3.2), see (b) of the proposition, has an immediate implication: the autocovariances tend to zero with growing lag:

$$\gamma(h) \to 0 \quad \text{as } h \to \infty \, .$$

This means that the correlation between x_t and x_{t-h} tends to decrease with growing lag h.

Note that x_t from (3.2) or Proposition 3.2 is defined as a linear combination of ε_{t-j}, $j \geq 0$. Therefore, one sometimes speaks of a **linear process**. Other authors reserve this label for the more restricted case of iid innovations, where all temporal dependence of $\{x_t\}$ arises exclusively from the MA coefficients $\{c_j\}$. The results of Proposition 3.2, however, hold under the weaker assumption that $\{\varepsilon_t\}$ is white noise.

Impulse Responses

The MA(∞) coefficients are often called **impulse responses**, since they measure the effect of a shock j periods ago on x_t:

$$\frac{\partial x_t}{\partial \varepsilon_{t-j}} = c_j \,.$$

Square summability implies that shocks are transient in that $c_j \to 0$ as $j \to \infty$. The speed at which the impulse responses converge to zero characterizes the dynamics. In particular, Campbell and Mankiw (1987) popularized the cumulated impulse responses as measure of **persistence**, where the cumulated effect is defined as

$$CIR(J) = \sum_{j=0}^{J} c_j \,.$$

This measure quantifies the total effect up to J periods back, if there occurred a unit shock in each past period, including the present period at time t. Asymptotically, one obtains the so-called long-run effect, often called total multiplier in economics. This measure is defined as

$$CIR := \lim_{J\to\infty} CIR(J) \,, \tag{3.3}$$

provided that this quantity exists. Clearly, under absolute summability of the impulse responses, CIR is well defined. Andrews and Chen (1994) advocated CIR as being superior to alternative measures of persistence. We will return to the measurement and interpretation of persistence in the next chapter.

The MA(∞) process is of considerable generality. In fact, Wold (1938) showed that every stationary process with expected value zero can be decomposed into a square summable, purely non-deterministic MA(∞) component and an uncorrelated component, say $\{\delta_t\}$, which is deterministic in the sense that it is perfectly predictable from its past:[2]

$$\sum_{j=0}^{\infty} c_j\, \varepsilon_{t-j} + \delta_t \,, \quad \sum_{j=0}^{\infty} c_j^2 < \infty \,, \; \mathrm{E}(\varepsilon_{t-j}\delta_t) = 0 \,, \tag{3.4}$$

[2]Typically, one assumes that δ_t is identically equal to zero for all t, since "perfectly predictable" only allows for trivial processes like e.g. $\delta_t = (-1)^t A$ or $\delta_t = A$, where A is some random variable, such that $\delta_t = -\delta_{t-1}$ or $\delta_t = \delta_{t-1}$, respectively. Of course, this does not rule out the case of a constant mean μ different from zero as assumed in (3.2).

where $E(\varepsilon_t) = E(\delta_t) = 0$. For details on this **Wold decomposition** see Fuller (1996, Theorem 2.10.2) or Brockwell and Davis (1991, Theorem 5.7.1).

In practice it is not reasonable to construct a model with infinitely many parameters c_j as they cannot be estimated from a finite number of observations without further restrictions. In the fourth section of this chapter, however, we will learn that very simple, so-called autoregressive processes with a finite number of parameters possess an MA(∞) representation. In order to discuss autoregressive processes rigorously, we need to concern ourselves with polynomials in the lag operator and their invertibility.

3.3 Lag Polynomials and Invertibility

Frequently, time series models are written by means of the **lag operator**[3] L. The lag operator shifts the process $\{x_t\}$ by one unit in time: $Lx_t = x_{t-1}$. By the inverse operator L^{-1} the shift is just reversed, $L^{-1}Lx_t = x_t$, or $L^{-1}x_t = x_{t+1}$. Successive use of the operator is denoted by its power, $L^j x_t = x_{t-j}, j \in \mathbb{Z}$. The identity is described with L^0, $L^0 x_t = x_t$. Applied to a constant c, the operator leaves the value unchanged, $Lc = c$.

Causal Linear Filters

Let us consider an input process $\{x_t\}$, $t \in \mathbb{T}$, which is transferred into an output process $\{y_t\}$ by linear filtering,

$$y_t = \sum_{j=0}^{p} w_j\, x_{t-j}\,, \tag{3.5}$$

with the real filter coefficients $\{w_j\}$ (which need not to add up to one). Therefore, y_t is a linear combination of values of x_{t-j} where we assume constant filters, i.e. the weights do not depend on t. In particular, the **filter** is called causal as y_t is defined by past and contemporaneous values of $\{x_t\}$ only.

The general linear causal filter from (3.5) can be formulated as **polynomial in the lag operator**:

$$F(L) = \sum_{j=0}^{p} w_j\, L^j \quad \text{with} \quad y_t = F(L)\, x_t\,.$$

[3]Many authors also speak of the backshift operator and write B.

Occasionally, two filters are put in a row:

$$y_t = F_1(L)\, x_t\,, \quad z_t = F_2(L)\, y_t = F_2(L)\, F_1(L)\, x_t\,,$$

$$F_1(L) = \sum_{j=0}^{p_1} w_{1,j}\, L^j\,, \quad F_2(L) = \sum_{j=0}^{p_2} w_{2,j}\, L^j\,.$$

Then, the filter $F(L)$ transforming $\{x_t\}$ into $\{z_t\}$ is defined by multiplying the filters, $F(L) = F_2(L)F_1(L)$, which is called a "convolution":

$$F_2(L)\, F_1(L) = \sum_{k=0}^{p_1+p_2} v_k L^k\,, \quad v_k = \sum_{j=0}^{k} w_{2,j} w_{1,k-j} = \sum_{j=0}^{k} w_{2,k-j} w_{1,j}\,.$$

This convolution is commutative:

$$F_1(L)\, F_2(L) = F_2(L)\, F_1(L)\,.$$

By expressing filters by means of the lag operator, we can manipulate them just as ordinary (complex-valued) polynomials. As an example we consider so-called difference filters.

Example 3.2 (Difference Filters) By means of the lag operator, filters can be constructed, for example the difference filter $\Delta = 1 - L$ or the difference of the previous year for quarterly data $\Delta_4 = 1 - L^4$:

$$\Delta x_t = x_t - x_{t-1}\,, \quad \Delta_4 x_t = x_t - x_{t-4}\,.$$

The seasonal difference filter for monthly observations ($S = 12$) as well as for daily observations ($S = 5$) are defined analogously:

$$\Delta_S = 1 - L^S\,.$$

Instead of extensively calculating the double difference,

$$\begin{aligned}\Delta(\Delta x_t) &= \Delta(x_t - x_{t-1}) = (x_t - x_{t-1}) - L(x_t - x_{t-1}) \\ &= (x_t - x_{t-1}) - (x_{t-1} - x_{t-2}) = x_t - 2\,x_{t-1} + x_{t-2}\,,\end{aligned}$$

we write in short by expanding $(1 - L)^2$:

$$\Delta^2 x_t = (1 - L)^2 x_t = (1 - 2\,L + L^2) x_t = x_t - 2\,x_{t-1} + x_{t-2}\,.$$

While the ordinary difference operator Δ eliminates a linear time trend from a time series, the second differences naturally remove a quadratic time trend:

$$\Delta^2(a + bt + ct^2) = c(t^2 - 2(t-1)^2 + (t-2)^2) = 2c.$$

The fact that the order of filtering is exchangeable is well demonstrated by means of the example of seasonal and ordinary differences:

$$\begin{aligned}\Delta\,\Delta_S &= (1-L)\,(1-L^S) = 1 - L - L^S + L^{S+1} \\ &= (1-L^S)\,(1-L) = \Delta_S\,\Delta\,. \quad \blacksquare\end{aligned}$$

Invertibility of Lag Polynomials

We define as a polynomial of degree p (also of order p) in the lag operator

$$P(L) = 1 + b_1\,L + \cdots + b_p\,L^p\,,\ b_p \neq 0\,, \tag{3.6}$$

with the real coefficients b_1 to b_p. For brevity, $P(L)$ is also called lag polynomial.

Consider a first degree polynomial as a special case of (3.6),

$$A_1(L) = 1 - a\,L\,,$$

where the reason for the negative sign will immediately be obvious. When and how can this polynomial be inverted? A **comparison of coefficients** ("method of undetermined coefficients") results in the following series expansion (see Problem 3.4):

$$(1-aL)^{-1} = \frac{1}{A_1(L)} = \sum_{j=0}^{\infty} a^j L^j\,.$$

As is well known, it holds that (infinite geometric series, see Problem 3.2)

$$\sum_{j=0}^{\infty} a^j = \frac{1}{1-a} < \infty \quad \Longleftrightarrow \quad |a| < 1\,. \tag{3.7}$$

Hence, it holds that

$$(1-aL)^{-1} = \frac{1}{A_1(L)} = \sum_{j=0}^{\infty} a^j L^j \quad \text{with} \sum_{j=0}^{\infty} |a^j| = \frac{1}{1-|a|} < \infty$$

if and only if $|a| < 1$. This condition of invertibility is frequently reformulated. In order to do this, we determine the so-called z-transform of the lag polynomial with z being an element of the complex numbers ($z \in \mathbb{C}$): $A_1(z) = 1 - az$. Now, $|a| < 1$

implies that the z-transform $A_1(z) = 1 - az$ exhibits only roots outside the unit circle, i.e. roots being greater than one in absolute value:

$$|a| < 1 \quad \Longleftrightarrow \quad \left[A_1(z) = 0 \Rightarrow |z| = \frac{1}{|a|} > 1 \right]. \tag{3.8}$$

Causally Invertible Polynomials

This condition of invertibility for $A_1(L)$ from (3.8) is easily conveyed to a polynomial $P(L)$ of the order p. We say $P(L)$ is **causally invertible** if there exists a power series expansion with non-negative powers and absolutely summable coefficients:

$$(P(L))^{-1} = \frac{1}{P(L)} = \sum_{j=0}^{\infty} \alpha_j L^j \quad \text{with} \sum_{j=0}^{\infty} |\alpha_j| < \infty .$$

The invertibility depends on the z-transform

$$P(z) = 1 + b_1 z + \cdots + b_p z^p , \quad z \in \mathbb{C} , \tag{3.9}$$

or rather on the absolute value of its roots. The following condition of invertibility is adopted from Brockwell and Davis (1991, Thm. 3.1.1), and it is discussed as an exercise (Problem 3.5).

Proposition 3.3

(a) The polynomial $1 - aL$ is causally invertible,

$$\frac{1}{1 - aL} = \sum_{j=0}^{\infty} a^j L^j \quad \textit{with} \sum_{j=0}^{\infty} |a^j| < \infty ,$$

if and only if $|a| < 1$.

(b) The polynomial $P(L)$ from (3.6) is causally invertible, i.e. for $(P(L))^{-1}$ there exits the absolutely summable series expansion,

$$(P(L))^{-1} = \frac{1}{P(L)} = \sum_{j=0}^{\infty} \alpha_j L^j \quad \textit{with} \sum_{j=0}^{\infty} |\alpha_j| < \infty ,$$

if and only if it holds for all roots of $P(z)$ that they are greater than one in absolute value:

$$P(z) = 0 \quad \Rightarrow \quad |z| > 1 . \tag{3.10}$$

Example 3.3 (Invertible MA Processes) The MA(1) process, $x_t = \varepsilon_t + b\varepsilon_{t-1}$, can now be formulated alternatively by applying the MA(1) polynomial $B(L) = 1 + b\,L$. Although not required for stationarity, it is usually assumed that $|b| < 1$. What for? This implies for the MA polynomial that:

$$B(z) = 0 \quad \Rightarrow \quad |z| = \frac{1}{|b|} > 1.$$

According to Proposition 3.3 the condition of invertibility is fulfilled and there exists

$$\frac{1}{B(L)} = \frac{1}{1 + b\,L} = \sum_{j=0}^{\infty} \alpha_j\, L^j,$$

where the coefficients $\{\alpha_j\}$ are absolutely summable. The $\{\alpha_j\}$ are obtained explicitly by comparison of coefficients in

$$\begin{aligned} 1 &= (1 + b\,L) \sum_{j=0}^{\infty} \alpha_j\, L^j \\ &= \alpha_0 + \alpha_1\, L + \alpha_2\, L^2 + \alpha_3\, L^3 + \cdots \\ &\quad + b\,(\alpha_0\, L + \alpha_1\, L^2 + \alpha_2\, L^3 + \cdots), \end{aligned}$$

yielding

$$1 = \alpha_0, \quad 0 = \alpha_1 + b\alpha_0, \quad 0 = \alpha_2 + b\alpha_1, \quad \text{etc.},$$

or

$$\alpha_0 = 1,\ \alpha_1 = -b,\ \alpha_2 = b^2,\ \text{and}\ \alpha_j = (-1)^j\, b^j,\ j \geq 0.$$

Hence, the MA(1) process $x_t = B(L)\,\varepsilon_t$ can be reformulated as follows:

$$\varepsilon_t = \frac{x_t}{1 + b\,L} = x_t - b\,x_{t-1} + b^2\,x_{t-2} - b^3\,x_{t-3} \pm \ldots,$$

or

$$\begin{aligned} x_t &= b\,x_{t-1} - b^2\,x_{t-2} + b^3\,x_{t-3} \pm \ldots + \varepsilon_t \\ &= \sum_{j=1}^{\infty} (-1)^{j-1}\, b^j\, x_{t-j} + \varepsilon_t\,, \quad \sum_{j=0}^{\infty} |b^j| < \infty. \end{aligned}$$

In other words: The invertible MA(1) process (for $|b| < 1$) can be expressed (in an absolutely summable manner) as a process depending on its own infinitely many lags. Such a process is called autoregressive (of infinite order). ■

Frequently, one is interested in the special case of a quadratic polynomial, $p = 2$. In this case, the so-called **Schur criterion** provides an equivalent reformulation of the condition of invertibility, rephrasing (3.10) in terms of the polynomial coefficients b_i directly. These have to fulfill three conditions simultaneously. We take the corresponding statement from e.g. Sydsæter, Strøm, and Berck (1999, p. 58).

Corollary 3.1 *For $p = 2$ the polynomial from (3.6),*

$$P(L) = 1 + b_1 L + b_2 L^2 ,$$

is causally invertible with absolutely summable series expansion $(P(L))^{-1}$ if and only if:

$$\begin{aligned} &(i) & 1 - b_2 \quad &> 0 , \\ \text{and } &(ii) & 1 - b_1 + b_2 &> 0 , \\ \text{and } &(iii) & 1 + b_1 + b_2 &> 0 . \end{aligned}$$

Instead of checking $|z_{1,2}| > 1$ for

$$z_{1,2} = \frac{-b_1 \pm \sqrt{b_1^2 - 4b_2}}{2b_2} ,$$

it may in practice be simpler to check (i) through (iii) from Corollary 3.1.

3.4 Autoregressive and Mixed Processes

Let $\{x_t\}$ be given by the following stochastic difference equation,

$$x_t = \nu + a_1 x_{t-1} + \cdots + a_p x_{t-p} + \varepsilon_t , \quad a_p \neq 0 , \quad t \in \mathbb{T} ,$$

defining an **autoregressive process** of the order p, AR(p). The properties of the general AR process can be illustrated well at the example $p = 1$.

AR(1)

Particularly, let $p = 1$:

$$x_t = \nu + a x_{t-1} + \varepsilon_t , \quad t \in \mathbb{T} . \tag{3.11}$$

When replacing x_{t-1} by this defining equation, one obtains:

$$\begin{aligned} x_t &= \nu + a(\nu + a\, x_{t-2} + \varepsilon_{t-1}) + \varepsilon_t \\ &= \nu + a\,\nu + a^2\, x_{t-2} + a\,\varepsilon_{t-1} + \varepsilon_t . \end{aligned}$$

Hence, x_t and x_{t-2} prove to be correlated. Continued substitution yields

$$x_t = \nu + a\,\nu + a^2\,\nu + a^3\, x_{t-3} + a^2\,\varepsilon_{t-2} + a\,\varepsilon_{t-1} + \varepsilon_t\,,$$

or for any $h \geq 0$:

$$x_t = (1 + a + \ldots + a^{h-1})\,\nu + a^h\, x_{t-h} + a^{h-1}\,\varepsilon_{t-h+1} + \ldots + a\,\varepsilon_{t-1} + \varepsilon_t .$$

Now, let us suppose that the index set $\mathbb{T}$ does not have a lower bound at zero but includes an infinite past, then h can be arbitrarily large and the substitution can be repeated ad infinitum. If it furthermore holds that

$$|a| < 1,$$

then the geometric series yields ($h \to \infty$)

$$1 + a + \ldots + a^{h-1} = \frac{1 - a^h}{1 - a} \to \frac{1}{1 - a},$$

and $a^h\, x_{t-h} \to 0$ in a sense that can be made rigoros, see Brockwell and Davis (1991, p. 71) or Fuller (1996, p. 39). In this manner, it follows for $h \to \infty$ under the aforementioned conditions that:

$$x_t = \frac{\nu}{1 - a} + 0 + \sum_{j=0}^{\infty} a^j\, \varepsilon_{t-j} .$$

This way one obtains an infinite MA representation with geometrically decaying coefficients, $c_j = a^j$ in (3.2). In fact, this representation can formally be obtained by inverting $1 - aL$, see Proposition 3.3 (a). The process is therefore stationary with (see Proposition 3.2)

$$\begin{aligned} \mathrm{E}(x_t) &= \mu = \frac{\nu}{1 - a}, \\ \mathrm{Var}(x_t) &= \sigma^2 \sum_{j=0}^{\infty} a^{2j} = \frac{\sigma^2}{1 - a^2} = \gamma(0) \end{aligned}$$

and

$$\begin{aligned}\mathrm{Cov}(x_t\,, x_{t+h}) &= \sigma^2 \sum_{j=0}^{\infty} a^j a^{j+h} \\ &= a^h\,\gamma(0).\end{aligned}$$

It follows

$$\rho(h) = a^h.$$

Note that these results are obtained for $|a| < 1$. Furthermore, stationarity also depends on the index set. That is to say, if it holds that $\mathbb{T} = \{0, 1, 2, \ldots\}$ then the above-mentioned repeated substitution cannot be performed infinitely often. From

$$x_t = (1 + a + \ldots + a^{t-1})\,\nu + a^t\,x_0 + a^{t-1}\,\varepsilon_1 + \cdots + a\,\varepsilon_{t-1} + \varepsilon_t$$

the expected value follows as time-dependent:

$$\mathrm{E}(x_t) = (1 + a + \ldots + a^{t-1})\,\nu + a^t \mathrm{E}(x_0), \quad t \in \{0, 1, \ldots\}.$$

In particular, this example shows that the stationarity behavior of a process can depend on the index set $\mathbb{T}$. Therefore, in general a stochastic process is not completely characterized without specifying $\mathbb{T}$.

Example 3.4 (AR(1)) Figure 3.2 displays 50 realizations each of AR(1) processes obtained by simulation. However, in the first case $a_1 = a = 0$ such that the graph depicts the realizations of $\{\varepsilon_t\}$. On the right the theoretical autocorrelogram, $\rho(h) = 0$ for $h > 0$, is shown. In the second panel, the case of a positive autocorrelation ($a_1 = a = 0.75$) is illustrated: Positive values tend to be followed by positive values (and vice versa for negative values), such that phases of positive realizations tend to alternate with phases of negative values. The corresponding autocorrelogram shows the geometrically decaying positive autocorrelations up to the order 10. In the last case, there is a negative autocorrelation ($a_1 = a = -0.75$); consistently, negative values tend to be followed by positive ones and vice versa positive values tend to be followed by negative ones. Therefore, the zero line is more often crossed than in the first case of no serial correlation. The corresponding autocorrelogram is alternating: $\rho(h) = |a|^h(-1)^h$ for $a = -0.75$. Qualitatively different patterns of autocorrelation cannot be generated by the simple AR(1) model. Note that the impulse responses or MA(∞) coefficients of the AR(1) model are $c_j = a^j$. Consequently, the cumulated effect defined in (3.3) becomes for $|a| < 1$:

$$CIR = \sum_{j=0}^{\infty} a^j = \frac{1}{1-a}\,.$$

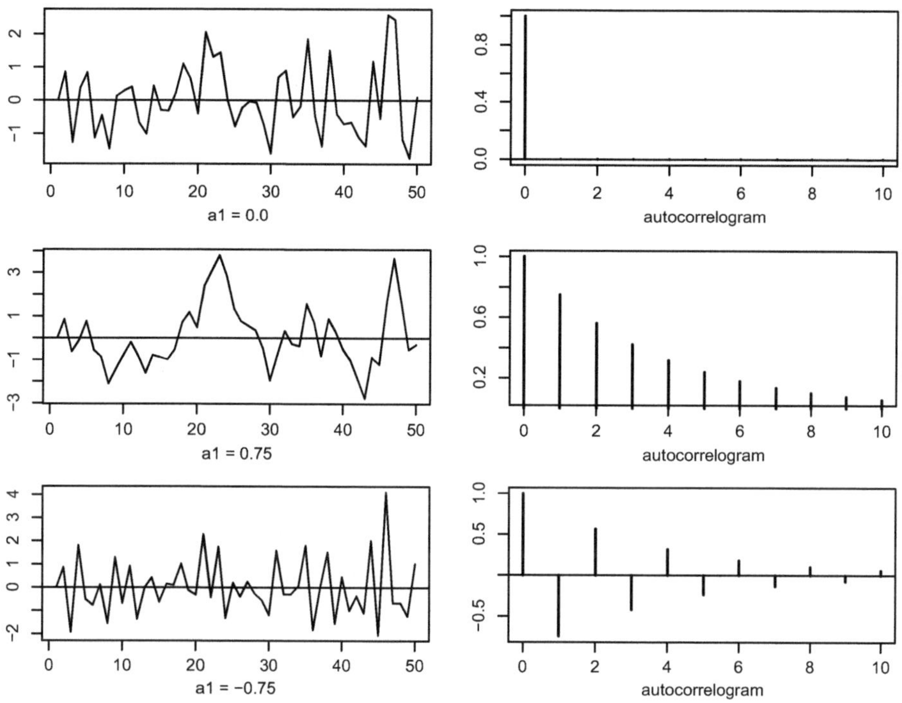

Fig. 3.2 Simulated AR(1) processes with $\sigma^2 = 1$ and $\nu = 0$

The larger a, the larger is *CIR*, which hence quantifies the persistence described above for positive a. ■

We have seen that the stationarity of an AR(1) process depends essentially on the absolute value of a. For the Markov property, however, this value is irrelevant. We talk about a Markov process if the entire past information $\mathcal{I}_t$ up to time t is concentrated in the last observation x_t:

$$\mathrm{P}\left(x_{t+s} \leq x | \mathcal{I}_t\right) = \mathrm{P}\left(x_{t+s} \leq x | x_t\right)$$

for all $s > 0$ and $x \in \mathbb{R}$. Assuming a normal distribution, we show in Problem 3.7 that every AR(1) process is a Markov process.

AR(p)

In general, the AR(p) process can be formulated equivalently by means of a lag polynomial:

$$A(L)\, x_t = \nu + \varepsilon_t\,, \quad A(L) = 1 - a_1 L - \cdots - a_p L^p\,, \quad t \in \mathbb{T}\,. \tag{3.12}$$

Under the condition

$$A(z) = 0 \quad \Rightarrow \quad |z| > 1$$

the autoregressive polynomial $A(L)$ is causally invertible with absolutely summable coefficients according to Proposition 3.3:

$$\frac{1}{A(L)} = \sum_{j=0}^{\infty} \alpha_j L^j, \quad \sum_{j=0}^{\infty} |\alpha_j| < \infty.$$

Therefore, under the condition of invertibility it holds that

$$x_t = \frac{\nu + \varepsilon_t}{A(L)} = \frac{\nu}{A(1)} + \sum_{j=0}^{\infty} \alpha_j \varepsilon_{t-j},$$

and $\{x_t\}$ is a stationary MA(∞) process, see Proposition 3.2. Hence, the autocovariance sequence is absolutely summable. Furthermore, in this case it holds for $h > 0$ (w.l.o.g.[4] we set $\nu = 0$ for simplification),

$$\begin{aligned}
\gamma(h) &= \mathrm{Cov}(x_t, x_{t+h}) \\
&= \mathrm{E}(x_t x_{t+h}) \\
&= \mathrm{E}\left(x_t(a_1 x_{t+h-1} + \ldots + a_p x_{t+h-p} + \varepsilon_{t+h})\right) \\
&= a_1\gamma(h-1) + \ldots + a_p\gamma(h-p) + 0.
\end{aligned}$$

Dividing by $\gamma(0)$ yields the recursive relation from the subsequent proposition: The autocorrelations are given by a deterministic difference equation of order p. The still missing *necessary* condition of stationarity from Proposition 3.4 ($A(1) > 0$) will be derived in Problem 3.6.

Proposition 3.4 (AR(p)) *Let $\{x_t\}$ be an AR(p) process from (3.12) with index set*[5] $\mathbb{T} = \{-\infty, \ldots, T\}$.

[4]The abbreviation stands for "without loss of generality". It is frequently used for assumptions that are substantially not necessary and that are only made to simplify the argument or the notation. In the example at hand, generally it would have to be written $(x_{t-j} - \mu)$ for all j; just as well, one can set $\mu = \nu/A(1)$ equal to zero and simply write x_{t-j}.

[5]Also in the following, the notation $\{-\infty, \ldots, T\}$ is always to denote the set of all integers without $\{T+1, T+2, \ldots\}$.

(a) *The process has an absolutely summable MA(∞) representation according to (3.2) if and only if it holds that*

$$A(z) = 0 \quad \Rightarrow \quad |z| > 1 .$$

Then the process is stationary with expectation $\mu = \nu/A(1)$. *The condition* $A(1) = 1 - \sum_{j=1}^{p} a_j > 0$ *is necessary for this.*

(b) *For stationary processes the autocorrelation sequence is absolutely summable where it holds that* $\rho(h) \neq 0$ *for all integer numbers h and*

$$\rho(h) = a_1\, \rho(h-1) + \ldots + a_p\, \rho(h-p)\,, \quad h > 0 .$$

Again note that the absolute summability of $\rho(h)$ implies: $\rho(h) \to 0$ for $h \to \infty$. The farther x_t and x_{t+h} are apart from each other the weaker tends to be their correlation.

Certain properties of the AR(1) process are lost for $p > 1$. In Problem 3.8 we show for a special case ($p = 2$) that the AR(p) process, $p > 1$, is not a Markov process in general.

AR(2)

Let $p = 2$,

$$x_t = \nu + a_1\, x_{t-1} + a_2\, x_{t-2} + \varepsilon_t\,,$$

with the autoregressive polynomial

$$A(L) = 1 - a_1 L - a_2 L^2 .$$

From Corollary 3.1, we know the conditions under which $(A(L))^{-1}$ can be expanded as an absolutely summable filter:

$$\begin{aligned} &(i) \quad 1 + a_2 \quad > 0\,, \\ \text{and } &(ii)\ 1 + a_1 - a_2 > 0\,, \\ \text{and } &(iii)\ 1 - a_1 - a_2 > 0\,. \end{aligned}$$

Consequently, under these three parameter restrictions the AR(2) process is stationary. The restrictions become even more obvious if they are solved for a_2 and depicted in a coordinate system:

$$\begin{aligned} &(i)\ a_2 > \ -1\,, \\ \text{and } &(ii)\ a_2 < 1 + a_1\,, \\ \text{and } &(iii)\ a_2 < 1 - a_1\,. \end{aligned}$$

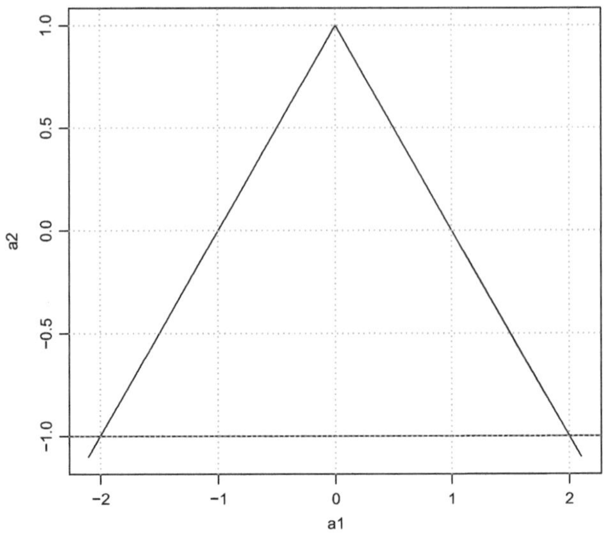

Fig. 3.3 Stationarity triangle for AR(2) processes

So, three lines are given, above or below which a_2 has to be, respectively. This is why one also talks about the stability or stationarity triangle, see Fig. 3.3. Within the triangle lies the stationarity region.

The autocorrelation series of the AR(2) case is determined from Proposition 3.4. For $h = 0$, it naturally holds that $\rho(0) = 1$. For $h = 1$, one obtains

$$\rho(1) = a_1\,\rho(0) + a_2\,\rho(-1).$$

Because of the symmetry, $\rho(-h) = \rho(h)$, it follows that

$$\rho(1) = \frac{a_1}{1 - a_2}.$$

Similarly, it follows that

$$\begin{aligned}\rho(2) &= a_1\,\rho(1) + a_2\,\rho(0) \\ &= \frac{a_1^2}{1 - a_2} + a_2.\end{aligned}$$

By repeated insertion into the second order difference equation,

$$\rho(h) = a_1\,\rho(h-1) + a_2\,\rho(h-2)\,, \quad h \geq 2\,,$$

the entire autocorrelation sequence is determined. Next, four numerical examples will be considered.

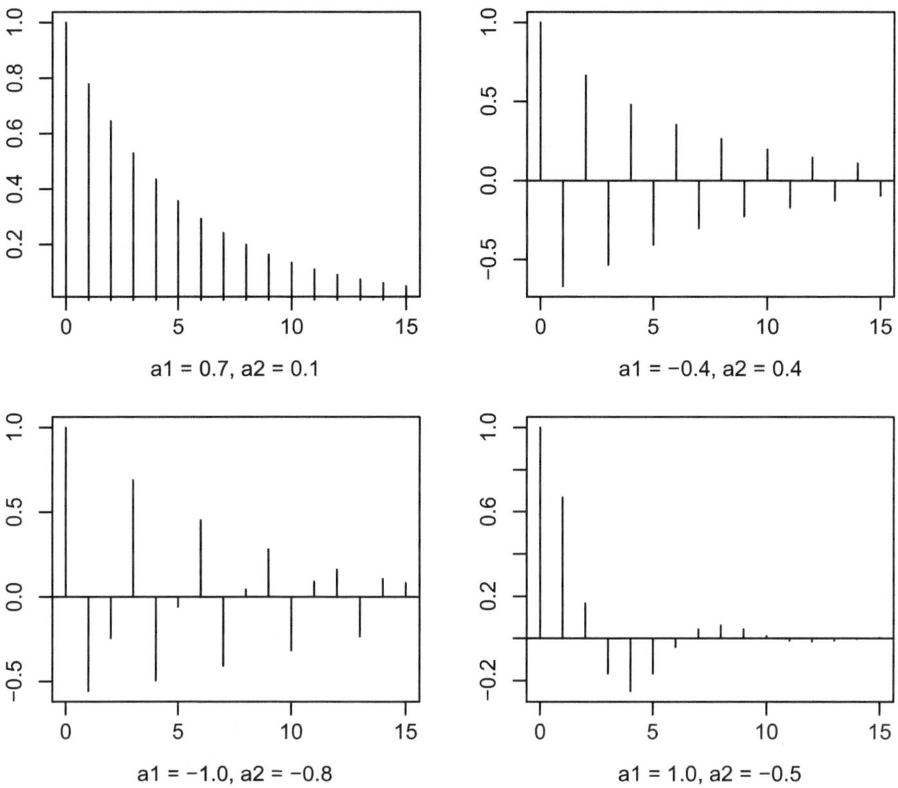

Fig. 3.4 Autocorrelograms for AR(2) processes

Example 3.5 (AR(2)) We now consider four numerical examples for AR(2) processes in order to characterize typical patterns of autocorrelation. In all cases, the stability conditions can be proven to be fulfilled. The corresponding autocorrelograms can be found in Fig. 3.4.

(i) $a_1 = 0.7$, $a_2 = 0.1$: In this case, all the autocorrelations are positive and they converge to zero with h; their behavior is similar to the autocorrelogram of a AR(1) process with $a_1 > 0$, see Fig. 3.2.

(ii) $a_1 = -0.4$, $a_2 = 0.4$: Starting with $\rho(1) < 0$, the autocorrelations alternate similarly to an AR(1) process with $a_1 < 0$, cf. Fig. 3.2.

(iii) $a_1 = -1.0$, $a_2 = -0.8$: In this case, we find a dynamic which cannot be generated by an AR(1) model; two negative autocorrelations of the first and second order are followed by a seemingly irregularly alternating pattern.

(iv) $a_1 = 1.0$, $a_2 = -0.5$: In the last case, the autocorrelations swing from the positive area to the negative area, then to the positive one and again to the negative one whereas the last-mentioned can hardly be perceived because of the small absolute values.

Therefore, the cases (iii) and (iv) show that the AR(2) process allows for richer dynamics and dependence structures than the simpler AR(1) process. ■

Autoregressive Moving Average Processes

Now, we consider a combination of AR and MA processes. Again, $\{x_t\}$ is given by a stochastic difference equation of order p, only that ε_t from (3.12) is replaced by an MA(q) process:

$$x_t = \nu + a_1 x_{t-1} + \cdots + a_p x_{t-p} + \varepsilon_t + b_1 \varepsilon_{t-1} + \cdots + b_q \varepsilon_{t-q}, \quad t \in \mathbb{T}.$$

Abbreviating, we talk about **ARMA(p, q) processes** assuming $a_p \neq 0$ and $b_q \neq 0$. Again, a more compact representation follows by using lag polynomials,

$$A(L)\, x_t = \nu + B(L)\, \varepsilon_t, \quad t \in \mathbb{T}, \tag{3.13}$$

where it is assumed that both polynomials

$$A(L) = 1 - a_1 L - \cdots - a_p L^p \text{ and } B(L) = 1 + b_1 L + \cdots + b_q L^q$$

do not have common roots.

A stationary MA(∞) representation hinges on the autoregressive polynomial such that the stationarity condition can be adopted from the pure AR(p) case in Proposition 3.4. It amounts to an absolutely summable expansion of $(A(L))^{-1}$ such that the process possesses an absolutely summable representation as MA(∞) process, see (3.2). If stationarity is given, the absolutely summable autocorrelation sequence can again be determined from a stable difference equation (see e.g. Brockwell & Davis, 1991, p. 93)

Proposition 3.5 (ARMA(p, q)) *Let $\{x_t\}$ be an ARMA(p, q) process from (3.13) with $\mathbb{T} = \{-\infty, \ldots, T\}$.*

(a) The process has an absolutely summable MA(∞) representation according to (3.2) if and only if it holds that

$$A(z) = 0 \quad \Rightarrow \quad |z| > 1.$$

Then the process is stationary with expectation $\mu = \nu / A(1)$. The condition $A(1) = 1 - \sum_{j=1}^{p} a_j > 0$ is necessary for this.

(b) For stationary processes the autocorrelation sequence is absolutely summable where it holds that $\rho(h) \neq 0$ for all integer numbers h and

$$\rho(h) = a_1 \rho(h-1) + \ldots + a_p \rho(h-p), \quad h \geq \max(p, q+1).$$

For $h \geq \max(p, q+1)$ the autocorrelations satisfy the same difference equation as in the pure AR(p) case. It is known that the solution to such a difference equation is bounded exponentially. Consequently, we have the following result, see e.g. Brockwell and Davis (1991, Prob. 3.11): For a stationary ARMA process there exist positive constants c and g with $0 < g < 1$, such that

$$|\rho(h)| \leq c\, g^h \tag{3.14}$$

for all $h = 1, 2, \ldots$. Hence, the decay rate of the autocorrelations is bounded exponentially. This shows again that stationary ARMA process are characterized be absolutely summable autocorrelations, see Problem 3.2. Proposition 3.5 will be discussed below for $p = q = 1$.

Before turning to the ARMA(1,1) case, we note that an analogous result to Proposition 3.5 (a) is available for an AR(∞) representation according to Proposition 3.3: The ARMA process has an absolutely summable AR(∞) representation (see Example 3.3) if and only if

$$B(z) = 0 \quad \Rightarrow \quad |z| > 1\,.$$

In this case the ARMA process is called invertible.

ARMA(1,1)

We now wish to obtain the autocorrelation structure of the ARMA(1,1) process,

$$x_t = a\,x_{t-1} + \varepsilon_t + b\,\varepsilon_{t-1}, \qquad |a| < 1,\ |b| < 1\,,$$

where $|a| < 1$ for stationarity, and $|b| < 1$ to ensure invertibility (see Example 3.3). The condition of no common roots of $1 - aL$ and $1 + bL$ is given for $a \neq -b$. In the case of common roots, the lag polynomials could be reduced and one would obtain

$$x_t = \frac{1 + bL}{1 + bL}\,\varepsilon_t = \varepsilon_t \quad \text{for} \quad a = -b\,.$$

Due to the invertibility of $1 - aL$, the process can be formulated as an infinite MA process,

$$\begin{aligned} x_t &= \frac{\varepsilon_t + b\,\varepsilon_{t-1}}{1 - aL} \\ &= \sum_{j=0}^{\infty} a^j\,\varepsilon_{t-j} + b\sum_{j=0}^{\infty} a^j\,\varepsilon_{t-1-j} \\ &= \varepsilon_t + \sum_{j=1}^{\infty}\left(a^j + b\,a^{j-1}\right)\varepsilon_{t-j}\,, \end{aligned}$$

where a shift of subscripts was carried out. In the notation of (3.2) the MA(∞) coefficients result as

$$c_j = a^{j-1}(a+b)\,, \quad j \geq 1\,.$$

In this way Proposition 3.2 yields for the variance:

$$\gamma(0) = \sigma^2 \left(1 + \sum_{j=1}^{\infty} a^{2j-2}\,(a+b)^2 \right) = \ \ldots \ = \sigma^2 \, \frac{(1+b^2+2ab)}{1-a^2}\,.$$

The autocovariance at lag one follows in the same way,

$$\gamma(1) = \sigma^2 \frac{(a+b)(1+ab)}{1-a^2},$$

such that it holds:

$$\rho(1) = \frac{(a+b)(1+ab)}{1+b^2+2ab}\,.$$

Furthermore we learn from Proposition 3.5:

$$\rho(h) = a\,\rho(h-1), \quad h \geq 2.$$

Hence, the MA(1) component (that is b) influences directly only $\rho(1)$ having only an indirect effect beyond the autocorrelation of the first order: For $h \geq 2$ a recursive relation between $\rho(h)$ and $\rho(h-1)$ holds true just as it applies to the pure AR(1) process. Thus, this yields four typical patterns. In order to identify these, it suffices entirely to concentrate on the numerator of $\rho(1)$ as well as on the sign of a as the denominator of $\rho(1)$ is always positive because it is a multiple of the variance. As $1+ab$ is positive due to the stationarity and the invertibility of the MA polynomial, the behavior of the autocorrelogram depends on the signs of $a+b$ and a only. The exponential bound for the autocorrelations is easily verified:

$$\rho(h) = a\,\rho(h-1) = \cdots = a^{h-1}\rho(1)\,. \quad h \geq 2.$$

Therefore, (3.14) applies with $g = |a|$ and $c = |\rho(1)|/|a|$.

Example 3.6 (ARMA(1,1)) The four possible patterns of the autocorrelogram of a stationary and invertible ARMA(1,1) model will be discussed and illustrated by numerical examples, cf. Fig. 3.5.

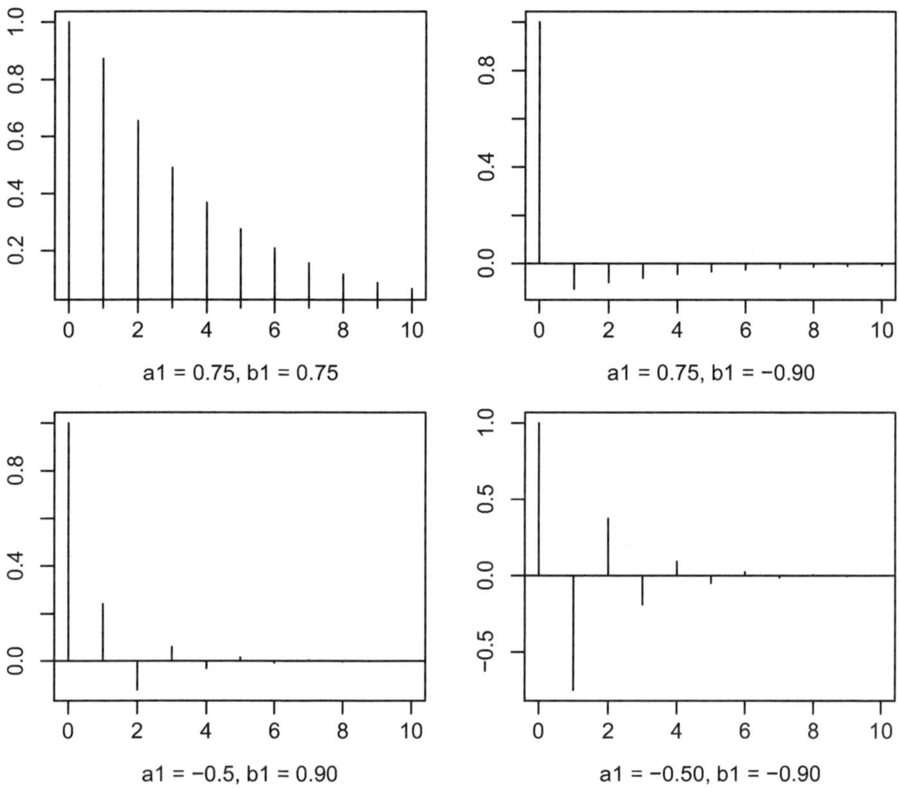

Fig. 3.5 Autocorrelograms for ARMA(1,1) processes

Case 1: Obviously, for $a + b > 0$ it holds that $\rho(1) > 0$. If, furthermore, $a > 0$ then the entire autocorrelogram proceeds above the zero line.

Case 2: For $a + b < 0$ and $a > 0$ one obtains exclusively negative autocorrelations.

Case 3: An alternating pattern, starting with $\rho(1) > 0$, is obtained for $a < 0$ and $a > -b$.

Case 4: For $a < 0$ and $a < -b$ the autocorrelation series is as well alternating but starts with a negative value.

Note that cases 1 and 4 can be generated qualitatively by a pure AR(1) process as well. For cases 2 and 3, however, there occur patterns which cannot be produced by an AR(1) process. Therefore, the ARMA(1,1) process allows for richer dynamic modeling than the AR(1) model does. Also when comparing with the AR(2) case, we find that the ARMA(1,1) model allows for additional dynamics. ■

3.5 Problems and Solutions

Problems

3.1 Where does the first order autocorrelation of an MA(1) process ($\rho(1) = \frac{b}{1+b^2}$) have its maximum and minimum?

3.2 Show for $g \in \mathbb{R} \setminus \{1\}$ (geometric series):

$$\sum_{i=0}^{n} g^i = \frac{1 - g^{n+1}}{1 - g}.$$

Conclusion: For $|g| < 1$ and $n \to \infty$ it holds that:

$$\sum_{i=0}^{\infty} g^i = \frac{1}{1 - g}.$$

3.3 Prove part (b) from Proposition 3.2.

3.4 Derive the series expansion

$$(1 - aL)^{-1} = \sum_{j=0}^{\infty} a^j L^j$$

for real a with $|a| < 1$.

3.5 Prove Proposition 3.3 (b).

3.6 Prove the necessary condition of causal invertibility from Proposition 3.4, that is:

$$\{A(z) = 0 \Rightarrow |z| > 1\} \Longrightarrow \{A(1) > 0\} ,$$

where $A(z) = 1 - a_1 z - \ldots - a_p z^p$.

3.7 Let $\{\varepsilon_t\} \sim \text{WN}(0, \sigma^2)$ be a Gaussian process. Show that $\{x_t\}$ with $x_t = a_1 x_{t-1} + \varepsilon_t$ is a Markov process.

3.8 Let $\{\varepsilon_t\} \sim \text{WN}(0, \sigma^2)$. Show that $\{x_t\}$ with $x_t = a_2 x_{t-2} + \varepsilon_t$ is not a Markov process ($a_2 \neq 0$).

Solutions

3.1 The traditional way to solve the problem is curve sketching. We consider the first order autocorrelation to be a function of b:

$$f(b) = \rho(1) = \frac{b}{1+b^2}.$$

Then the quotient rule for the first order derivative yields

$$f'(b) = \frac{1+b^2-2b^2}{(1+b^2)^2} = \frac{(1-b)(1+b)}{(1+b^2)^2}.$$

The roots of the derivative are given by $|b| = 1$. In $b = -1$ there is a change of sign of $f'(b)$, namely from a negative to a positive slope. Hence, in $b = -1$ there is a relative (and also an absolute) minimum. Because of $f(b)$ being an odd function (symmetric about the origin), there is a maximum in $b = 1$. Therefore, the maximum possible correlation in absolute value is

$$|f(-1)| = f(1) = \frac{1}{2}.$$

One may also tackle the problem by more elementary means. Note that for $b \neq 0$

$$\frac{1}{|b|} - 2 + |b| = \left(\frac{1}{\sqrt{|b|}} - \sqrt{|b|}\right)^2 \geq 0,$$

which is equivalent to

$$\frac{1}{2} \geq \frac{1}{\frac{1}{|b|} + |b|} = \frac{|b|}{1+b^2} = |f(b)|,$$

with $f(b) = \rho(1)$ defined above. Since $|f(-1)| = f(1) = \frac{1}{2}$, this solves the problem.

3.2 By S_n we denote the following sum for finite n:

$$S_n = \sum_{i=0}^{n} g^i = 1 + g + \ldots + g^{n-1} + g^n.$$

Multiplication by g yields

$$g\, S_n = g + g^2 + \ldots + g^n + g^{n+1}.$$

Therefore, it holds that

$$S_n - g\, S_n = 1 - g^{n+1}.$$

By ordinary factorization the formula

$$S_n = \frac{1 - g^{n+1}}{1 - g}$$

and therefore the claim is verified.

3.3 The absolute summability of $\gamma(h)$ follows from the absolute summability of the linear coefficients $\{c_j\}$ allowing for a change of the order of summation. In order to do so, we first apply the triangle inequality:

$$\begin{aligned}
\frac{1}{\sigma^2}\sum_{h=0}^{\infty}|\gamma(h)| &= \sum_{h=0}^{\infty}\left|\sum_{j=0}^{\infty} c_j\, c_{j+h}\right| \\
&\le \sum_{h=0}^{\infty}\sum_{j=0}^{\infty}\left|c_j\, c_{j+h}\right| = \sum_{h=0}^{\infty}\sum_{j=0}^{\infty}\left|c_j\right|\,\left|c_{j+h}\right| \\
&= \sum_{j=0}^{\infty}\left|c_j\right|\left(\sum_{h=0}^{\infty}\left|c_{j+h}\right|\right),
\end{aligned}$$

where at the end round brackets were placed for reasons of clarity. The final term is further bounded by enlarging the expression in brackets:

$$\sum_{j=0}^{\infty}\left|c_j\right|\left(\sum_{h=0}^{\infty}\left|c_{j+h}\right|\right) \le \sum_{j=0}^{\infty}\left|c_j\right|\left(\sum_{h=0}^{\infty}\left|c_h\right|\right).$$

Therefore, the claim follows indeed from the absolute summability of $\{c_j\}$.

3.4 For the proof we denote $(1 - aL)^{-1}$ as $\sum_{j=0}^{\infty}\alpha_j L^j$,

$$\frac{1}{1 - aL} = \sum_{j=0}^{\infty}\alpha_j L^j\,,$$

and determine the coefficients α_j. By multiplying this equation with $1 - aL$, we obtain

$$\begin{aligned}
1 &= (1 - aL) \sum_{j=0}^{\infty} \alpha_j L^j \\
&= \alpha_0 + \alpha_1 L^1 + \alpha_2 L^2 + \ldots \\
&\quad - a\,\alpha_0 L^1 - a\,\alpha_1 L^2 - a\,\alpha_2 L^3 - \ldots .
\end{aligned}$$

Now, we compare the coefficients associated with L^j on the left- and on the right-hand side:

$$\begin{aligned}
1 &= \alpha_0 \,, \\
0 &= \alpha_1 - a\,\alpha_0 \,, \\
0 &= \alpha_2 - a\,\alpha_1 \,, \\
&\vdots \\
0 &= \alpha_j - a\,\alpha_{j-1} \,, \quad j \geq 1 \,.
\end{aligned}$$

As claimed, the solution of the difference equation obtained in this way, ($\alpha_j = a\,\alpha_{j-1}$), is obviously $\alpha_j = a^j$.

3.5 We factorize $P(z) = 1 + b_1 z + \ldots + b_p z^p$ with roots $z_1, \ldots, z_p$ of this polynomial (fundamental theorem of algebra):

$$P(z) = b_p\,(z - z_1) \ldots (z - z_p)\,.$$

From each bracket we factorize $-z_j$ out such that

$$P(z) = b_p\,(-1)^p\,z_1\,\ldots z_p \left(1 - \frac{z}{z_1}\right) \ldots \left(1 - \frac{z}{z_p}\right).$$

Because of $P(0) = 1$, we obtain $b_p\,(-1)^p\,z_1\,\ldots z_p = 1$. Therefore the factorization simplifies to

$$\begin{aligned}
P(z) &= \left(1 - \frac{z}{z_1}\right) \ldots \left(1 - \frac{z}{z_p}\right) \\
&= P_1(z) \cdots P_p(z)\,,
\end{aligned}$$

with

$$P_k(z) = 1 - \frac{z}{z_k} = 1 - \zeta_k z\,, \quad k = 1, \ldots, p\,,$$

where $\zeta_k = 1/z_k$. From part a) we know that

$$\frac{1}{P_k(L)} = \sum_{j=0}^{\infty} \zeta_k^j L^j \quad \text{with} \sum_{j=0}^{\infty} |\zeta_k^j| < \infty$$

if and only if

$$|z_k| = \frac{1}{|\zeta_k|} > 1 .$$

Now, consider the convolution (sometimes called Cauchy product) for $k \neq \ell$:

$$\frac{1}{P_k(L)} \frac{1}{P_\ell(L)} = \sum_{j=0}^{\infty} c_j L^j$$

with

$$c_j := \sum_{i=0}^{j} \zeta_k^i \zeta_\ell^{j-i} .$$

We have $\sum_{j=0}^{\infty} |c_j| < \infty$ if and only if both $P_k^{-1}(L)$ and $P_\ell^{-1}(L)$ are absolutely summable, which holds true if and only if

$$|z_k| > 1 \quad \text{and} \quad |z_\ell| > 1 .$$

Repeating this argument we obtain that

$$\frac{1}{P(L)} = \frac{1}{P_1(L)} \cdots \frac{1}{P_p(L)} = \sum_{j=0}^{\infty} c_j L^j \quad \text{with} \sum_{j=0}^{\infty} |c_j| < \infty$$

if and only if (3.10) holds. Quod erat demonstrandum.

3.6 At first we reformulate the autoregressive polynomial $A(z) = 1 - a_1 z - \ldots - a_p z^p$ in its factorized form with roots $z_1, \ldots, z_p$ (again by the fundamental theorem of algebra):

$$A(z) = -a_p(z - z_1) \ldots (z - z_p) .$$

For $z = 1$ this amounts to

$$A(1) = -a_p(1 - z_1) \ldots (1 - z_p) . \tag{3.15}$$

Because of $A(0) = 1$ we obtain as well:

$$1 = -a_p(-1)^p z_1 \ldots z_p \,. \tag{3.16}$$

Now we proceed in two steps, treating the cases of complex and real roots separately.

(A) Complex roots: Note that for a root $z_1 \in \mathbb{C}$ it holds that the complex conjugate, $z_2 = \overline{z_1}$, is a root as well. Then calculating with complex numbers yields for the product

$$\begin{aligned}(1-z_1)(1-z_2) &= (1-z_1)(1-\overline{z_1}) \\ &= (1-z_1)(\overline{1-z_1}) \\ &= |1-z_1|^2 > 0\,.\end{aligned}$$

Hence, for $p > 2$, complex roots contribute positively to $A(1)$ in (3.15). If $p = 2$, the roots are only complex if $a_2 < 0$, since the discriminant is $a_1^2 + 4a_2$; hence, $A(1) > 0$ by (3.15).

(B) Since the effect of complex roots is positive, we now concentrate on real roots z_i, for which it holds that $|z_i| > 1$ by assumption. So, we assume without loss of generality that the polynomial has no complex roots, or that all complex roots have been factored out. Two sub-cases have to be distinguished. (1) Even degree: For an even p we again distinguish between two cases. Case 1, $a_p > 0$: Because of (3.16) there has to be an odd number of negative roots and therefore there has to be an odd number of positive roots as well. For the latter it holds that $(1 - z_i) < 0$ while the first naturally fulfill $(1 - z_i) > 0$. Hence, as claimed, it follows from (3.15) that $A(1)$ is positive. Case 2, $a_p < 0$: In this case one argues quite analogously. Because of (3.16) there is an even number of positive and negative roots such that the requested claim follows from (3.15) as well. (2) Odd degree: For an odd p one obtains the requested result as well by distinction of the two cases for the sign of a_p. We omit details.

Hence, the proof is complete.

3.7 The normality of $\{\varepsilon_t\}$ implies a multivariate Gaussian distribution of

$$\begin{pmatrix}\varepsilon_{t+1}\\ \vdots \\ \varepsilon_{t+s}\end{pmatrix} \sim \text{ii}\,\mathcal{N}_s\left(\begin{pmatrix}0\\ \vdots \\ 0\end{pmatrix}, \sigma^2 I_s\right)$$

with the identity matrix I_s of dimension s. The s-fold substitution yields

$$\begin{aligned}x_{t+s} &= a_1^s x_t + a_1^{s-1}\varepsilon_{t+1} + \ldots + a_1\varepsilon_{t+s-1} + \varepsilon_{t+s} \\ &= a_1^s x_t + \sum_{i=0}^{s-1} a_1^i \varepsilon_{t+s-i}.\end{aligned}$$

The sum over the white noise process has the moments

$$\mathrm{E}\left(\sum_{i=0}^{s-1} a_1^i \varepsilon_{t+s-i}\right) = 0, \quad \mathrm{Var}\left(\sum_{i=0}^{s-1} a_1^i \varepsilon_{t+s-i}\right) = \sigma^2 \sum_{i=0}^{s-1} a_1^{2i},$$

and, furthermore, it is normally distributed:

$$\sum_{i=0}^{s-1} a_1^i \varepsilon_{t+s-i} \sim \mathcal{N}\left(0, \sigma^2 \sum_{i=0}^{s-1} a_1^{2i}\right).$$

Hence, x_{t+s} given x_t follows a Gaussian distribution with the corresponding moments:

$$x_{t+s} | x_t \sim \mathcal{N}\left(a_1^s x_t, \sigma^2 \sum_{i=0}^{s-1} a_1^{2i}\right).$$

As x_{t+s} can be expressed as a function of x_t and $\varepsilon_{t+1}, \ldots, \varepsilon_{t+s}$ alone, the further past of the process does not matter for the conditional distribution of x_{t+s}. Therefore, for the entire information $\mathcal{I}_t$ up to time t it holds that:

$$x_{t+s} | \mathcal{I}_t \sim \mathcal{N}\left(a_1^s x_t, \sigma^2 \sum_{i=0}^{s-1} a_1^{2i}\right).$$

Hence, the Markov property (2.9) has been shown. It holds independently of the concrete value of a_1.

3.8 For

$$x_t = a_2 x_{t-2} + \varepsilon_t,$$

we obtain for $s = 1$ the conditional expectations $\mathrm{E}(x_{t+1} | \mathcal{I}_t) = a_2 x_{t-1}$, and

$$\mathrm{E}(x_{t+1} | x_t) = \mathrm{E}(a_2 x_{t-1} + \varepsilon_{t+1} | x_t) = a_2 \mathrm{E}(x_{t-1} | x_t),$$

with $\mathcal{I}_t = \sigma(x_t, x_{t-1}, \ldots, x_1)$. As the conditional expectations are not equivalent, the conditional distributions are not the same. Hence, it generally holds that

$$\mathrm{P}(x_{t+1} \leq x | x_t) \neq \mathrm{P}(x_{t+1} \leq x | \mathcal{I}_t),$$

which proves that $\{x_t\}$ is not a Markov process.

References

Andrews, D. W. K., & Chen, H.-Y. (1994). Approximately median-unbiased estimation of autoregressive models. *Journal of Business & Economic Statistics, 12*, 187–204.

Brockwell, P. J., & Davis, R. A. (1991). *Time series: Theory and methods* (2nd ed.). New York: Springer.

Campbell, J. Y., & Mankiw, N. G. (1987). Are output fluctuations transitory? *Quarterly Journal of Economics, 102*, 857–880.

Fuller, W. A. (1996). *Introduction to statistical time series* (2nd ed.). New York: Wiley.

Sydsæter, K., Strøm, A., & Berck, P. (1999). *Economists' mathematical manual* (3rd ed.). Berlin/New York: Springer.

Wold, H. O. A. (1938). *A study in the analysis of stationary time series*. Stockholm: Almquist & Wiksell.

References

Spectra of Stationary Processes

4

4.1 Summary

Spectral analysis (or analysis in the frequency domain) aims at detecting cyclical movements in a time series. These may originate from seasonality, a trend component or from a business cycle. The theoretical spectrum of a stationary process is the quantity measuring how strongly cycles with a certain period, or frequency, account for total variance. Typically, elaborations on spectral analysis are formally demanding requiring e.g. knowledge of complex numbers and Fourier transformations. In this textbook we have tried for a way of presenting and deriving the relevant results being less elegant but in return managing with less mathematical burden. The next section provides the definitions and intuition behind spectral analysis. Section 4.3 is analytically more demanding containing some general theory. This theory is exemplified with the discussion of spectra from particular ARMA processes, hence building on the previous chapter.

4.2 Definition and Interpretation

In this chapter we assume the most general case considered previously, i.e. the infinite MA process that is only square summable, $\{x_t\}_{t\in\mathbb{T}}$, $\mathbb{T} \subseteq \mathbb{Z}$,

$$x_t = \mu + \sum_{j=0}^{\infty} c_j \, \varepsilon_{t-j} \,, \quad \sum_{j=0}^{\infty} c_j^2 < \infty \,, \; c_0 = 1 \,, \tag{4.1}$$

with $\{\varepsilon_t\} \sim \mathrm{WN}(0, \sigma^2)$. The autocovariances,

$$\gamma(h) = \mathrm{Cov}(x_t, x_{t+h}) = \gamma(-h) \,, \quad h \in \mathbb{Z} \,,$$

U. Hassler, *Stochastic Processes and Calculus*, Springer Texts in Business and Economics, DOI 10.1007/978-3-319-23428-1_4

are given in Proposition 3.2 (a). We do not assume that $\{c_j\}$ and hence $\{\gamma(h)\}$ are absolutely summable,[1] simply because this will not hold under long memory treated in the next chapter. We wish to construct a function f that allows to express the autocovariances as weighted cosine waves of different periodicity,[2]

$$\gamma(h) = \int_{-\pi}^{\pi} \cos(\lambda h) f(\lambda)\, d\lambda \,.$$

The basic ingredient of an analysis of periodicity is the **cosine cycle** whose properties we want to recall as an introduction.

Periodic Cycles

By $c_\lambda(t)$ we denote the cycle based on the cosine,[3]

$$c_\lambda(t) = \cos(\lambda t) \;, \quad t \in \mathbb{R},$$

where λ with $\lambda > 0$ is called **frequency**. The frequency is inversely related to the **period** P,

$$P = \frac{2\pi}{\lambda}.$$

For $\lambda = 1$ one obtains the cosine function which is 2π–periodic and even (symmetric about the ordinate):

$$c_1(t) = \cos(t) = \cos(t + 2\pi) = c_1(t + 2\pi) \,,$$

$$c_1(-t) = \cos(-t) = \cos(t) = c_1(t).$$

More generally, it holds with $P = 2\pi/\lambda$ that:

$$c_\lambda(t) = \cos(\lambda t) = \cos(\lambda t + 2\pi) = \cos(\lambda(t + P)) = c_\lambda(t + P) \,.$$

Therefore the cosine cycle $c_\lambda(t)$ with frequency λ has the period $P = 2\pi/\lambda$. Of course, the symmetry of $c_1(t)$ carries over:

$$c_\lambda(t) = c_\lambda(-t) \,.$$

[1] The assumption of absolute summability underlies most textbooks when it comes to spectral analysis, see e.g. Hamilton (1994) or Fuller (1996).

[2] From Brockwell and Davis (1991, Coro. 4.3.1) in connection with Brockwell and Davis (Thm. 5.7.2) one knows that such an expression exists.

[3] Here, the so-called amplitude is equal to one ($|c_\lambda(t)| \leq 1$), and the phase shift is zero ($c_\lambda(0) = 1$).

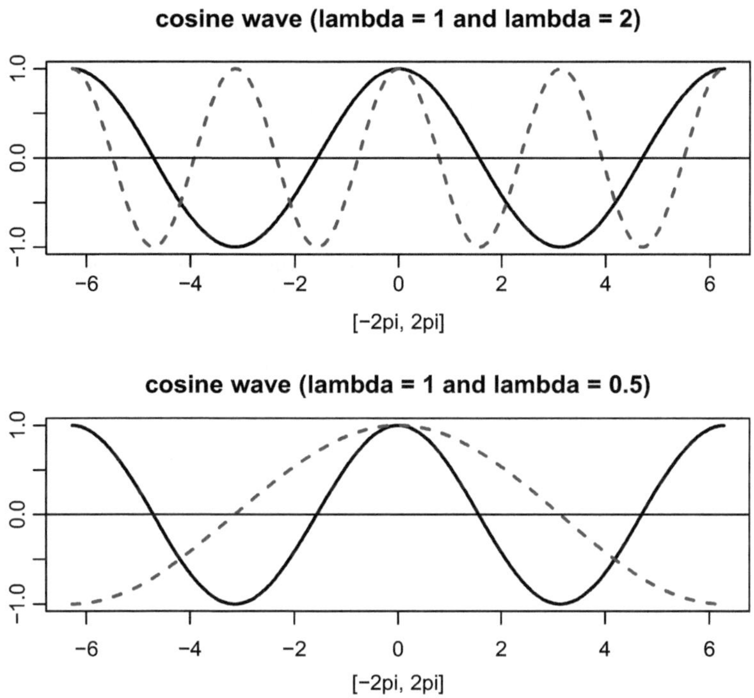

Fig. 4.1 Cosine cycle with different frequencies

For $\lambda = 1$, $\lambda = 2$ and $\lambda = 0.5$ these properties are graphically illustrated in Fig. 4.1.
Finally, remember the derivative of the cosine,

$$\frac{d\,c_\lambda(t)}{d\,t} = c'_\lambda(t) = -\lambda\ \sin(\lambda t)\,,$$

which we will use repeatedly.

Definition

For convenience, we now rephrase the MA(∞) process in terms of the lag polynomial $C(L)$ of infinite order,

$$x_t = \mu + C(L)\,\varepsilon_t \quad \text{with} \quad C(L) = \sum_{j=0}^{\infty} c_j L^j\,.$$

Next, we define the so-called **power transfer function** $T_C(\lambda)$ of this polynomial:[4]

$$T_C(\lambda) = \sum_{j=0}^{\infty} c_j^2 + 2\sum_{h=1}^{\infty}\sum_{j=0}^{\infty} c_j c_{j+h} \cos(\lambda h), \quad \lambda \in \mathbb{R} \setminus \{\lambda^*\}. \tag{4.2}$$

Note that T_C may not exist everywhere, there may be singularities at some frequency λ^* such that $T_C(\lambda)$ goes off to infinity as $\lambda \to \lambda^*$; but at least the power transfer function is integrable. The key result in Proposition 4.1 (e) is from from Brockwell and Davis (1991, Coro. 4.3.1, Thm. 5.7.2); it will be proved explicitly in Problem 4.1 under the simplifying assumption of absolute summability. The first four statements in the following proposition are rather straightforward and will be justified below.

Proposition 4.1 (Spectrum) *Define for $\{x_t\}$ from (4.1) the spectrum*

$$f(\lambda) = T_C(\lambda)\,\frac{\sigma^2}{2\pi}.$$

It has the following properties:

(a) $f(-\lambda) = f(\lambda)$,
(b) $f(\lambda) = f(\lambda + 2\pi)$,
(c) $f(\lambda) \geq 0$,
(d) $f(\lambda)$ is continuous in λ under absolute summability, $\sum_j |c_j| < \infty$.
(e) For all $h \in \mathbb{Z}$:

$$\gamma(h) = \int_{-\pi}^{\pi} f(\lambda)\cos(\lambda h)\,d\lambda = 2\int_0^{\pi} f(\lambda)\cos(\lambda h)\,d\lambda.$$

Substituting the autocovariance expression from Proposition 3.2 into (4.2), the following representation of the spectrum exists:

$$f(\lambda) = \frac{\gamma(0)}{2\pi} + \frac{2}{2\pi}\sum_{h=1}^{\infty}\gamma(h)\cos(\lambda h) = \frac{1}{2\pi}\sum_{h=-\infty}^{\infty}\gamma(h)\cos(\lambda h). \tag{4.3}$$

The symmetry of the spectrum in Proposition 4.1 (a) immediately follows from the symmetry of the cosine function. From the periodicity of the cosine, (b) follows as well. Both results jointly explain why the spectrum is normally considered on the restricted domain $[0, \pi]$ only. Property (c) follows from the definition of the power transfer function, see Footnote 6 below. Finally, the continuity of $f(\lambda)$ claimed in

[4] A more detailed and technical exposition is reserved for the next section. Our expression in (4.2) can be derived from the expression in Brockwell and Davis (1991, eq. 5.7.9), which is given in terms of complex numbers.

(d) under absolute summability results from uniform convergence, see Fuller (1996, Thm. 3.1.9).

We call the function f (or f_x, if we want to emphasize that $\{x_t\}$ is the underlying process) the **spectrum** of $\{x_t\}$. Frequently, one also talks about spectral density or spectral density function as f is a non-negative function which could be standardized in such a way that the area beneath it would be equal to one.

Interpretation

The usual interpretation of the spectrum is based on Proposition 4.1. Result (e) and (4.3) jointly show the spectrum and the autocovariance series to result from each other. In a sense, spectrum and autocovariances are two sides of the same coin. The spectrum can be determined from the autocovariances by definition and having the spectrum, Proposition 4.1 provides the autocovariances. The case $h = 0$ with

$$\text{Var}(x_t) = \gamma(0) = \int_{-\pi}^{\pi} f(\lambda)\,d\lambda = 2\int_{0}^{\pi} f(\lambda)\,d\lambda$$

is particularly interesting. This equation implies: The spectrum at λ_0 measures how strongly the cycle with frequency λ_0 and therefore of period $P_0 = 2\pi/\lambda_0$ adds to the variance of the process. If f has a maximum at λ_0, then the dynamics of $\{x_t\}$ is dominated by the corresponding cycle or period; inversely, if the spectrum has a minimum at λ_0, then the corresponding cycle is of less relevance for the behavior of $\{x_t\}$ than all other cycles. For $\lambda \to 0$, period P converges to infinity. A cycle with an infinitely long period is interpreted as a trend or a long-run component. Hence, $f(0)$ indicates how strongly the process is dominated by a **trend component**.

Frequently, the analysis of the autocovariance structure or the autocorrelation structure of a process is called "analysis in the time domain" as $\gamma(h)$ measures the direct temporary dependence between x_t and x_{t+h}. Correspondingly, the spectral analysis is often referred to as "analysis in the frequency domain". Proposition 4.1 and the definition in (4.3) show how to move back and forth between time and frequency domain.

Examples

Example 4.1 (White Noise) Let us consider the white noise process $x_t = \varepsilon_t$ being free from serial correlation. By definition it immediately follows that the spectrum is constant:

$$f_\varepsilon(\lambda) = \sigma^2/2\pi\,, \quad \lambda \in [0, \pi]\,.$$

According to Proposition 4.1 all frequencies account equally strongly for the variance of the process. Analogously to the perspective in optics that the "color" white results if all frequencies are present equally strongly, serially uncorrelated processes are also often called "white noise". ■

Example 4.2 (Season) Let us consider the ordinary seasonal MA process from Example 3.1,

$$x_t = \varepsilon_t + b\varepsilon_{t-S}$$

with

$$\gamma(0) = \sigma^2 \left(1 + b^2\right), \ \gamma(S) = \sigma^2 b$$

and $\gamma(h) = 0$ else. By definition we obtain for the spectrum from (4.3)

$$2\pi f(\lambda) = \gamma(0) + 2\gamma(S)\cos(\lambda S)$$

or

$$f(\lambda) = \left(1 + b^2 + 2b\cos(\lambda S)\right)\sigma^2/2\pi .$$

In Problem 4.2 we determine that there are extrema at

$$0, \frac{\pi}{S}, \frac{2\pi}{S}, \ldots, \frac{(S-1)\pi}{S}, \pi .$$

The corresponding values are

$$f(0) = f\left(\frac{2\pi}{S}\right) = \ldots = (1+b)^2\sigma^2/2\pi ,$$
$$f\left(\frac{\pi}{S}\right) = f\left(\frac{3\pi}{S}\right) = \ldots = (1-b)^2\sigma^2/2\pi .$$

Depending on the sign of b, maxima and minima are followed by each other, respectively. In Fig. 4.2 we find two typical shapes of the spectrum of the seasonal MA process for[5] $S = 4$ (quarterly data) with $b = 0.7$ and $b = -0.5$. First, let us interpret the case $b > 0$. There are maxima at the frequencies $0, \pi/2$ and π. Corresponding cycles are of the period

$$P_0 = \frac{2\pi}{0} = \infty, \quad P_1 = \frac{2\pi}{\pi/2} = 4, \quad P_2 = \frac{2\pi}{\pi} = 2.$$

[5]The variance of the white noise is set to one , $\sigma^2 = 1$. This is also true for all spectra of this chapter depicted in the following.

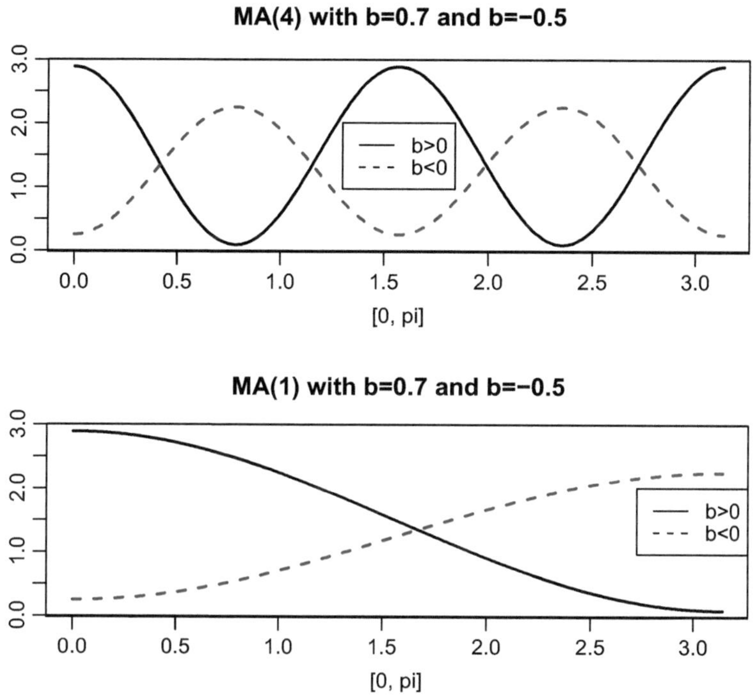

Fig. 4.2 Spectra ($2\pi f(\lambda)$) of the MA(S) process from Example 4.2

The trend is the first infinitely long "period". The second cycle has the period $P_1 = 4$, i.e. four quarters which is why this is the annual cycle. The third cycle with $P_2 = 2$ is the semi-annual cycle with only two quarters. These three cycles dominate the process for $b > 0$. Inversely, for $b < 0$ it holds that these very cycles add particularly little to the variance of the process. ■

Example 4.3 (MA(1)) Specifically for $S = 1$ the seasonal MA process passes into the MA(1) process. Accordingly, one obtains two extrema at zero and π:

$$f(0) = (1+b)^2 \sigma^2/2\pi\,, \quad f(\pi) = (1-b)^2 \sigma^2/2\pi.$$

In between the spectrum reads

$$f(\lambda) = \left(1 + b^2 + 2b\cos(\lambda)\right) \sigma^2/2\pi.$$

For $b = 0.7$ and $b = -0.5$, respectively, the spectra were calculated, see Fig. 4.2. For $b < 0$ one spots the relative absence of a trend (frequency zero matters least) while for $b > 0$ precisely the long-run component as a trend dominates the process. ■

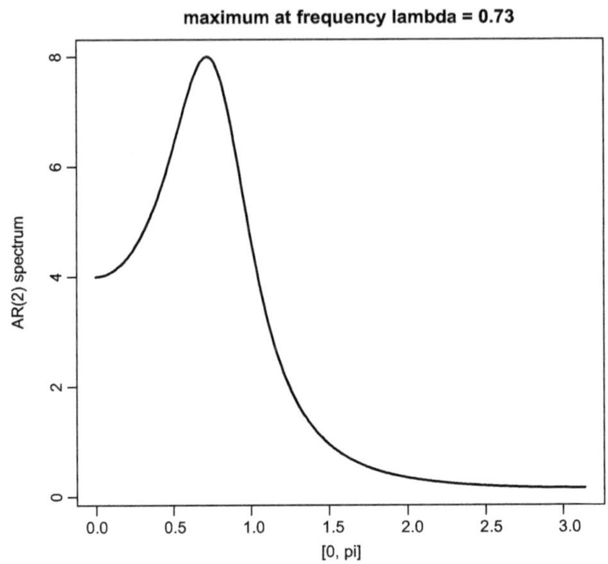

Fig. 4.3 Spectrum ($2\pi f(\lambda)$) of business cycle with a period of 8.6 years

Example 4.4 (Business Cycle) The spectrum is not only used for modelling seasonal patterns but as well for determining the length of a typical business cycle. Let us assume a process with annual observations having the spectrum depicted in Fig. 4.3. The maximum is at $\lambda = 0.73$. How do we interpret this fact with regard to contents? The dominating frequency $\lambda = 0.73$ corresponds to a period of about 8.6 (years). A frequency of this magnitude is often called "business cycle frequency" being interpreted as the frequency which corresponds to the business cycle. In fact, Fig. 4.3 does not comprise an empirical spectrum. Rather, one detects the theoretical spectrum of the AR(2) model whose autocorrelogram is depicted in Fig. 3.4 down to the right. The cycle, which can be seen in the autocorrelogram there, translates into the spectral maximum from Fig. 4.3. ■

4.3 Filtered Processes

The ARMA process or more generally the infinite MA process have been defined as filtered white noise. In order to systematically derive a formal expression for their spectra, we start quite generally with the relation between input and output of a filter in the frequency domain.

Filtered Processes

As in the previous chapter, we consider the causal, time-invariant, linear filter $F(L)$,

$$F(L) = \sum_{j=0}^{p} w_j L^j,$$

where L again denotes the lag operator and $p = \infty$ is allowed for. The filter is assumed to be absolutely summable, which trivially holds true for finite order p. Let the process $\{x_t\}$ be generated by filtering of the stationary process $\{e_t\}$,

$$x_t = F(L)\, e_t.$$

Then, how does the corresponding spectrum of $\{x_t\}$ for a given spectrum f_e of $\{e_t\}$ read? The answer is based on the **power transfer function** $T_F(\lambda)$ that we briefly touched upon in the previous section[6]:

$$T_F(\lambda) = \sum_{j=0}^{\infty} w_j^2 + 2 \sum_{h=1}^{\infty} \sum_{j=0}^{\infty} w_j w_{j+h} \cos(\lambda h)\ . \tag{4.4}$$

At a first glance, this expression appears cumbersome. However, in the next section we will see that for concrete ARMA processes it simplifies radically. If $F(L)$ is a finite filter (i.e. with finite p), then the sums of $T_F(\lambda)$ are truncated accordingly, see (4.8) in the next section. With $T_F(\lambda)$ the following proposition considerably simplifies the calculation of theoretical spectra (for a proof of an even more general result see Brockwell and Davis (1991, Thm. 4.4.1), while Fuller (1996, Thm. 4.3.1) covers our case where $\{e_t\}$ has absolutely summable autocovariances).

[6] The mathematically experienced reader will find the expression in (4.4) to be unnecessarily complicated as the transformation $T_F(\lambda)$ can be written considerably more compactly by using the exponential function in the complex space. It holds that

$$T_F(\lambda) = \left|F(e^{-i\lambda})\right|^2 = F(e^{i\lambda})F(e^{-i\lambda}),$$

where Euler's formula allows for expressing the complex-valued exponential function by sine and cosine,

$$e^{i\lambda} = \cos\lambda + i \sin\lambda, \quad i^2 = -1,$$

with the conjugate complex number $e^{-i\lambda} = \cos\lambda - i\sin\lambda$, where i denotes the imaginary unit. Instead of burdening the reader with complex numbers and functions, we rather expect him or her to handle the more cumbersome definition from (4.4). By the way, the term "power transfer function" stems from calling $F(e^{-i\lambda})$ alone transfer function of the filter $F(L)$, and $T_F(\lambda) = \left|F(e^{-i\lambda})\right|^2$ being the power thereof.

Proposition 4.2 (Spectra of Filtered Processes) *Let $\{e_t\}$ be a stationary process with spectrum $f_e(\lambda)$. The filter*

$$F(L) = \sum_{j=0}^{\infty} w_j L^j$$

be absolutely summable, $\sum_{j=0}^{\infty} | w_j | < \infty$, and $\{x_t\}$ be

$$x_t = F(L)\, e_t .$$

Then, $\{x_t\}$ is stationary with spectrum

$$f_x(\lambda) = T_F(\lambda) f_e(\lambda) \,, \quad \lambda \in [0, \pi] \,,$$

where $T_F(\lambda)$ is defined in (4.4).

Example 4.5 (Infinite MA) Let $e_t = \varepsilon_t$ from Proposition 4.2 be white noise with

$$f_\varepsilon(\lambda) = \sigma^2/2\pi \,,$$

and consider an absolutely summable MA(∞) process,

$$x_t = C(L)\, \varepsilon_t = \sum_{j=0}^{\infty} c_j \varepsilon_{t-j} \,.$$

Then Proposition 4.2 kicks in:

$$f_x(\lambda) = \left(\sum_{j=0}^{\infty} c_j^2 + 2 \sum_{h=1}^{\infty} \sum_{j=0}^{\infty} c_j c_{j+h} \cos(\lambda h) \right) \frac{\sigma^2}{2\pi} \,, \quad \lambda \in [0, \pi] \,. \tag{4.5}$$

This special case of Proposition 4.2 will be verified in Problem 4.3. Note that the spectrum given in (4.5) equals of course the result from Proposition 4.1 with (4.2), which continues to hold without absolute summability. ■

Persistence

We now return more systematically to the issue of **persistence** that we have touched upon in the example of the AR(1) process in the previous chapter. Loosely speaking, we understand by persistence the degree of (positive) autocorrelation such that subsequent observations form clusters: positive observations tend to be followed by positive ones, while negative observations tend to induce negative ones. With persistence we try to capture the strength of such a tendency, which depends not only on the autocorrelation coefficient at lag one but also on higher order lags. In the previous chapter we mentioned that it has been suggested to measure persistence by means of the cumulated impulse responses *CIR* defined in (3.3). This quantity shows up in the spectrum at frequency zero by Proposition 4.2. Assume that $\{x_t\}$ is an MA(∞) process with absolutely summable impulse response sequence $\{c_j\}$. We then have:

$$f_x(0) = T_C(0)\frac{\sigma^2}{2\pi} = \left(\sum_{j=0}^{\infty} c_j^2 + 2\sum_{h=1}^{\infty}\sum_{j=0}^{\infty} c_j c_{j+h}\right)\frac{\sigma^2}{2\pi} = \left(\sum_{j=0}^{\infty} c_j\right)^2 \frac{\sigma^2}{2\pi},$$

or

$$f_x(0) = (CIR)^2\,\frac{\sigma^2}{2\pi}.$$

Hence, the larger *CIR*, the stronger is the contribution of the trend component at frequency zero to the variance of the process, which formalizes our concept of persistence. Cogley and Sargent (2005) applied as relative spectral measure for persistence the ratio of $2\pi f_x(0)/\gamma_x(0)$ pioneered previously by Cochrane (1988); it can be interpreted as a variance ratio and is hence abbreviated as *VR*:

$$VR := \frac{2\pi f_x(0)}{\gamma_x(0)} = \frac{\left(\sum_{j=0}^{\infty} c_j\right)^2}{\sum_{j=0}^{\infty} c_j^2}. \tag{4.6}$$

In the case of a stationary AR(1) process, $x_t = a_1 x_{t-1} + \varepsilon_t$, it holds that (see Problem 4.6)

$$VR = \frac{1-a_1^2}{(1-a_1)^2} = \frac{1+a_1}{1-a_1}\begin{cases} > 1 \text{ if } a_1 > 0 \\ = 1 \text{ if } a_1 = 0 \\ < 1 \text{ if } a_1 < 0 \end{cases}. \tag{4.7}$$

In the case of $a_1 = 0$ (white noise) we have no persistence, and $VR = 1$. For $a_1 > 0$ the process is all the more persistent the larger a_1 is. Following Hassler (2014), one may say that a process has negative persistence if $VR < 1$. The plot of a series under negative persistence will typically display a zigzag pattern as observed in the last plot in Fig. 3.2. The limiting cases of $VR = 0$ (also called antipersistent) and $VR = \infty$ (also called strongly persistent) will be dealt with in Chap. 5.

ARMA Spectra

As a consequence of the previous proposition, we can derive what the spectrum of a stationary ARMA process $\{x_t\}$ looks like. Remember the definition from (3.13),

$$A(L)\, x_t = B(L)\, \varepsilon_t \, .$$

Now, define

$$y_t = A(L)\, x_t = B(L)\, \varepsilon_t \, .$$

By Proposition 4.2 one obtains for the spectra

$$f_y(\lambda) = T_A(\lambda)\, f_x(\lambda) = T_B(\lambda)\, \sigma^2/2\pi \, .$$

The assumption of a stationary MA(∞) representation[7] implies $T_A(\lambda) > 0$. Consequently, one may solve for f_x rendering the following corollary.

Corollary 4.1 (ARMA Spectra) *Let $\{x_t\}$ be a stationary ARMA(p, q) process*

$$A(L)\, x_t = \nu + B(L)\, \varepsilon_t.$$

Its spectrum is given by

$$f_x(\lambda) = \frac{T_B(\lambda)}{T_A(\lambda)} \frac{\sigma^2}{2\pi}, \qquad \lambda \in [0, \pi] \, ,$$

where $T_B(\lambda)$ and $T_A(\lambda)$ are the power transfer functions of $B(L)$ and $A(L)$.

Often, we restrict the class of stationary ARMA processes to the **invertible** ones, meaning we assume that the moving average polynomial $B(L)$ satisfies the invertibility condition of Proposition 3.3: All solutions of $B(z) = 0$ are larger than 1 in absolute value. This implies as in Footnote 7 that $T_B(\lambda) > 0$, such that the invertible ARMA spectrum is strictly positive for all λ.

---

[7] According to Proposition 3.5 we rule out autoregressive roots on the unit circle, such that $A(e^{-i\lambda}) \neq 0$, and $\left|A(e^{-i\lambda})\right|^2 > 0$. By assumption, $|z| = 1$ implies $A(z) \neq 0$, and here, $z = e^{-i\lambda}$ with

$$|e^{-i\lambda}|^2 = (\cos\lambda)^2 + (\sin\lambda)^2 = 1 \, .$$

In the next section we will learn that the calculation of the functions $T_A(\lambda)$ and $T_B(\lambda)$ and thereby the calculation of the spectra do not pose any problem, cf. Eq. (4.8).

4.4 Examples of ARMA Spectra

The ARMA filters $A(L)$ and $B(L)$ are assumed to be of finite order. In order to calculate the spectrum, the power transfer function is needed due to Corollary 4.1. Thus, next we will get to know a simple trick allowing for quickly calculating the power transfer function of a finite filter.

Summation over the Diagonal

We consider for finite p the filter $F(L)$ with the coefficients $w_0, w_1, \ldots, w_p$ being collected in a vector:

$$w = \begin{pmatrix} w_0 \\ w_1 \\ \vdots \\ w_{p-1} \\ w_p \end{pmatrix}.$$

The outer product yields a matrix where w' stands for the transposition of the column w:

$$ww' = \begin{pmatrix} w_0 \\ w_1 \\ \vdots \\ w_{p-1} \\ w_p \end{pmatrix} \left(w_0, w_1, \ldots, w_{p-1}, w_p\right)$$

$$= \begin{pmatrix} w_0^2 & w_0 w_1 & \ldots & w_0 w_{p-1} & w_0 w_p \\ w_1 w_0 & w_1^2 & \ldots & w_1 w_{p-1} & w_1 w_p \\ \vdots & \vdots & \cdots & \vdots & \vdots \\ w_{p-1} w_0 & w_{p-1} w_1 & \cdots & w_{p-1}^2 & w_{p-1} w_p \\ w_p w_0 & w_p w_1 & \cdots & w_p w_{p-1} & w_p^2 \end{pmatrix}.$$

Obviously, the matrix is symmetric. Now, we add the cosine as function of $|j - i|$, $\cos(\lambda \mid j - i \mid)$, to the entries $w_i w_j$. Let the resulting matrix be called $M_F(\lambda)$. It becomes:

$$\begin{pmatrix} w_0^2 \cos(0) & w_0 w_1 \cos(\lambda) & \dots & w_0 w_p \cos(\lambda p) \\ w_0 w_1 \cos(\lambda) & w_1^2 \cos(0) & \dots & w_1 w_p \cos(\lambda (p-1)) \\ \vdots & \vdots & \dots & \vdots \\ w_0 w_{p-1} \cos(\lambda (p-1)) & w_1 w_{p-1} \cos(\lambda (p-2)) & \cdots & w_{p-1} w_p \cos(\lambda) \\ w_0 w_p \cos(\lambda p) & w_1 w_p \cos(\lambda (p-1)) & \cdots & w_p^2 \cos(0) \end{pmatrix}$$

The rule for calculating $T_F(\lambda)$ reads in words: "Add up the sums over all diagonals of $M_F(\lambda)$":

$$\left[w_0^2 + \dots w_p^2\right] + 2\left[w_0 w_1 + \cdots w_{p-1} w_p\right] \cos(\lambda) + \cdots + 2\left[w_0 w_p\right] \cos(\lambda p).$$

This corresponds exactly to (4.4) for finite p:

$$T_F(\lambda) = \sum_{j=0}^{p} w_j^2 + 2 \sum_{h=1}^{p} \left[\sum_{j=0}^{p-h} w_j w_{j+h} \right] \cos(\lambda h). \tag{4.8}$$

AR(1) Spectra

The autoregressive polynomial of order one reads

$$A(L) = 1 - a_1 L,$$

i.e. the filter coefficients are

$$w_0 = 1 \text{ and } w_1 = -a_1.$$

Hence, for the power transfer function, (4.8) provides us with

$$T_A(\lambda) = 1 + a_1^2 - 2a_1 \cos(\lambda),$$

and Corollary 4.1 yields for the spectrum

$$2\pi f(\lambda) = \frac{\sigma^2}{1 + a_1^2 - 2a_1 \cos(\lambda)}.$$

In Problem 4.4 we will show that there are extrema at $\lambda = 0$ and $\lambda = \pi$, where the slope of the spectrum is zero. For $a_1 > 0$ the spectrum decreases on $[0, \pi]$, i.e. the most significant frequency is $\lambda = 0$: The process is dominated by trending behavior. Figure 4.4 shows that this is the more true the greater a_1 is: The greater a_1, the steeper and higher grows the spectrum in the area around zero. Mirror-inversely,

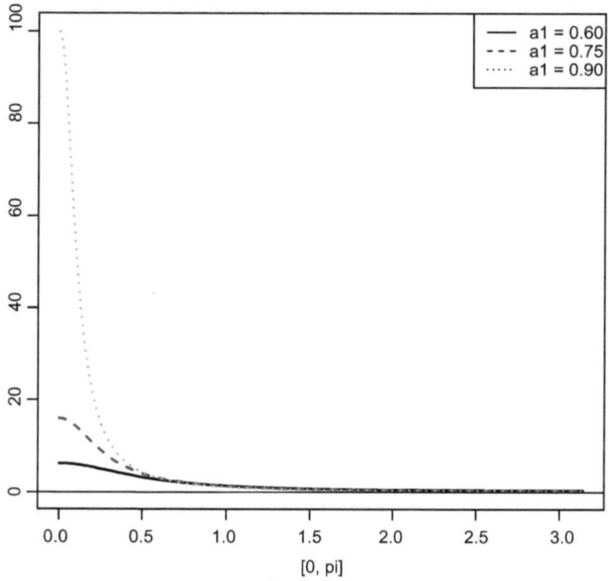

Fig. 4.4 AR(1) spectra ($2\pi f(\lambda)$) with positive autocorrelation

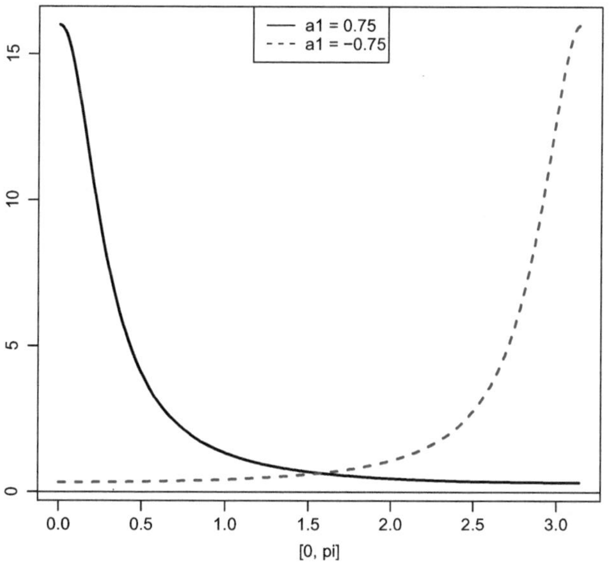

Fig. 4.5 AR(1) spectra ($2\pi f(\lambda)$), cf. Fig. 3.2

for $a_1 < 0$ it holds that the trend component matters least, see Fig. 4.5. The direct comparison to the time domain in Fig. 3.2 is also interesting. The case in which $a_1 > 0$ with the spectral maximum at $\lambda = 0$ translates in persistence of the process:

Observations temporarily lying close together have similar numerical values, i.e. the autocorrelation function is positive. For $a_1 < 0$, however, observations following each other have the tendency to change their sign as, in this case, there is just no trending behavior.

AR(2) Spectra

The AR(2) process is given by

$$x_t = \frac{\varepsilon_t}{A(L)} \quad \text{with} \quad A(L) = 1 - a_1 L - a_2 L^2.$$

In Problem 4.5 we recapitulate the principle of the "summation over the diagonal" and thus we show

$$T_A(\lambda) = 1 + a_1^2 + a_2^2 + 2\left[a_1 (a_2 - 1)\cos(\lambda) - a_2 \cos(2\lambda)\right].$$

Therefore, due to Corollary 4.1, the corresponding spectrum reads

$$2\pi f(\lambda) = \frac{\sigma^2}{T_A(\lambda)}.$$

For $a_2 = 0$ one obtains the AR(1) case.

In Fig. 4.6 spectra for four parameter constellations are depicted; these are exactly the four cases for which autocorrelograms are given in Fig. 3.4. The top left case could be well approximated by an AR(1) process. This is also roughly true for the top right case; however, closer inspection reveals that the AR(2) spectrum is not minimal at frequency zero. Both the lower spectra entirely burst the AR(1) scheme. On the bottom right we have the example of the business cycle, see Fig. 4.3. The spectrum on the bottom left is even more extreme: Except for a rather small area around $\lambda = 2$, it is zero almost everywhere which is why there is no trend component. The process is determined by almost only one cycle which can be seen in the autocorrelogram as well.

ARMA(1,1) Spectra

Now, let us consider the two filters

$$A(L) = 1 - a_1 L \quad \text{and} \quad B(L) = 1 + b_1 L.$$

We know the filter transfer function of $B(L)$ from Example 4.3:

$$T_B(\lambda) = 1 + b_1^2 + 2\, b_1 \cos(\lambda).$$

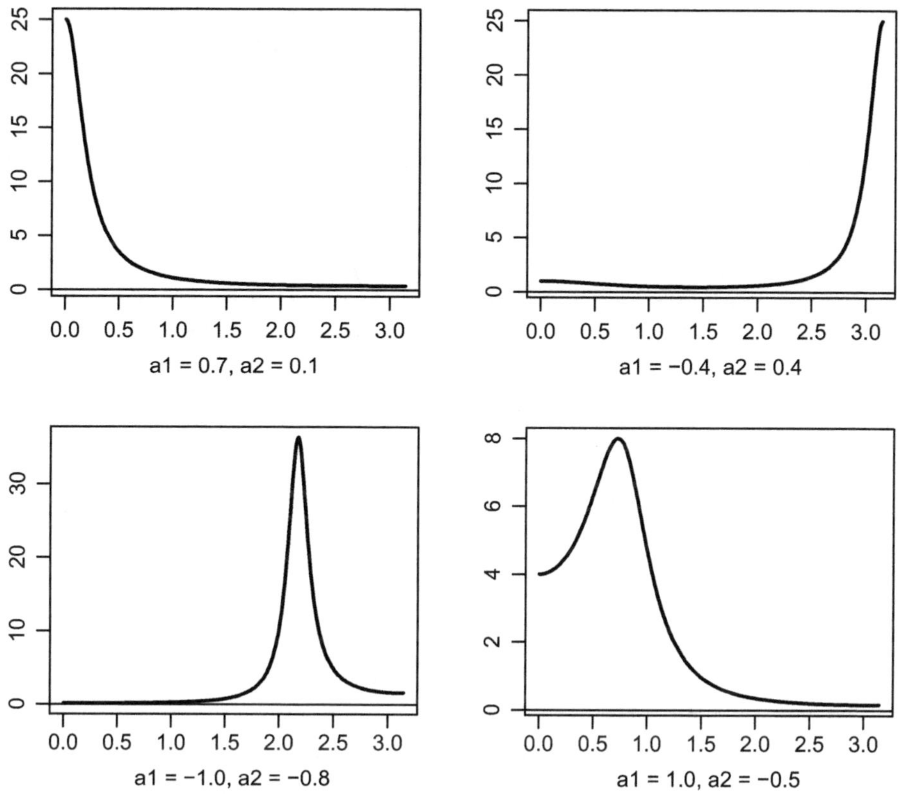

Fig. 4.6 AR(2) spectra ($2\pi f(\lambda)$), cf. Fig. 3.4

The transformation of $A(L)$ was determined at the beginning of this section. Due to Corollary 4.1, we put the spectrum together as follows:

$$2\pi f(\lambda) = \frac{1 + b_1^2 + 2\, b_1\, \cos(\lambda)}{1 + a_1^2 - 2\, a_1\, \cos(\lambda)}\, \sigma^2\,, \quad \lambda \in [0, \pi]\,.$$

In order to have this illustrated, consider the examples from Fig. 4.7. The cases correspond in their graphical arrangement to the autocorrelograms from Fig. 3.5. The cases top right and bottom left are interesting. At the top on the right, the entire absence of a trend is reflected in a negative autocorrelogram close to zero. At the bottom on the left, beside the trend, cycles of higher frequencies add to the process as well, the process consequently being positively autocorrelated of the first order and then exhibiting an alternating pattern of autocorrelation.

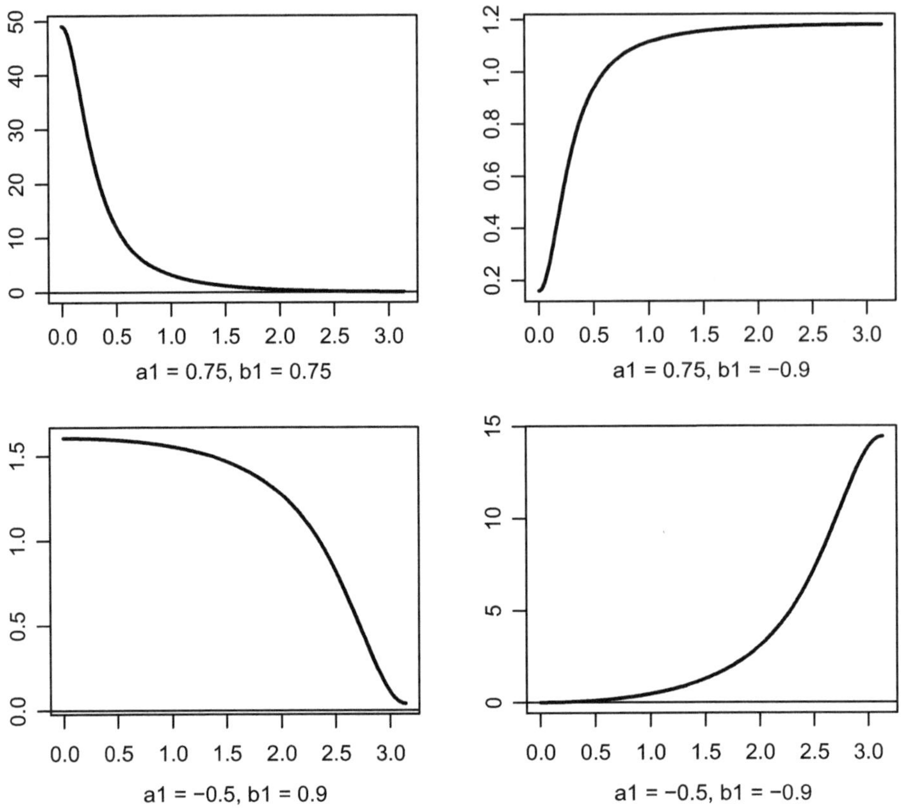

Fig. 4.7 ARMA(1,1) spectra ($2\pi f(\lambda)$), cf. Fig. 3.5

Multiplicative Seasonal AR Process

If one wants to have a decaying autocorrelation function not dropping to zero, then one does not choose a pure MA model as in Example 4.2. The most basic seasonal autoregressive model is based on the filter $(1-a_S L^S)$. Frequently, the trend component is to have an additional weight which is why one adds the AR(1) factor $(1-a_1 L)$:

$$\begin{aligned} A(L) &= (1-a_1 L)\,(1-a_S L^S) \\ &= 1-a_1 L - a_S L^S + a_1\, a_S L^{S+1} \, . \end{aligned}$$

Therefore, we have an AR($S+1$) model with parameter restrictions. The spectrum is adopted from Problem 4.6 in which $T_A(\lambda)$ is given:

$$2\pi f(\lambda) = \frac{\sigma^2}{T_A(\lambda)} \, .$$

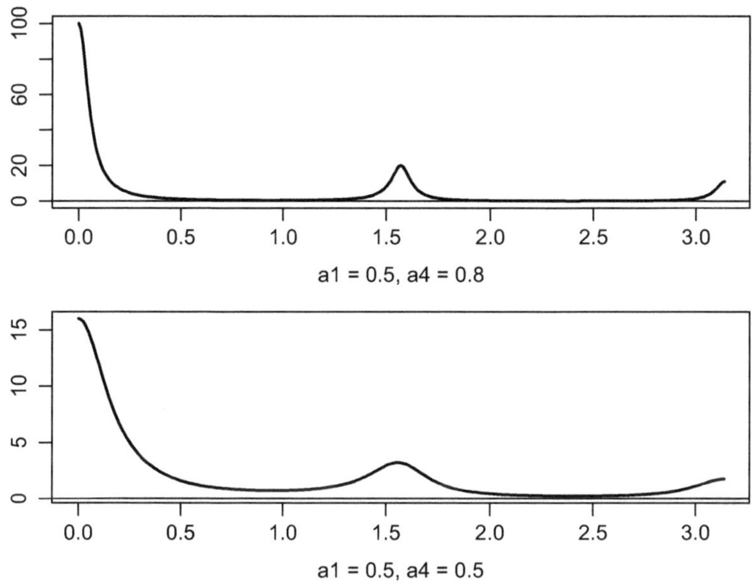

Fig. 4.8 Spectra ($2\pi f(\lambda)$) of multiplicative seasonal AR processes ($S = 4$)

In Fig. 4.8, we show two examples for the quarterly case ($S = 4$). With the frequencies $\lambda = \pi$ and $\lambda = \pi/2$ the semi-annual cycles with $P = 2$ quarters period and the annual cycles with $P = 4$ quarters length are modelled (one also talks about seasonal cycles). As $a_1 = 0.5$ is positive in both the spectra, the trend (at frequency zero) dominates the seasonal cycles. The annual and semi-annual cycles add both equally strongly to the variance of the process. However, in the case $a_4 = 0.8$, the seasonal component is more pronounced than in the case $a_4 = 0.5$ as in the upper spectrum both the seasonal peaks are not only higher than in the lower one (note the scale on the ordinate) but most of all steeper: In the upper graph, the area beneath the spectrum substantially concentrates on the three frequencies $0, \pi/2$ and π, whereas it is more spread over all frequencies in the lower one.

4.5 Problems and Solutions

Problems

4.1 Prove Proposition 4.1 (e) under the additional assumption of absolute summability.

4.2 Determine the extrema in the spectrum of the seasonal MA process from Example 4.2.

4.3 Prove the structure of the spectrum (4.5) for absolutely summable MA(∞) processes.

4.4 Determine the extrema of the AR(1) spectrum.

4.5 Determine the power transfer function $T_A(\lambda)$ of the filter $A(L) = 1 - a_1 L - a_2 L^2$.

4.6 Determine the power transfer function $T_A(\lambda)$ of the multiplicative quarterly AR filter $A(L) = (1 - a_1 L)(1 - a_4 L^4) = 1 - a_1 L - a_4 L^4 + a_1 a_4 L^5$.

4.7 Determine the persistence measure *VR* from (4.6) for a stationary and invertible ARMA(1,1) process. Discuss its behavior in particular for the MA(1) model (in comparison with the AR(1) case given in (4.7)).

Solutions

4.1 We define the entity A_h and will show that it equals $\gamma(h)$. Due to the symmetry of the cosine function and of the even spectrum it holds by definition that:

$$\begin{aligned} A_h &:= 2 \int_0^\pi f(\lambda) \cos(\lambda h)\, d\lambda \\ &= \int_{-\pi}^{\pi} f(\lambda) \cos(\lambda h)\, d\lambda \\ &= \frac{1}{2\pi} \int_{-\pi}^{\pi} \sum_{l=-\infty}^{\infty} \gamma(l) \cos(\lambda l) \cos(\lambda h)\, d\lambda . \end{aligned}$$

Because of the absolute summability, the order of summation and integration is interchangeable:

$$\begin{aligned} 2\pi A_h &= \sum_{l=-\infty}^{\infty} \gamma(l) \int_{-\pi}^{\pi} \cos(\lambda l) \cos(\lambda h)\, d\lambda \\ &= \gamma(0) \int_{-\pi}^{\pi} \cos(\lambda h)\, d\lambda + 2 \sum_{l=1}^{\infty} \gamma(l) \int_{-\pi}^{\pi} \cos(\lambda l) \cos(\lambda h)\, d\lambda . \end{aligned}$$

For $h = 0$ it holds that[8]

$$
\begin{aligned}
2\pi A_0 &= \gamma(0)\, 2\pi + 2\sum_{l=1}^{\infty} \gamma(l)\, \frac{\sin(\pi l) - \sin(-\pi l)}{l} \\
&= 2\pi\, \gamma(0) + 0.
\end{aligned}
$$

Accordingly, for $h \neq 0$ it holds that

$$
2\pi A_h = 0 + 2\sum_{l=1}^{\infty} \gamma(l) \int_{-\pi}^{\pi} \frac{\cos(\lambda(l-h)) + \cos(\lambda(l+h))}{2}\, d\lambda,
$$

where the trigonometric formula

$$
2\cos x \cos y = \cos(x-y) + \cos(x+y)
$$

was used. By this we obtain

$$
2\pi A_h = \gamma(h) \int_{-\pi}^{\pi} (1 + \cos(2\lambda h))\, d\lambda
$$

as one can see that for $k \in \mathbb{Z}\backslash\{0\}$ the integral is

$$
\int_{-\pi}^{\pi} \cos(\lambda k)\, d\lambda = \frac{\sin(\pi k) - \sin(-\pi k)}{k} = 0\,.
$$

So, we finally obtain

$$
2\pi A_h = \gamma(h)(2\pi + 0) = 2\pi\, \gamma(h)
$$

for $h \neq 0$ as well. Hence, $A_h = \gamma(h)$ for all h, and the proof is complete.

4.2 The spectrum

$$
f(\lambda) = \left(1 + b^2 + 2b\cos(\lambda S)\right) \sigma^2/2\pi
$$

[8]We use

$$
\int \cos(\lambda \ell) d\lambda = \frac{\sin(\lambda \ell)}{\ell},
$$

and $\sin(\pi k) = 0$ for $k \in \mathbb{Z}$.

is given. In order to determine the extrema, we consider the derivative,

$$f'(\lambda) = -2bS \sin(\lambda S)\, \sigma^2/2\pi,$$

with $(S+1)$ zeros

$$0, \frac{\pi}{S}, \frac{2\pi}{S}, \ldots, \frac{(S-1)\pi}{S}, \pi$$

on the interval $[0, \pi]$. The sign of the second derivative depends on b:

$$f''(\lambda) = -2bS^2 \cos(\lambda S)\, \sigma^2/2\pi.$$

One obtains

$$f''(0) = f''\left(\frac{2\pi}{S}\right) = \ldots = -2bS^2\, \sigma^2/2\pi\,,$$

$$f''\left(\frac{\pi}{S}\right) = f''\left(\frac{3\pi}{S}\right) = \ldots = +2bS^2\, \sigma^2/2\pi.$$

Accordingly, maxima and minima follow each other. For $b > 0$, the sequence of extrema begins with a maximum at zero; for $b < 0$, one obtains a minimum, inversely.

4.3 The autocovariances of

$$x_t = \sum_{j=0}^{\infty} c_j \varepsilon_{t-j}$$

are known from Proposition 3.2:

$$\gamma_x(h) = \sigma^2 \sum_{j=0}^{\infty} c_j c_{j+h}.$$

For the spectrum $f_x(\lambda)$ it follows:

$$2\pi \frac{f_x(\lambda)}{\sigma^2} = \frac{\gamma_x(0)}{\sigma^2} + 2\sum_{h=1}^{\infty} \frac{\gamma_x(h)}{\sigma^2} \cos(\lambda h)$$

$$= \sum_{j=0}^{\infty} c_j^2 + 2\sum_{h=1}^{\infty}\sum_{j=0}^{\infty} c_j c_{j+h} \cos(\lambda h)\,.$$

Hence, the claim is verified.

4.4 For

$$f(\lambda) = \left(1 + a_1^2 - 2\,a_1\,\cos(\lambda)\right)^{-1}\,\sigma^2/2\pi$$

one obtains by differentiation

$$f'(\lambda) = -\left(1 + a_1^2 - 2\,a_1\,\cos(\lambda)\right)^{-2}(-2\,a_1)\,(-\sin(\lambda))\,\sigma^2/2\pi\,.$$

Obviously, candidates for extrema are $\lambda = 0$ and $\lambda = \pi$:

$$f'(0) = f'(\pi) = 0\,.$$

The sign of the derivative depends on a_1 only:

$$f'(\lambda) < 0\,,\ \lambda \in [0, \pi] \quad \Longleftrightarrow \quad a_1 > 0\,.$$

Accordingly,

$$f(0) = (1 - a_1)^{-2}\,\sigma^2/2\pi \quad \text{and} \quad f(\pi) = (1 + a_1)^{-2}\,\sigma^2/2\pi$$

are maxima and minima, depending on the sign of a_1.

4.5 With the vector of coefficients

$$a = \begin{pmatrix} 1 \\ -a_1 \\ -a_2 \end{pmatrix}$$

we obtain as outer product

$$a\,a' = \begin{pmatrix} 1 & -a_1 & -a_2 \\ -a_1 & a_1^2 & a_1 a_2 \\ -a_2 & a_1 a_2 & a_2^2 \end{pmatrix}.$$

Adding the cosine, it follows that

$$M_A(\lambda) = \begin{pmatrix} 1 & -a_1\,\cos(\lambda) & -a_2\,\cos(2\,\lambda) \\ -a_1\,\cos(\lambda) & a_1^2 & a_1 a_2\,\cos(\lambda) \\ -a_2\,\cos(2\,\lambda) & a_1 a_2\,\cos(\lambda) & a_2^2 \end{pmatrix}.$$

By summation over the diagonal we obtain due to symmetry

$$T_A(\lambda) = 1 + a_1^2 + a_2^2 + 2\,[-a_1 + a_1\,a_2]\,\cos(\lambda) + 2\,[-a_2]\,\cos(2\lambda)\,,$$

which results from (4.8) as well. This is in accordance with the result in the text.

4.6 Using (4.8) with $p = 5$ yields the following expression:

$$\begin{aligned} T_A(\lambda) &= 1 + a_1^2 + a_4^2 + a_1^2 a_4^2 \\ &\quad +2\left[-a_1 - a_1 a_4^2\right] \cos(\lambda) + 2\left[a_1 a_4\right] \cos(3\lambda) \\ &\quad +2\left[-a_4 - a_1^2 a_4\right] \cos(4\lambda) + 2\left[a_1 a_4\right] \cos(5\lambda) . \end{aligned}$$

This is simply an exercise in concentration and is simplified by the following equalities:

$$w_0 = 1, \quad w_1 = -a_1, \quad w_2 = w_3 = 0, \quad w_4 = -a_4, \quad w_5 = a_1 a_4 .$$

4.7 In the previous section we discussed the ARMA(1,1) process with the polynomials

$$A(L) = 1 - a_1 L \quad \text{and} \quad B(L) = 1 + b_1 L .$$

Evaluating the spectrum given there we have:

$$2\pi f(0) = \frac{1 + b_1^2 + 2\,b_1}{1 + a_1^2 - 2\,a_1}\,\sigma^2 = \frac{(1 + b_1)^2}{(1 - a_1)^2}\,\sigma^2 .$$

The variance we copy from Chap. 3:

$$\gamma(0) = \frac{(1 + b_1^2 + 2a_1 b_1)}{1 - a_1^2}\,\sigma^2 .$$

By (4.6) we obtain

$$VR = \frac{1 + a_1}{1 - a_1}\,\frac{(1 + b_1)^2}{1 + b_1^2 + 2a_1 b_1} .$$

If $b_1 = 0$, the AR(1) case from (4.7) is of course reproduced. If $a_1 = 0$, the MA(1) case results as

$$VR = \frac{(1 + b_1)^2}{1 + b_1^2} \begin{cases} > 1 \text{ if } b_1 > 0 \\ = 1 \text{ if } b_1 = 0 \\ < 1 \text{ if } b_1 < 0 \end{cases} .$$

We hence have negative persistence for $b_1 < 0$, which reflects the negative autocorrelation. For $b_1 > 0$, it is straightforward to verify that VR is growing with b_1, reaching a maximum value of $VR = 2$ for $b_1 = 1$. This corresponds to the persistence of an AR(1) process with $a_1 = 1/3$. Hence, the invertible MA(1) process with $|b_1| < 1$ can only capture very moderate persistence in comparison with the AR(1) case where VR grows with a_1 beyond any limit.

References

Brockwell, P. J., & Davis, R. A. (1991). *Time series: Theory and methods* (2nd ed.). New York: Springer.
Cochrane, J. H. (1988). How big is the random walk in GNP? *Journal of Political Economy, 96*, 893–920.
Cogley, T., & Sargent T. S. (2005). Drifts and volatilities: Monetary policies and outcomes in the post WWII US. *Review of Economic Dynamics, 8*, 262–302.
Fuller, W. A. (1996). *Introduction to statistical time series* (2nd ed.). New York: Wiley.
Hamilton, J. (1994). *Time series analysis*. Princeton: Princeton University Press.
Hassler, U. (2014). Persistence under temporal aggregation and differencing. *Economics Letters, 124*, 318–322.

Long Memory and Fractional Integration 5

5.1 Summary

Below Proposition 3.5 we saw that the autocorrelation sequence of any stationary ARMA process dies out at exponential rate: $|\rho(h)| \leq c\, g^h$, see (3.14). This is too restrictive for many time series of stronger persistence, which display long memory in that the autocovariance sequence vanishes at a slower rate. In some fields of economics and finance long memory is treated as an empirical stylized fact.[1] Fractional integration as a model for long memory will be presented in this chapter. In the same paper where Granger (1981) introduced the Nobel prize winning concept of cointegration (see Chap. 16) he addressed the idea of fractional integration, too. For an early survey on fractional integration and applications see Baillie (1996).

5.2 Persistence and Long Memory

We have already briefly touched upon the so-called random walk, see Eq. (1.9). In terms of the difference operator this can be written as $\Delta x_t = \varepsilon_t$, i.e. the process has to be differenced once to obtain stationarity. Alternatively, the process is given by a cumulation or summation over the shocks, $x_t = \sum_{j=1}^{t} \varepsilon_j$, see (1.8), which is the reason to call the process $\{x_t\}$ integrated of order 1, see also Chap. 14. In this section, differencing or integration of order 1 will be extended by introducing non-integer orders of differencing and integration.

[1] See e.g. the special issue edited by Maasoumi and McAleer (2008) in *Econometric Reviews* on "Realized Volatility and Long Memory".

U. Hassler, *Stochastic Processes and Calculus*, Springer Texts in Business and Economics, DOI 10.1007/978-3-319-23428-1_5

Persistence

By persistence we understand how strongly a past shock affects the presence of a stochastic process. We stick to the MA(∞) representation behind the Wold decomposition of a stationary process that we briefly touched upon in Chap. 3,

$$x_t = \sum_{j=0}^{\infty} c_j \varepsilon_{t-j}, \quad \sum_{j=0}^{\infty} c_j^2 < \infty,$$

with $\{\varepsilon_t\}$ forming a white noise sequence. The impulse responses coefficients $\{c_j\}$ measures the response of x_t on a shock j periods ago. With stationary processes, the shocks are transitory in that $\lim_{j\to\infty} c_j = 0$. In particular, for stationary ARMA processes we know that the impulse responses die out so fast that they are summable in absolute value, see (3.2). To model a stronger degree of persistence and **long memory**, we require a slower convergence to zero. The model of fractional integration of order d will impose the so-called hyperbolic decay rate,

$$c_j = c\, j^{d-1}, \quad c \neq 0.$$

Under $d < 1$, the sequence $\{c_j\}$ converges to zero. Clearly, the larger d, the stronger is the persistence in that j^{d-1} dies out more slowly. Hence, the parameter d measures the strength of persistence. Contrary to the exponential case characteristic of ARMA processes, hyperbolic decay is so slow for positive $d > 0$, that the impulse responses are not summable. In Problem 5.1 we will establish the following convergence result concerning the so-called (generalized) harmonic series, often also called p-series:

$$\sum_{j=1}^{\infty} j^{-p} < \infty \quad \text{if and only if } p > 1. \tag{5.1}$$

Moreover, we will show in Problem 5.2 that exponential decay to zero is faster than hyperbolic one:

$$\lim_{j\to\infty} \frac{g^j}{j^{d-1}} = 0, \quad 0 < g < 1, \ |d| < 1.$$

In order to illustrate the different decay rates, we display in Figs. 5.1 and 5.2 sequences j^{d-1} and g^{j-1}, respectively; by construction they all have the value 1 at $j = 1$.

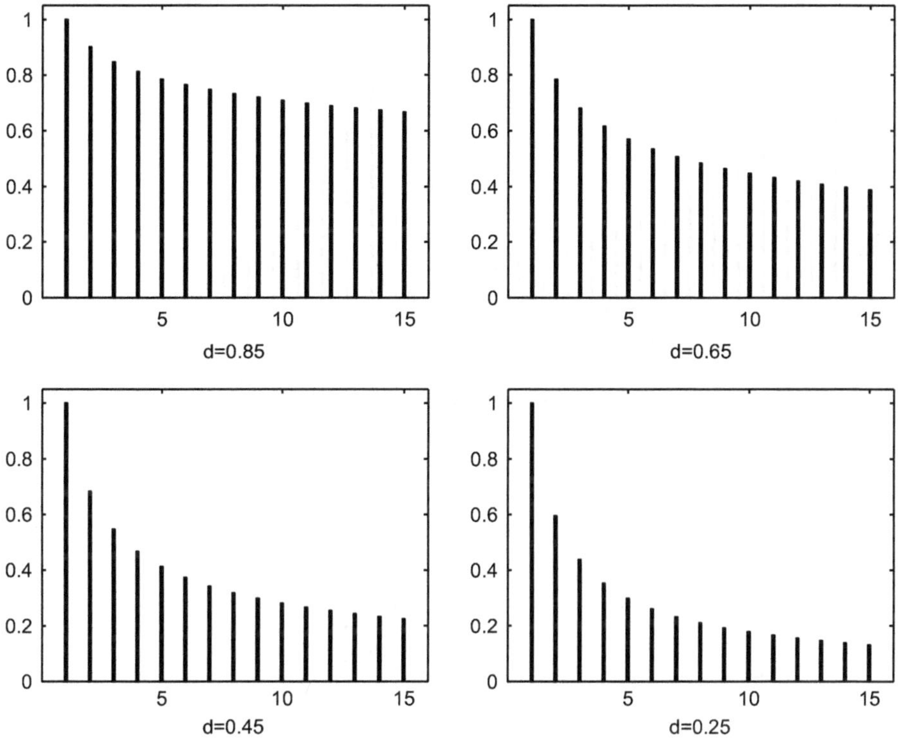

Fig. 5.1 j^{d-1} for $d = 0.85, 0.65, 0.45, 0.25$

For a process to be stationary, we require the impulse response sequence to be square summable, see Proposition 3.2. From (5.1) we learn that $\sum_{j=1}^{\infty} j^{2d-2}$ is finite if and only if $d < 0.5$, which hence turns out to be the stationarity condition for processes with impulse responses $\{j^{d-1}\}$.The model of fractional integration, however, does not directly assume $c_j = c\, j^{d-1}$; rather this power law will hold only true for large j,

$$c_j \sim c\, j^{d-1}\,, \quad j \to \infty\,,$$

where "$\sim$ for $j \to \infty$" is to be understood as a proper limit in the following way:

$$a_j \sim b_j \quad \Longleftrightarrow \quad \lim_{j\to\infty} \frac{a_j}{b_j} = 1\,, \quad b_j \neq 0\,. \tag{5.2}$$

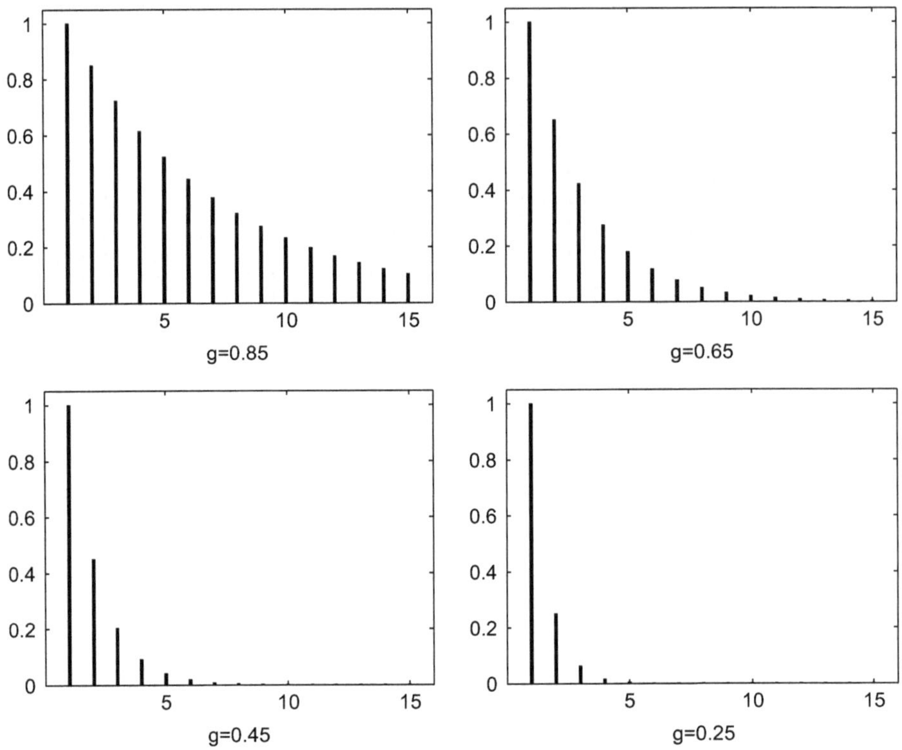

Fig. 5.2 g^{j-1} for $g = 0.85, 0.65, 0.45, 0.25$

Fractional Differencing and Integration

With the usual difference operator $\Delta = (1 - L)$, see Example 3.2, we define **fractional differences** by binomial expansion[2]:

$$\Delta^d = (1-L)^d = 1 - dL - \frac{d\,(1-d)}{2} L^2 - \frac{d\,(1-d)\,(2-d)}{6} L^3 - \dots$$
$$= \sum_{j=0}^{\infty} \pi_j L^j \,, \quad d > -1 \,.$$

[2]For the rest of this chapter we maintain $d > -1$, which guarantees that $\{\pi_j\}$ converges to 0 with growing j, making the infinite expansion meaningful.

Readers not be familiar with binomial series for $d \notin \mathbb{N}$ may wish to consult e.g. Trench (2013, Sect. 4.5). The binomial series results from a Taylor expansion of $(1-z)^d$ about $z = 0$, hence also called Maclaurin series. The coefficients $\{\pi_j\}$ are given in terms of binomial coefficients

$$\binom{d}{j} = \frac{d\,(d-1)\,\cdots\,(d-j+1)}{j!}\,,$$

yielding the recursion

$$\pi_j = \binom{d}{j}(-1)^j = \frac{j-1-d}{j}\,\pi_{j-1}\,, \quad j \geq 1\,,\ \pi_0 = 1\,. \tag{5.3}$$

For natural numbers d one has the more familiar finite expansions,

$$(1-L)^1 = 1 - L\,, \quad (1-L)^2 = 1 - 2L + L^2\,,$$

while the expansion in (5.3) holds for non-integer (or fractional) values of d, too. In Problem 5.3 we derive the behavior for large j,

$$\pi_j \sim \frac{j^{-d-1}}{\Gamma(-d)}\,, \quad j \to \infty\,,\ d \neq 0, \tag{5.4}$$

where $\Gamma(\cdot)$ is the so-called **Gamma function** introduced at greater detail below its definition in (5.18) in the Problem section.

Similarly to fractional differences, we may define the **fractional integration** operator upon inversion,

$$\Delta^{-d} = (1-L)^{-d} = \sum_{j=0}^{\infty} \psi_j L^j\,, \quad \psi_0 = 1\,,$$

where the coefficients are given by simply replacing d by $-d$ in (5.3):

$$\psi_j = \binom{-d}{j}(-1)^j = \frac{j-1+d}{j}\,\psi_{j-1}\,, \quad j \geq 1\,. \tag{5.5}$$

The same arguments establishing (5.4) hence show

$$\psi_j \sim \frac{j^{d-1}}{\Gamma(d)}\,, \quad j \to \infty\,, \quad d \neq 0\,. \tag{5.6}$$

Fractional integration thus imposes the hyperbolic decay rate discussed above, where the speed of convergence varies with d. From (5.1) we observe with (5.6) that $\{\psi_j\}$ is summable if and only if $d < 0$, in which case

$$\sum_{j=0}^{\infty} \psi_j = (1-z)^{-d}\big|_{z=1} = 0\,, \quad d < 0\,. \tag{5.7}$$

Further, $\{\psi_j\}$ is square summable if and only if $d < 0.5$. These are the ingredients to define a fractionally integrated process.

5.3 Fractionally Integrated Noise

We now apply the above findings to the simplest case of fractional noise (which is short for: fractionally integrated noise), define long memory in the time domain in (5.9), and translate it into the frequency domain.

Fractional Noise and Long Memory

In case of fractionally integrated noise the fractional differencing filter Δ^d has to be applied to $\{x_t\}$ in order to obtain white noise $\{\varepsilon_t\}$ with variance σ^2: $\Delta^d x_t = \varepsilon_t$. Equivalently, we write after inverting the differences

$$\begin{aligned} x_t &= (1-L)^{-d}\,\varepsilon_t \\ &= \sum_{j=0}^{\infty} \psi_j \varepsilon_{t-j}\,, \quad t \in \mathbb{Z}\,, \quad d < 0.5\,, \end{aligned} \tag{5.8}$$

with $\{\psi_j\}$ from (5.5) being the sequence of impulse response coefficients measuring the effect of a past shock on the presence. As we have discussed above the impulse responses die out the more slowly the larger d is. In that sense we interpret d as measure of persistence or memory. The impulse responses die out so slowly that they are not absolutely summable for positive memory parameter d. Consequently for $d > 0$, $\{x_t\}$ from (5.8) does not belong to the class of processes with absolutely summable autocovariances characterized in (3.2), while all stationary ARMA processes belong to this class. Hence, fractional integration models for $d > 0$ a feature that is not captured by traditional ARMA processes, which we call **strong persistence**; it is defined in the time domain by

$$\sum_{j=1}^{J} \psi_j \to \infty\,,\ J \to \infty\,, \qquad \text{if } d > 0\,.$$

Consequently, *CIR* or *VR* defined in (3.3) and (4.6), respectively, do not exist. Even though the MA coefficients are not summable for positive d, the process is still stationary as long as $d < 0.5$ because $\sum_j \psi_j^2 < \infty$ due to (5.6) and (5.1). Further, the process is often called invertible for $d > -0.5$ since an autoregressive representation exists that is square summable[3]:

$$\sum_{j=0}^{\infty} \pi_j x_{t-j} = \varepsilon_t \quad \text{with} \sum_{j=0}^{\infty} \pi_j^2 < \infty .$$

Note that the existence of an autoregressive representation in the mean square sense does not require square summability; in fact, Bondon and Palma (2007) extend the range of invertibility in the mean square to $d > -1$. Given the existence of the MA(∞) representation the following properties of fractionally integrated noise are proven in the Problem section.

Proposition 5.1 (Fractional noise, time domain) *For fractionally integrated noise from (5.8) it holds with $-1 < d < 0.5$ that*

(a) the variance equals

$$\gamma(0) = \gamma(0; d) = \sigma^2 \frac{\Gamma(1-2d)}{(\Gamma(1-d))^2} ,$$

with $\Gamma(\cdot)$ being defined in (5.18), and $\gamma(0; d)$ achieves its minimum for $d = 0$;

(b) the autocovariances equal

$$\begin{aligned} \gamma(h) &= \frac{h-1+d}{h-d} \gamma(h-1) , \quad h = 1, 2, \ldots , \\ &\sim \gamma_d \sigma^2 h^{2d-1}, \quad h \to \infty , \; d \neq 0 , \end{aligned}$$

with

$$\gamma_d = \frac{\Gamma(1-2d)}{\Gamma(d)\Gamma(1-d)} ,$$

where $\gamma_d < 0$ if and only if $d < 0$;

(c) the autocorrelations $\rho(h) = \rho(h; d)$ grow with d for $d > 0$.

Let us briefly comment those results. First, since $\Gamma(1) = 1$, the minimum variance obtained in the white noise case ($d = 0$) is of course $\gamma(0; 0) = \sigma^2$. Second,

[3]A more technical exposition can be found in Brockwell and Davis (1991, Thm. 13.2.1) or Giraitis, Koul, and Surgailis (2012, Thm. 7.2.1), although they consider only the range $|d| < 0.5$.

from the hyperbolic decay of the autocovariance sequence we observe that $\gamma(h)$ converges to zero with h as long as $d < 0.5$, but for $d > 0$ so slowly, that we have long memory defined as

$$\sum_{h=0}^{H} |\, \gamma\,(h)\, | \to \infty\,,\; H \to \infty \quad \text{if}\, d > 0\,. \tag{5.9}$$

In particular, the autocovariances die out the more slowly the larger the memory parameter d is. Obviously, the same feature can be rephrased in terms of autocorrelations. The recursion carries over to the autocorrelations, and Proposition 5.1 (b) yields

$$\rho(h) \sim \frac{\gamma_d\,\sigma^2}{\gamma(0)}\, h^{2d-1}, \quad h \to \infty\,.$$

For a numerical and graphical illustration see Fig. 5.3. The asymptotic constant γ_d has the same sign as d, meaning that in case of long memory the autocovariances converge to zero from above, and vice versa from below zero for $d < 0$, see again

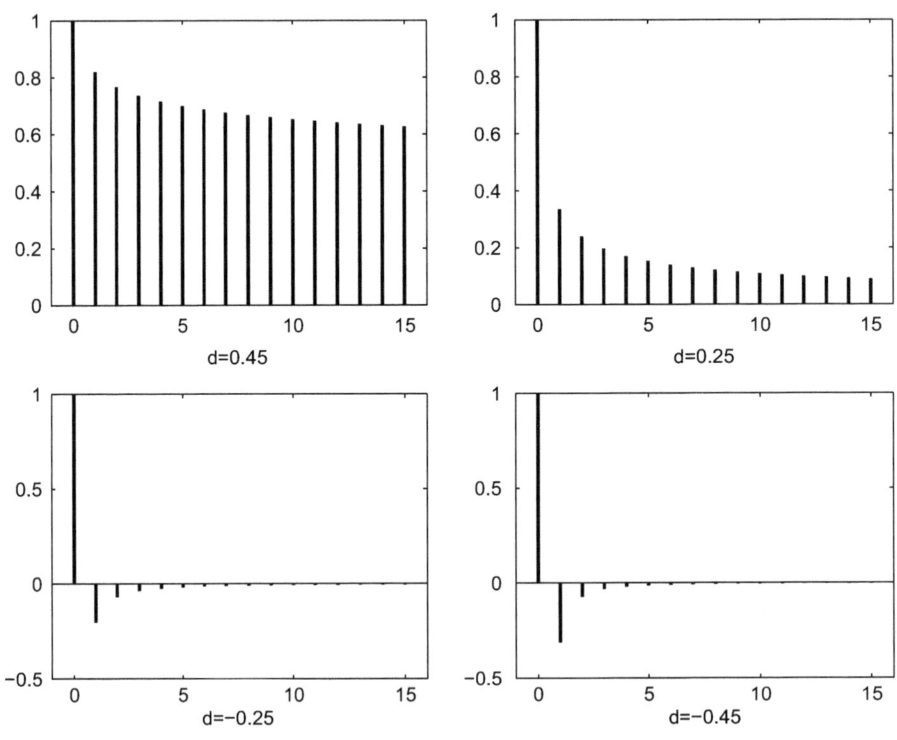

Fig. 5.3 $\rho(h)$ from Proposition 5.1 for $d = 0.45, 0.25, -0.25, -0.45$

Fig. 5.3. Note, however, that γ_d collapses to zero as $d \to 0$, simply meaning that the hyperbolic decay rate does not hold for $d = 0$. Third, a similar effect that d has on $\rho(h)$ at long lags, holds true for finite h. More precisely, Proposition 5.1 (c) says for each finite h that the autocorrelation grows with d (for $d > 0$), which reinforces the interpretation of d as measure of persistence or long memory.[4]

The case of negative d results in short memory in that the autocovariances are absolutely summable, which is clear again from the p-series in (5.1). This case is sometimes called antipersistent, the reason for that being

$$\sum_{j=1}^{\infty} \psi_j = 0\,, \qquad \text{if } d < 0\,.$$

This property translates into a special case of short memory, namely

$$\sum_{h=-\infty}^{\infty} \rho(h) = 0\,, \quad \text{if } d < 0\,,$$

as we will become obvious from the spectrum at frequency zero.

Long Memory in the Frequency Domain

It is obvious from the definition of the spectrum in (4.3) that it does not exist at the origin under long memory ($d > 0$), because the autocovariances are not summable. Still, the previous chapter has been set up sufficiently general to cover long memory, see (4.1). Given a singularity at frequency $\lambda = 0$, one still may determine the rate at which $f(\lambda)$ goes off to infinity as λ approaches 0. To determine f, we have to evaluate the power transfer function of $(1-L)^{-d}$ from Proposition 4.1 and obtain[5]

$$T_{(1-L)^{-d}}(\lambda) = (2 - 2\cos(\lambda))^{-d} = \left(4\sin^2\left(\frac{\lambda}{2}\right)\right)^{-d},$$

[4]More complicated is the effect of changes in d if $d < 0$, see Hassler (2014).

[5]Readers not familiar with complex numbers, $i^2 = -1$, may skip the following equation, see also Footnote 6 in Chap. 4:

$$\begin{aligned} T_{(1-L)^{-d}}(\lambda) &= (1 - e^{i\lambda})^{-d}(1 - e^{-i\lambda})^{-d} \\ &= (1 - e^{i\lambda} - e^{-i\lambda} + 1)^{-d} \\ &= (2 - 2\cos(\lambda))^{-d}. \end{aligned}$$

where the trigonometric half-angle formula was used for the second equality:

$$2\,\sin^2(x) = 1 - \cos(2x)\,. \tag{5.10}$$

We hence have the following result.

Proposition 5.2 (Fractional noise, frequency domain) *Under the assumptions of Proposition 5.1 it holds for the spectrum of fractional noise* $x_t = (1-L)^{-d}\varepsilon_t$ *that*

$$f(\lambda) = \left(4\sin^2\left(\frac{\lambda}{2}\right)\right)^{-d}\frac{\sigma^2}{2\pi}\,,\quad \lambda > 0\,, \tag{5.11}$$

and

$$f(\lambda) \sim \lambda^{-2d}\frac{\sigma^2}{2\pi}\,,\quad \lambda \to 0\,. \tag{5.12}$$

The second statement in Proposition 5.2 is again understood to be asymptotic: Similarly to (5.2) we denote for two function $a(x)$ and $b(x) \neq 0$:

$$a(x) \sim b(x) \text{ for } x \to 0 \quad \Longleftrightarrow \quad \lim_{x\to 0}\frac{a(x)}{b(x)} = 1\,. \tag{5.13}$$

Since $\lim_{x\to 0}\sin(x)/x = 1$ we write $\sin(x) \sim x$ for $x \to 0$. Consequently, (5.12) arises from (5.11).

From Proposition 5.2 we learn that long memory ($d > 0$) translates into a spectral singularity at frequency zero, and the negative slope is the steeper the larger d is. In other words: the longer the memory, the stronger is the contribution of the long-run trend to the variance of the process. The antipersistent case in contrast is characterized by the opposite extreme: $f(0) = 0$. For an illustration, have a look at Fig. 5.4.

Example 5.1 (Fractionally Integrated Noise) Although the fractional noise is dominated by the trend component at frequency zero (strongly persistent) for $d > 0$, the process is stationary as long as $d < 0.5$. Consequently, a typical trajectory can not drift off but displays somehow reversing trends. In Fig. 5.5 we see from simulated data that the deviations from the zero line are stronger for $d = 0.45$ than for $d = 0.25$. The antipersistent series ($d = -0.45$), in contrast, displays an oscillating behavior due to the negative autocorrelation. ■

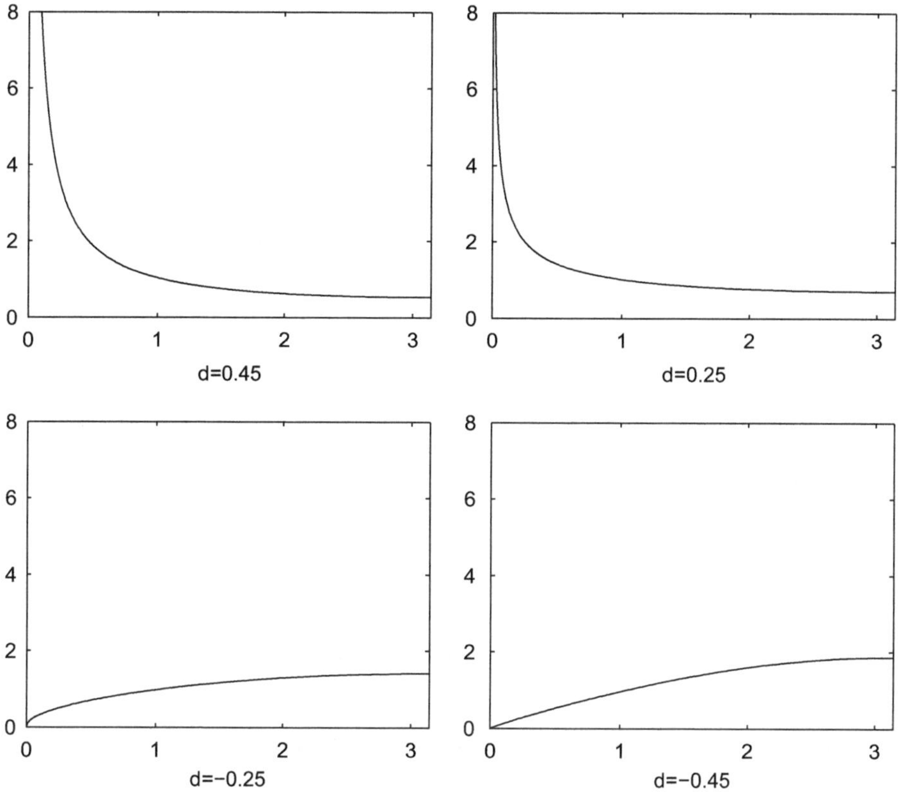

Fig. 5.4 $2\pi f(\lambda)$ from Proposition 5.2 for $d = 0.45, 0.25, -0.25, -0.45$

5.4 Generalizations

On top of long memory as implied by fractional integration for $0 < d < 0.5$, we now want to allow for additional short memory. We assume that Δ^d has to be applied to $\{x_t\}$ in order to obtain a short memory process $\{e_t\}$: $\Delta^d x_t = e_t$. At the end of this section, the order of integration d will be extended beyond $d = 0.5$ to cover nonstationary processes, too. Thus we define general fractionally integrated processes of order d, in short $x_t \sim I(d)$.[6]

[6]The use of '$\sim$' with a differing meaning from that one in (5.2) should not be a source for confusion.

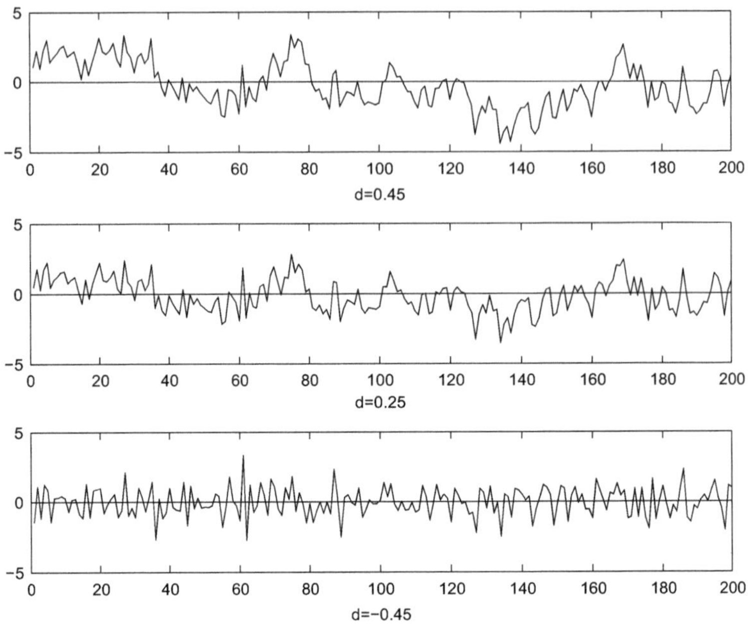

Fig. 5.5 Simulated fractional noise for $d = 0.45, 0.25, -0.45$

Fractionally Integrated ARMA Processes (ARFIMA)

Since the papers by Granger and Joyeux (1980) and Hosking (1981), it is often assumed that $\{e_t\}$ is a stationary and invertible ARMA(p, q) process, $A(L)e_t = B(L)\varepsilon_t$, with spectrum

$$f_e(\lambda) = \frac{T_B(\lambda)}{T_A(\lambda)} \frac{\sigma^2}{2\pi},$$

see Corollary 4.1. An ARFIMA(p,d,q) process is defined by replacing ε_t in (5.8) by e_t, such that

$$A(L)\Delta^d x_t = B(L)\varepsilon_t .$$

With the expansion from (5.5) one obtains

$$x_t = (1-L)^{-d} e_t = \sum_{j=0}^{\infty} \psi_j e_{t-j}, \quad t \in \mathbb{Z}, \quad d < 0.5 . \tag{5.14}$$

Under stationarity and invertibility of the ARMA process, the spectrum f_e of $\{e_t\}$ is bounded and bounded away from zero everywhere:

$$0 < f_e(\lambda) < \infty, \quad \lambda \in [0, \pi]. \tag{5.15}$$

The results from Propositions 5.1 and 5.2 carry over to the ARFIMA(p,d,q) process, see also Brockwell and Davis (1991, Sect. 13.2).

Proposition 5.3 (ARFIMA) *Let the ARMA process with* $A(L)e_t = B(L)\varepsilon_t$ *be stationary and invertible. Then the ARFIMA process with* $A(L)\Delta^d x_t = B(L)\varepsilon_t$ *and* $-1 < d < 0.5$ *is stationary, and it holds that*

(a) the spectral density $f(\lambda)$ *is given as*

$$f(\lambda) = 4^{-d} \sin^{-2d}\left(\frac{\lambda}{2}\right) f_e(\lambda), \quad \lambda > 0$$

$$\sim \lambda^{-2d} f_e(0), \quad \lambda \to 0;$$

(b) the autocovariances satisfy

$$\gamma(h) \sim \gamma_d f_e(0)\, 2\pi\, h^{2d-1}, \quad h \to \infty, \; d \neq 0,$$

with γ_d *from Proposition 5.1.*

Hosking (1981, Thm. 2) and Brockwell and Davis (1991, Thm. 13.2.2) cover only $|d| < 0.5$, but their proof carries over to $-1 < d \leq -0.5$. Further, they state only $\gamma(h) \sim C\, h^{2d-1}$ for some constant $C \neq 0$; looking at the details of the proof of Brockwell and Davis (Thm. 13.2.2), however, it turns out that $C = \gamma_d f_e(0)\, 2\pi$, which of course covers the case of Proposition 5.1, too. In particular, we find again that long memory defined by a non-summable autocovariance sequence translates into a spectral peak at $\lambda = 0$. This feature occurs for $d > 0$.

Semiparametric Models

The parametric assumption that $\{e_t\}$ is an ARMA process is by no means essential for Proposition 5.3 to hold. More generally, we now define a stationary process $\{e_t\}$ to be integrated of order 0, $e_t \sim I(0)$, if

$$e_t = \sum_{k=0}^{\infty} b_k \varepsilon_{t-k}, \quad \text{with} \sum_{k=0}^{\infty} |b_k| < \infty \text{ and } \sum_{k=0}^{\infty} b_k \neq 0, \tag{5.16}$$

where $b_0 = 1$ and $\{\varepsilon_t\} \sim \text{WN}(0, \sigma^2)$. The absolute summability of $\{b_k\}$ rules out long memory (or $d > 0$) in that the autocovariances of $\{e_t\}$ are absolutely summable, see Proposition 3.2; the second condition that the sequence $\{b_k\}$ does not sum up to zero rules out that $\{e_t\}$ is integrated of order d with $d < 0$, see (5.7). This motivates to call processes from (5.16) integrated of order 0. Consequently, $\{x_t\}$ from (5.14) with $\{e_t\}$ from (5.16) is called integrated of order d, $x_t \sim I(d)$. The spectrum of $\{e_t\}$ is of course given by Proposition 4.2, see (4.5). With f_e being the spectrum of the $I(0)$ process, Proposition 5.3 continues to hold without changes, provided $0 < d$ (see Giraitis et al., 2012, Prop. 3.1.1).

A further question is the behavior of the impulse responses of a general $I(d)$ process without parametric model: Does the decay rate j^{d-1} of $\{\psi_j\}$ from Δ^{-d} carry over? The answer is almost yes, but mild additional assumptions have to be imposed on $\{b_k\}$ from (5.16). Denote

$$x_t = \Delta^{-d} e_t = \sum_{j=0}^{\infty} \psi_j e_{t-j} = \sum_{j=0}^{\infty} c_j \varepsilon_{t-j},$$

where the MA coefficients are given by convolution:

$$c_j = \sum_{k=0}^{j} b_k \psi_{j-k}, \quad j \geq 0.$$

Hassler and Kokoszka (2010) prove that a necessary and sufficient condition for

$$c_j \sim \frac{\sum_{k=0}^{\infty} b_k}{\Gamma(d)} j^{d-1}, \quad d > 0,$$

is under long memory

$$k^{1-d} b_k \to 0, \quad k \to \infty. \tag{5.17}$$

This is a very weak condition satisfied by all stationary ARMA models and most other processes of practical interest. Hassler (2012) proves that this condition remains necessary in the case of antipersistence, $d < 0$, and establishes a mildly stronger sufficient condition.

The statistical literature often refrains from a fractionally integrated model of the type $x_t = (1 - L)^{-d} e_t$, and directly assumes for the corresponding spectral behavior in a vicinity of the origin:

$$f(\lambda) \sim \lambda^{-2d} g(\lambda), \quad \lambda \to 0.$$

When it comes to estimation of d, technical smoothness restrictions are imposed on g, including of course the minimum assumption

$$0 < g(0) < \infty\,,$$

which is required to identify d.

Nonstationary Processes

The simplest way to define a process that is integrated of a higher order than $d < 0.5$ is as follows. Consider a sequence $\{y_t\}$ that has to be differenced once in order to obtain an $I(\delta)$ process, $\Delta y_t = x_t$, $x_t \sim I(\delta)$ or $y_t = y_{t-1} + x_t$. Given a starting value $y_0 = 0$, the solution of this difference equation for $t \in \{1, 2, \ldots, n\}$ is

$$y_t = \sum_{j=1}^{t} x_j\,, \quad t = 1, 2, \ldots, n\,.$$

Since $\{y_t\}$ is given by integration over an $I(\delta)$ process, we say that $\{y_t\}$ is integrated of order d, $y_t \sim I(d)$, with $d = \delta + 1$. For $d \geq 0.5$, i.e. $\delta \geq -0.5$, the process $\{y_t\}$ is necessarily nonstationary. We illustrate this type of nonstationarity or drift by means of an example.

Example 5.2 (Nonstationary Fractional Noise) The middle graph in Fig. 5.6 displays a realization of a random walk ($d = 1$). It drifts off from the zero line for very long time spans and crosses only a few times. The $I(1.45)$ process drifts even more pronouncedly displaying a much smoother trajectory than the random walk. The $I(0.55)$ process does not drift as strongly, hitting the zero line much more often. In fact, comparing the $I(0.55)$ series with the $I(0.45)$ case from Fig. 5.5, one can imagine that it may be hard to tell apart stationarity and nonstationarity in finite samples. ■

The case of $\delta = 0$ or $y_t \sim I(1)$ is of particular interest in many financial and economic applications. Hence, one may wish to test whether a process is $I(1)$ or not,

$$H_0 : d = 1 \quad \text{vs.} \quad H_1 : d \neq 1\,.$$

One method to discriminate more specifically between $d = 1$ and $d = 0$ is the celebrated test by Dickey and Fuller (1979), see Chap. 15. In a fractionally integrated framework it is more generally possible to decide e.g. whether a process is nonstationary or not,

$$H_0 : d \geq 0.5 \quad \text{vs.} \quad H_1 : d < 0.5\,,$$

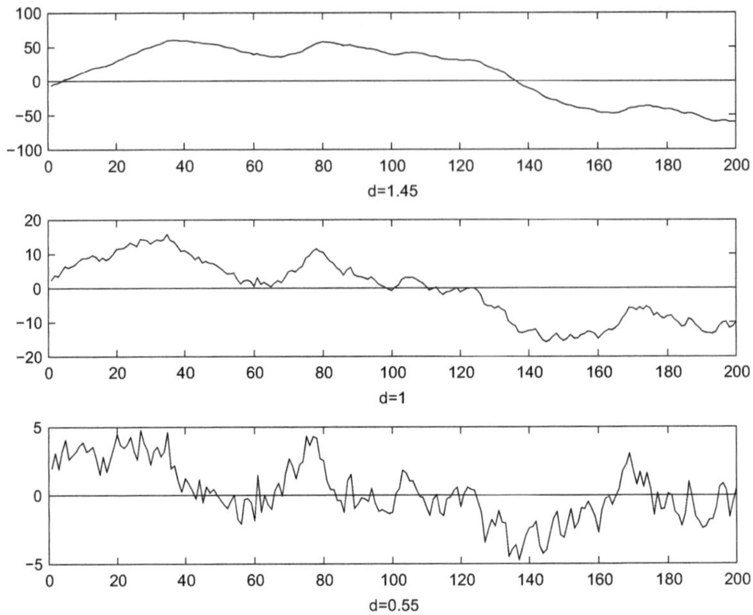

Fig. 5.6 Nonstationary fractional noise for $d = 1.45, 1.0, 0.55$

or whether a process has short memory or not,

$$H_0 : d \leq 0 \quad \text{vs.} \quad H_1 : d > 0 \, .$$

Demetrescu, Kuzin, and Hassler (2008) suggested a simple procedure similar to the Dickey-Fuller test, to test for arbitrary values d_0.

5.5 Problems and Solutions

Problems

5.1 Consider the p-series $\sum_{j=1}^{J} j^{-p}$ for $J \to \infty$. Show that the limit is finite if and only if $p > 1$, see (5.1).

5.2 Show that exponential decay is faster than hyperbolic decay, i.e.

$$\lim_{j\to\infty} \frac{g^j}{j^{d-1}} = 0 \quad \text{for } 0 < g < 1 \, , \; |d| < 1 \, .$$

5.3 Show (5.4).

Hint: Use properties of the Gamma function

$$\Gamma(x) = \begin{cases} \int_0^\infty t^{x-1} e^{-t} dt\,, & x > 0 \\ \infty\,, & x = 0 \\ \Gamma(x+1)/x\,, & x < 0\,, x \neq -1, -2, \ldots \end{cases} . \tag{5.18}$$

5.4 Establish the following expression for the autocovariance $\gamma(h)$ of fractionally integrated noise:

$$\gamma(h) = \sigma^2 \frac{\Gamma(1-2d)\Gamma(d+h)}{\Gamma(d)\Gamma(1-d)\Gamma(1-d+h)} . \tag{5.19}$$

Hint: Use Proposition 4.1 with (5.11), and apply the following identity from Gradshteyn and Ryzhik (2000, 3.631.8):

$$\int_0^\pi \sin^{\nu-1}(x)\cos(ax)dx = \frac{\pi\cos(\frac{a\pi}{2})\Gamma(\nu+1)}{2^{\nu-1}\nu\,\Gamma(\frac{\nu+a+1}{2})\Gamma(\frac{\nu-a+1}{2})}, \quad \text{where } \nu > 0 .$$

5.5 Show Proposition 5.1 (a).
Hint: Use (5.19).

5.6 Show Proposition 5.1 (b).
Hint: Use (5.19).

5.7 Show Proposition 5.1 (c).

5.8 Consider the ARFIMA(0,d,1) model $\Delta^d x_t = B(L)\,\varepsilon_t$, $B(L) = 1 + b\,L$. Show for this special case that the proportionality constant from Proposition 5.3 (b) is indeed $\gamma_d f_e(0)\,2\pi$.

Solutions

5.1 We define the p-series of the first J terms,

$$S_J(p) = \sum_{j=1}^{J} \frac{1}{j^p} .$$

First, we discuss the case separating the convergence region from the divergent one ($p = 1$). For convenience choose $J = 2^n$, such that

$$S_{2^n}(1) = \left(1 + \frac{1}{2}\right) + \left(\frac{1}{3} + \frac{1}{4}\right) + \cdots + \left(\frac{1}{2^{n-1}+1} + \cdots + \frac{1}{2^n}\right)$$
$$> \frac{1}{2} + \frac{2}{4} + \frac{4}{8} + \cdots + \frac{2^{n-1}}{2^n} = \frac{n}{2}.$$

Since the lower bound of $S_{2^n}(1)$ diverges to infinity with n, $S_J(1)$ must diverge, too.

Second, consider the case where $p < 1$, such that $j^p < j$, or $j^{-p} > j^{-1}$. Hence, $S_J(p) > S_J(1)$, and divergence of $S_J(1)$ implies divergence for $p < 1$.

Third, for $p > 1$, we group the terms for $J = 2^n - 1$ as follows.

$$S_{2^n-1}(p) = 1 + \left(\frac{1}{2^p} + \frac{1}{3^p}\right) + \cdots + \left(\frac{1}{(2^{n-1})^p} + \cdots + \frac{1}{(2^n-1)^p}\right)$$
$$< 1 + \frac{2}{2^p} + \frac{4}{4^p} + \cdots + \frac{2^{n-1}}{(2^{n-1})^p}$$
$$= 1 + \frac{1}{2^{p-1}} + \frac{1}{4^{p-1}} + \cdots + \frac{1}{(2^{n-1})^{p-1}}.$$

We now abbreviate $g = \frac{1}{2^{p-1}}$ with $0 < g < 1$ since $p > 1$. Consequently,

$$S_{2^n-1}(p) < 1 + g + g^2 + \ldots + g^{n-1} = \frac{1-g^n}{1-g}$$
$$< \sum_{i=0}^{\infty} g^i = \frac{1}{1-g} = \frac{2^{p-1}}{2^{p-1}-1},$$

where we use the geometric series, see Problem 3.2. Hence, $S_{2^n-1}(p)$ is bounded for every n, while growing monotonically at the same time, which establishes convergence for $p > 1$. Hence, the proof of (5.1) is complete.

We want to add a final remark. While convergence is ensured for $p > 1$, an explicit expression for the limit is by no means obvious, and indeed only known for selected values. For example, for $p = 2$ one has the famous result by Leonhard Euler:

$$\lim_{J\to\infty} S_J(2) = \sum_{j=1}^{\infty} \frac{1}{j^2} = \frac{\pi^2}{6}.$$

5.2 Since the limit of the ratio of interest is indeterminate, we apply L'Hospital's rule:

$$\lim_{j\to\infty}\frac{g^j}{j^{d-1}} = \lim_{j\to\infty}\frac{j^{1-d}}{g^{-j}} = \lim_{j\to\infty}\frac{(1-d)j^{-d}}{g^{-j}\log(g)(-1)}$$
$$= \frac{d-1}{\log(g)}\lim_{j\to\infty}\frac{g^j}{j^d} = 0\,, \quad d \geq 0\,.$$

If $-1 < d < 0$, the expression g^j/j^d is still indeterminate. Application of L'Hospital's rule twice, however, yields:

$$\lim_{j\to\infty}\frac{g^j}{j^d} = \lim_{j\to\infty}\frac{j^{-d}}{g^{-j}} = \lim_{j\to\infty}\frac{-dj^{-d-1}}{g^{-1}\log(g)(-1)}$$
$$= \frac{d}{\log(g)}\lim_{j\to\infty}\frac{g^j}{j^{1+d}} = 0.$$

This establishes the claim.

5.3 Prior to solving the problem, we review some useful properties of the Gamma function that is often employed to simplify manipulations with binomial expressions, see e.g. Sydsæter, Strøm, and Berck (1999, p.52), and in much greater detail Gradshteyn and Ryzhik (2000, Sect. 8.31), or Rudin (1976, Ch. 8) containing proofs. For integer numbers, Γ coincides with the factorial,

$$\Gamma(n+1) = n\,(n-1)\,\cdots\,2 = n!\,,$$

which implies a recursive relation holding in fact in general:

$$\Gamma(x+1) = x\,\Gamma(x)\,. \tag{5.20}$$

Hence, obviously $\Gamma(1) = \Gamma(2) = 1$, and a further value often encountered is $\Gamma(0.5) = \sqrt{\pi}$. The recursive relation further yields the rate of divergence at the origin,

$$\Gamma(x) \sim x^{-1}\,, \quad x \to 0\,,$$

which justifies the convention $\Gamma(0)/\Gamma(0) = 1$. Finally, we want to approximate the Gamma function for large arguments. Remember Stirling's formula for factorials,

$$n! \sim \sqrt{2\pi\,n}\left(\frac{n}{e}\right)^n\,.$$

It generalizes to

$$\Gamma(x) \sim \sqrt{\frac{2\pi}{x}} \left(\frac{x}{e}\right)^x$$

for $x \to \infty$, which again has to be read as

$$\lim_{x\to\infty} \Gamma(x) \Big/ \sqrt{\frac{2\pi}{x}} \left(\frac{x}{e}\right)^x = 1\,.$$

Consequently, we have for finite x and y and large integer n that[7]

$$\frac{\Gamma(n+x)}{\Gamma(n+y)} \sim n^{x-y}, \quad n \to \infty\,. \tag{5.21}$$

Now, we turn to establishing (5.4). Repeated application of (5.20) gives

$$\frac{\Gamma(j-d)}{\Gamma(-d)} = (j-d-1)\,(j-d-2)\cdots(-d)\,.$$

By definition of the binomial coefficients we conclude from (5.3) with $\Gamma(j+1) = j!$ that

$$\begin{aligned}\pi_j &= \frac{\Gamma\,(j-d)}{\Gamma\,(j+1)\,\Gamma\,(-d)}, \quad j \geq 0 \\ &\sim \frac{j^{-d-1}}{\Gamma\,(-d)}, \quad j \to \infty\,,\end{aligned}$$

where the approximation relies on (5.21). This is the required result.

5.4 With Proposition 4.1 and (5.11) we compute

$$\begin{aligned}\gamma(h) &= 2\int_0^{\pi} f(\lambda)\cos(\lambda h)d\lambda \\ &= \frac{4^{-d}}{\pi} 2\sigma^2 \int_0^{\frac{\pi}{2}} \sin^{-2d}(x)\cos(2hx)dx,\end{aligned}$$

[7]Use $(1+x/n)^n \to e^x$.

where we substituted $\frac{\lambda}{2} = x$. Both $\sin^2(x)$ and $\cos(2hx)$ are even symmetric around $\frac{\pi}{2}$, such that

$$\int_0^{\pi} \sin^{-2d}(x)\cos(2hx)dx = 2\int_0^{\frac{\pi}{2}} \sin^{-2d}(x)\cos(2hx)dx.$$

Hence, the integration formula 3.631.8. from Gradshteyn and Ryzhik (2000) can be applied with $\nu = 1 - 2d > 0$ as long as $d < 0.5$ and $a = 2h$:

$$\begin{aligned}
\gamma(h) &= \frac{4^{-d}\sigma^2}{\pi}\int_0^{\pi} \sin^{-2d}(x)\cos(2hx)dx \\
&= \frac{4^{-d}\sigma^2}{\pi}\frac{\pi\cos(h\pi)\Gamma(2-2d)}{2^{-2d}(1-2d)\Gamma(1-d+h)\Gamma(1-d-h)} \\
&= \sigma^2\frac{(-1)^h\Gamma(2-2d)}{(1-2d)\Gamma(1-d+h)\Gamma(1-d-h)} \\
&= \sigma^2\frac{(-1)^h\Gamma(1-2d)}{\Gamma(1-d+h)\Gamma(1-d-h)}.
\end{aligned}$$

Using (5.20) once more, one can show that

$$\frac{\Gamma(d+h)}{\Gamma(d)} = (-1)^h\frac{\Gamma(1-d)}{\Gamma(1-d-h)}.$$

Therefore, we finally have

$$\gamma(h) = \sigma^2\frac{\Gamma(1-2d)\Gamma(d+h)}{\Gamma(d)\Gamma(1-d)\Gamma(1-d+h)},$$

which is (5.19).

5.5 With (5.19) we obtain

$$\gamma(0; d) = \frac{\Gamma(1-2d)}{(\Gamma(1-d))^2},$$

where we assumed $\sigma^2 = 1$ without loss of generality. Instead of the variance, we will equivalently minimize the natural logarithm thereof. We determine as derivative

$$\begin{aligned}
\frac{\partial\log(\gamma(0;d))}{\partial d} &= -2\left(\frac{\Gamma'(1-2d)}{\Gamma(1-2d)} - \frac{\Gamma'(1-d)}{\Gamma(1-d)}\right) \\
&= -2\left(\psi(1-2d) - \psi(1-d)\right),
\end{aligned}$$

where the so-called psi function is defined as logarithmic derivative of the Gamma function,

$$\psi(x) = \frac{\partial \log(\Gamma(x))}{\partial x}.$$

The psi function is strictly increasing for $x > 0$, which can be shown in different ways, see e.g. Gradshteyn and Ryzhik (2000, 8.362.1). Consequently,

$$\psi(1-d) - \psi(1-2d) \begin{cases} > 0, & 0 < d < 0.5 \\ = 0, & d = 0 \\ < 0, & d < 0 \end{cases},$$

which proves that $\log(\gamma(0;d))$, and hence $\gamma(0;d)$, takes on its minimum for $d = 0$. This solves the problem.

5.6 The recursive relative for $\gamma(h)$ is obvious with (5.19) and (5.20) at hand.
For $h \to \infty$, the approximation in (5.21) yields

$$\gamma(h) \sim \sigma^2 \frac{\Gamma(1-2d)}{\Gamma(d)\Gamma(1-d)} h^{2d-1},$$

which defines the constant from Proposition 5.1 (b):

$$\gamma_d = \frac{\Gamma(1-2d)}{\Gamma(d)\Gamma(1-d)}.$$

The Gamma function is positive for positive arguments and negative on the interval $(-1, 0)$. Hence, the sign of γ_d equals the sign of $\Gamma(d)$ since $d < 0.5$, which completes the proof of Proposition 5.1 (b).

5.7 In terms of autocorrelations the recursion from Proposition 5.1 (b) becomes

$$\rho(h;d) = f(h;d)\,\rho(h-1;d), \quad h \geq 1,$$

where the factor $f(h;d)$ is positive for $d > 0$,

$$f(h;d) = \frac{h-1+d}{h-d} > 0 \quad \text{with} \quad \frac{\partial f(h;d)}{\partial d} > 0,$$

such that $\rho(h;d) > 0$ since $\rho(0;d) = 1$. Hence, we have

$$\frac{\partial \rho(h;d)}{\partial d} = \frac{\partial f(h;d)}{\partial d}\rho(h-1;d) + f(h;d)\frac{\partial \rho(h-1;d)}{\partial d},$$

which is positive by induction since

$$\frac{\partial \rho(1;d)}{\partial d} = \frac{\partial f(1;d)}{\partial d} > 0.$$

Hence, $\rho(h;d)$ is growing with d as required.

5.8 The spectral density of the MA(1) component $e_t = \varepsilon_t + b\varepsilon_{t-1}$ is known from Example 4.3:

$$f_e(0) = (1+b)^2\sigma^2/2\pi.$$

We now express x_t in terms of a fractional noise called $y_t = \Delta^{-d}\varepsilon_t$:

$$\begin{aligned} x_t &= \Delta^{-d}B(L)\varepsilon_t = B(L)\Delta^{-d}\varepsilon_t \\ &= (1+bL)\,y_t\,. \end{aligned}$$

Let $\gamma_x(h)$ and $\gamma_y(h)$ denote the autocovariances of $\{x_t\}$ and $\{y_t\}$, respectively. It holds that

$$\begin{aligned} \gamma_x(h) &= \mathrm{E}\left[(y_t + by_{t-1})\,(y_{t+h} + by_{t+h-1})\right] \\ &= (1+b^2)\gamma_y(h) + b\left(\gamma_y(h-1) + \gamma_y(h+1)\right)\,. \end{aligned}$$

With the behavior of $\gamma_y(h)$ from Proposition 5.1 it follows

$$\begin{aligned} \frac{\gamma_x(h)}{h^{2d-1}} &\to (1+b^2)\,\gamma_d\sigma^2 + 2\,b\,\gamma_d\sigma^2 \\ &= \gamma_d f_e(0)\,2\pi\,, \end{aligned}$$

as required.

References

Baillie, R. T. (1996). Long memory processes and fractional integration in econometrics. *Journal of Econometrics, 73*, 5–59.

Bondon, P., & Palma, W. (2007). A class of antipersistent processes. *Journal of Time Series Analysis, 28*, 261–273.

Brockwell, P. J., & Davis, R. A. (1991). *Time series: Theory and methods* (2nd ed.). New York: Springer.

Demetrescu, M., Kuzin, V., & Hassler, U. (2008). Long memory testing in the time domain. *Econometric Theory, 24*, 176–215.

Dickey, D. A., & Fuller, W. A. (1979). Distribution of the estimators for autoregressive time series with a unit root. *Journal of the American Statistical Association, 74*, 427–431.

Giraitis, L., Koul, H. L., & Surgailis, D. (2012). *Large sample inference for long memory processes*. London: Imperial College Press.

Gradshteyn, I. S., & Ryzhik, I. M. (2000). *Table of integrals, series, and products* (6th ed.). London/San Diego: Academic Press.

Granger, C. W. J. (1981). Some properties of time series data and their use in econometric model specification. *Journal of Econometrics, 16*, 121–130.

Granger, C. W. J., & Joyeux, R. (1980). An Introduction to long-memory time series models and fractional differencing. *Journal of Time Series Analysis, 1*, 15–29.

Hassler, U. (2012). Impulse responses of antipersistent processes. *Economics Letters, 116*, 454–456.

Hassler, U. (2014). Persistence under temporal aggregation and differencing. *Economics Letters, 124*, 318–322.

Hassler, U., & Kokoszka (2010). Impulse responses of fractionally integrated processes with long memory. *Econometric Theory, 26*, 1855–1861.

Hosking, J. R. M. (1981). Fractional differencing. *Biometrika, 68*, 165–176.

Maasoumi, E., & McAleer, M. (2008). Realized volatility and long memory: An overview. *Econometric Reviews, 27*, 1–9.

Rudin, W. (1976). *Principles of mathematical analyis* (3rd ed.). New York: McGraw-Hill.

Sydsæter, K., Strøm, A., & Berck, P. (1999). *Economists' mathematical manual* (3rd ed.). Berlin/New York: Springer.

Trench, W. F. (2013). *Introduction to real analysis*. Free Hyperlinked Edition 2.04 December 2013. Downloaded on 10th May 2014 from http://digitalcommons.trinity.edu/mono/7.

Processes with Autoregressive Conditional Heteroskedasticity (ARCH) 6

6.1 Summary

In particular in the case of financial time series one often observes a highly fluctuating volatility (or variance) of a series: Agitated periods with extreme amplitudes alternate with rather quiet periods being characterized by moderate observations. After some short preliminary considerations concerning models with time-dependent heteroskedasticity, we will discuss the model of autoregressive conditional heteroskedasticity (ARCH), for which Robert F. Engle was awarded the Nobel prize in the year 2003. After a generalization (GARCH), there will be a discussion on extensions relevant for practice. Throughout this chapter, the innovations or shocks $\{\varepsilon_t\}$ stand for a pure random process as defined in Example 2.7.

6.2 Time-Dependent Heteroskedasticity

The heteroskedasticity allowed for here is modeled as time-dependent **volatility** by[1]

$$x_t = \sigma_t \, \varepsilon_t \,, \quad \varepsilon_t \sim \text{iid}(0, 1) \,, \tag{6.1}$$

[1]The following equation could be extended by a mean function, e.g. of a regression-type,

$$x_t = \alpha + \beta z_t + \sigma_t \, \varepsilon_t \,,$$

or

$$x_t = a_1 x_{t-1} + \cdots + a_p x_{t-p} + \sigma_t \, \varepsilon_t \,.$$

We restrict our exposition and concentrate on modeling volatility exclusively, although in practice time-dependent heteroskedasticity is often found with regression errors.

U. Hassler, *Stochastic Processes and Calculus*, Springer Texts in Business and Economics, DOI 10.1007/978-3-319-23428-1_6

where $\{\sigma_t\}$ is the volatility process being stochastically independent of $\{\varepsilon_t\}$ by assumption, and $\{\varepsilon_t\}$ is a pure random process with unit variance. There exist two routes to model the volatility process. The first one, often labeled as processes with stochastic volatility, assumes an unobserved, or latent, process $\{h_t\}$ behind the volatility: $\sigma_t = \exp(h_t/2)$. This implies for the squared data

$$x_t^2 = e^{h_t}\varepsilon_t^2 \quad \text{or} \quad \log(x_t^2) = h_t + \log(\varepsilon_t^2)\,.$$

For an early survey on stochastic volatility processes see Taylor (1994). A second strand in the literature assumes that σ_t depends on observed data, in particular on past observations x_{t-j}. This class of models has been called autoregressive conditional heteroskedasticity (ARCH). ARCH processes are widely spread and successful in practice and will be the focus of attention in the present chapter.

Heteroskedasticity as a Function of the Past

In this chapter the variance function is modeled by the observed past of the process itself:

$$\sigma_t^2 = f\left(x_{t-1}, x_{t-2}, \cdots\right)\,. \tag{6.2}$$

By plugging in x_{t-j} from (6.1) one obtains:

$$\sigma_t^2 = f\left(\sigma_{t-1}\varepsilon_{t-1}, \sigma_{t-2}\varepsilon_{t-2}, \cdots\right)\,.$$

We will show that the process from (6.1) is a martingale difference sequence. Remember the definition of the information set $\mathcal{I}_{t-1}$ generated by the past of x_t up to x_{t-1}. Then it holds that[2]

$$\begin{aligned}
\mathrm{E}\left(x_t \,|\, \mathcal{I}_{t-1}\right) &= \mathrm{E}\left(x_t \,|\, x_{t-1}, x_{t-2}, \cdots\right)\\
&= \mathrm{E}\left(\sigma_t\varepsilon_t \,|\, x_{t-1}, x_{t-2}, \cdots\right)\\
&= \sigma_t \mathrm{E}\left(\varepsilon_t \,|\, x_{t-1}, x_{t-2}, \cdots\right)\\
&= \sigma_t \mathrm{E}\left(\varepsilon_t\right)\,,
\end{aligned}$$

as ε_t is independent of x_{t-j} for $j > 0$ by construction. With ε_t being zero on average, it follows that

$$\mathrm{E}\left(x_t \,|\, \mathcal{I}_{t-1}\right) = 0\,,$$

[2]When conditioning on $\mathcal{I}_{t-1}$, one often writes $\mathrm{E}\left(\cdot \,|\, x_{t-1}, x_{t-2}, \cdots\right)$ instead of $\mathrm{E}\left(\cdot \,|\, \mathcal{I}_{t-1}\right)$.

which proves (integrability of $\{x_t\}$ assumed) that $\{x_t\}$ is in fact a martingale difference sequence, see below (2.11). The variance of the martingale difference is determined in the following way (as x_t is zero on average,

$$\begin{aligned}\operatorname{Var}(x_t) &= \operatorname{E}\left(x_t^2\right)\\ &= \operatorname{E}\left(\sigma_t^2\varepsilon_t^2\right)\\ &= \operatorname{E}\left(\sigma_t^2\right)\operatorname{E}\left(\varepsilon_t^2\right)\\ &= \operatorname{E}\left(\sigma_t^2\right),\end{aligned}$$

as σ_t^2 from (6.2) and ε_t (with variance 1 from (6.1)) are stochastically independent. Hence, the following proposition is verified.

Proposition 6.1 (Heteroskedastic Martingale Differences) *Let $\{x_t\}$ be from (6.1) and $\{\sigma_t^2\}$ from (6.2) with $E(\sigma_t^2) < \infty$ independent of $\{\varepsilon_t\}$. Then $\{x_t\}$ is a martingale difference sequence with variance*

$$Var(x_t) = E(x_t^2) = E(\sigma_t^2).$$

Let us remember Proposition 2.2. Due to the martingale difference property it holds that

$$\operatorname{E}(x_t) = 0 \quad \text{and} \quad \gamma(h) = \operatorname{E}(x_t x_{t+h}) = 0\,,\ h \neq 0\,.$$

Hence, the process is serially uncorrelated with expectation zero which would be supposed e.g. for returns. However, the process is generally not independent over time. The (weak) stationarity of $\{x_t\}$ depends on the possibly variable variance; if the variance $\operatorname{Var}(x_t)$ is constant, then the entire process is stationary.

Heuristics

Now, the question is how the functional dependence in (6.2) should be specified and parameterized. Heteroskedasticity as an empirical phenomenon has been known to observers on financial markets for a long time. Before ARCH models were introduced, it had been measured by moving a window of width B through the data and averaging over the squares:

$$s_t^2 = \frac{1}{B}\sum_{i=1}^{B} x_{t-i}^2\,.$$

For every point in time t one averages over the past preceding B values in order to determine the variance in t. In doing so, we do not center x_{t-i} around the arithmetic mean as we think of returns with $\operatorname{E}(x_t) = 0$ when applying the procedure, cf.

Proposition 6.1. With daily observations (with five trading days a week) one chooses e.g. $B = 20$ which approximately corresponds to a time window of a month. A first improvement of the time-dependent volatility measurement is obtained by using a weighted average where the weights, $g_i \geq 0$, are not negative:

$$s_t^2 = \sum_{i=1}^{B} g_i x_{t-i}^2 \quad \text{with} \quad \sum_{i=1}^{B} g_i = 1. \tag{6.3}$$

Example 6.1 (Exponential Smoothing) For the weighting function g_i one often uses an exponential decay:

$$g_i = \frac{\lambda^{i-1}}{1 + \lambda + \ldots + \lambda^{B-1}} \quad \text{with} \quad 0 < \lambda < 1.$$

Note that the denominator is just defined such that it holds that $\sum_{i=1}^{B} g_i = 1$. With growing B one furthermore obtains

$$1 + \lambda + \ldots + \lambda^{B-1} = \frac{1 - \lambda^B}{1 - \lambda} \to \frac{1}{1 - \lambda}.$$

Inserting the exponentially decaying weights in (6.3), we get the following result for $B \to \infty$:

$$s_t^2(\lambda) = (1 - \lambda) \sum_{i=1}^{\infty} \lambda^{i-1} x_{t-i}^2.$$

Now it is an easy exercise to verify the following recursive relation:

$$s_t^2(\lambda) = (1 - \lambda) x_{t-1}^2 + \lambda s_{t-1}^2(\lambda). \tag{6.4}$$

We will call $s_t^2(\lambda)$ the exponentially smoothed volatility or variance. In order to be able to calculate it for $t = 2, \ldots, n$, we need a starting value. Typically, $s_1^2(\lambda) = x_1^2$ is chosen which leads to $s_2^2(\lambda) = x_1^2$. ■

The ARCH and GARCH processes which are subsequently introduced are models leading to volatility specifications which generalize s_t^2 and $s_t^2(\lambda)$ from (6.3) and (6.4), respectively.

6.3 ARCH Models

So-called autoregressive conditional heteroskedasticity models can be traced back to Engle (1982). We consider the case of the order q and specify σ_t^2 from (6.2) as follows:

$$\sigma_t^2 = \alpha_0 + \alpha_1 x_{t-1}^2 + \ldots + \alpha_q x_{t-q}^2, \tag{6.5}$$

where it is assumed that

$$\alpha_0 > 0 \quad \text{and} \quad \alpha_i \geq 0\,,\ i = 1, \ldots, q\,, \tag{6.6}$$

in order to guarantee $\sigma_t^2 > 0$. Note that this variance function corresponds to s_t^2 from (6.3). For $\alpha_1 = \cdots = \alpha_q = 0$, however, the case of homoskedasticity is modeled.

Conditional Moments

Given $x_{t-1}, \ldots, x_{t-q}$, one naturally obtains zero for the conditional expectation as ARCH processes are martingale differences, see Proposition 6.1. For the conditional variance it holds that

$$\begin{aligned}
\operatorname{Var}\left(x_t \,|\, x_{t-1}, \ldots, x_{t-q}\right) &= \mathrm{E}\left(x_t^2 \,|\, x_{t-1}, \ldots, x_{t-q}\right) \\
&= \sigma_t^2 \mathrm{E}\left(\varepsilon_t^2 \,|\, x_{t-1}, \cdots, x_{t-q}\right) \\
&= \sigma_t^2\,,
\end{aligned}$$

as ε_t is again independent of x_{t-j} and has a unit variance. Hence, for the variance it conditionally holds that:

$$\operatorname{Var}(x_t|x_{t-1}, \ldots, x_{t-q}) = \alpha_0 + \alpha_1\, x_{t-1}^2 + \ldots + \alpha_q\, x_{t-q}^2\,,$$

which explains the name of the models: The conditional variance is modeled autoregressively (where "autoregressive" means in this case: dependent on the past of the process). Thus, extreme amplitudes in the previous period are followed by high volatility in the present period resulting in so-called volatility clusters. If the assumption of normality of the innovations is added, then the conditional distribution of x_t given the past is normal as well:

$$x_t \,|\, x_{t-1}, \ldots x_{t-q} \sim \mathcal{N}\left(0,\, \alpha_0 + \alpha_1\, x_{t-1}^2 + \ldots + \alpha_q\, x_{t-q}^2\right).$$

But, although the original work by Engle (1982) assumed $\varepsilon_t \sim \mathrm{ii}\,\mathcal{N}(0,1)$, the assumption of normality is not crucial for ARCH effects.

Stationarity

By Proposition 6.1 it holds for the variance that

$$\operatorname{Var}(x_t) = \mathrm{E}(\sigma_t^2) = \alpha_0 + \alpha_1 \mathrm{E}(x_{t-1}^2) + \cdots + \alpha_q \mathrm{E}(x_{t-q}^2)\,.$$

If the process is stationary, $\operatorname{Var}(x_t) = \operatorname{Var}(x_{t-j})$, $j = 1, \ldots, q$, then its variance results as

$$\operatorname{Var}(x_t) = \frac{\alpha_0}{1 - \alpha_1 - \ldots - \alpha_q}\,.$$

For a positive variance expression, this requires necessarily (due to $\alpha_0 > 0$):

$$1 - \alpha_1 - \ldots - \alpha_q > 0\,.$$

In fact, this condition is sufficient for stationarity as well. In Problem 6.1 we therefore show the following result.

Proposition 6.2 (Stationary ARCH)
Let $\{x_t\}$ be from (6.1) and $\{\sigma_t^2\}$ from (6.5) with (6.6). The process is weakly stationary if and only if it holds that

$$\sum_{j=1}^{q} \alpha_j < 1\,.$$

Correlation of the Squares

We define $e_t = x_t^2 - \sigma_t^2$. Due to Proposition 6.1 the expected value is zero: $\mathrm{E}(e_t) = 0$. Adding x_t^2 to both sides of (6.5), one immediately obtains

$$x_t^2 = \alpha_0 + \alpha_1\, x_{t-1}^2 + \ldots + \alpha_q\, x_{t-q}^2 + e_t\,.$$

From this we learn that an ARCH(q) process implies an autoregressive structure for the squares $\{x_t^2\}$. The serial dependence of an ARCH process originates from the squares of the process. Because of $\alpha_i \geq 0$, x_t^2 and x_{t-i}^2 are positively correlated which again allows to capture volatility clusters.

In Figs. 6.1 and 6.2 ARCH(1) time series of the length 500 are simulated. For this purpose pseudo-random numbers $\{\varepsilon_t\}$ are generated as normally distributed and $\alpha_0 = 1$ is chosen. The effect of α_1 is now quite obvious. The larger the value of this parameter, the more obvious are the volatility clusters. Long periods with little movement are followed by shorter periods of vehement, extreme amplitudes which can be negative as well as positive. These volatility clusters become even more obvious in the respective lower panel of the figures, in which the squared observations $\{x_t^2\}$ are depicted. Because of the positive autocorrelation of the squares, small amplitudes tend again to be followed by small ones while extreme observations appear to follow each other.

Skewness and Kurtosis

In the first section of the chapter on basic concepts from probability theory we have defined the kurtosis by means of the fourth moment of a random variable and we have denoted the corresponding coefficient by γ_2. For $\gamma_2 > 3$ the density function is more "peaked" than the one of the normal distribution: On the one hand the

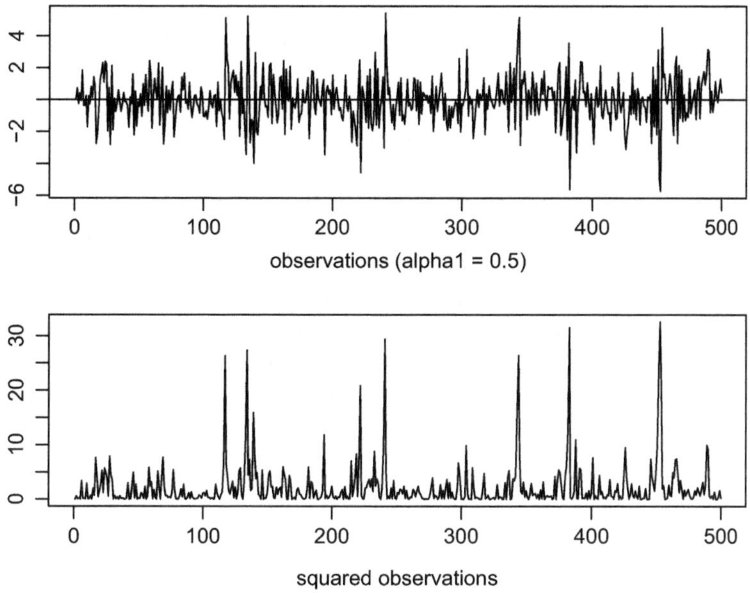

Fig. 6.1 ARCH(1) with $\alpha_0 = 1$ and $\alpha_1 = 0.5$

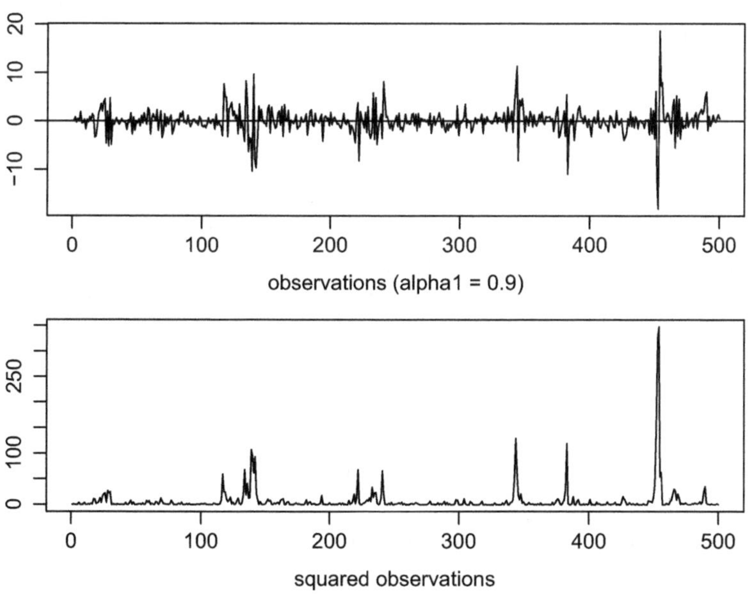

Fig. 6.2 ARCH(1) with $\alpha_0 = 1$ and $\alpha_1 = 0.9$

values are more concentrated around the expected value, on the other hand there occur extreme observations in the tail of the distribution with higher probability ("fat-tailed and highly peaked"). For stationary ARCH processes with Gaussian innovations ($\varepsilon_t \sim \text{ii}\,\mathcal{N}(0,1)$) it holds that the kurtosis exceeds 3 (provided it exists at all):

$$\gamma_2 > 3 .$$

The corresponding derivation can be found in Problem 6.2. Due to this excess kurtosis, ARCH is generally incompatible with the assumption of an *un*conditional Gaussian distribution.

We define the skewness coefficient γ_1 similarly to the kurtosis by the third moment of a standardized random variable. The skewness coefficient of ARCH models depends on the symmetry of ε_t. If this innovation is symmetric, then it follows that $\mathrm{E}(\varepsilon_t^3) = 0$. Hence, $\gamma_1 = 0$ follows for the corresponding ARCH process (due to independence of σ_t and ε_t):

$$\begin{aligned} \mathrm{E}(x_t^3) &= \mathrm{E}(\sigma_t^3) \cdot \mathrm{E}(\varepsilon_t^3) \\ &= \mathrm{E}(\sigma_t^3) \cdot 0 = 0. \end{aligned}$$

Thereby it was only used that ε_t is symmetrically distributed.

Example 6.2 (ARCH(1)) In particular for a stationary ARCH(1) process with $\alpha_1^2 < \frac{1}{3}$ and Gaussian innovations ε_t the kurtosis is finite and it results as (see Problem 6.3):

$$\gamma_2 = 3\,\frac{1-\alpha_1^2}{1-3\,\alpha_1^2} > 3 .$$

For $\alpha_1^2 \geq \frac{1}{3}$ there occur extreme observations with a high probability such that the kurtosis is no longer constant. Consider a stationary ARCH(1) process ($\alpha_1 < 1$) with $\alpha_1^2 = 1/3$. Under this condition one has for $\mu_{4,t} = \mathrm{E}(x_t^4)$ with $\mathrm{E}(x_{t-1}^2) = \alpha_0/(1-\alpha_1)$ assuming Gaussianity:

$$\begin{aligned} \mu_{4,t} &= \mathrm{E}(\varepsilon_t^4)\,\mathrm{E}(\sigma_t^4) = 3\,\mathrm{E}\left((\alpha_0 + \alpha_1 x_{t-1}^2)^2\right) \\ &= 3\left(\alpha_0^2 + 2\,\frac{\alpha_0^2\alpha_1}{1-\alpha_1} + \alpha_1^2\mu_{4,t-1}\right) = c + 3\,\alpha_1^2\mu_{4,t-1} = c + \mu_{4,t-1} , \end{aligned}$$

where the constant c is appropriately defined. Continued substitution yields thus

$$\mu_{4,t} = c\,t + \mu_{4,0} .$$

Hence, we observe that the kurtosis grows linearly over time if $3\,\alpha_1^2 = 1$. ■

6.4 Generalizations

Some extensions of the ARCH model having originated from empirical features of financial data will be covered in the following.

GARCH

In practice, with many financial series it can be observed that the correlation of the squares reaches far into the past. Therefore, for an adequate modeling a large q is needed, i.e. a large number of parameters. A very economical parametrization, however, is allowed for by the GARCH model.

Generalized ARCH processes of the order p and q were introduced by Bollerslev (1986) and are defined by their volatility function

$$\sigma_t^2 = \alpha_0 + \alpha_1 x_{t-1}^2 + \cdots + \alpha_q x_{t-q}^2 + \beta_1 \sigma_{t-1}^2 + \cdots + \beta_p \sigma_{t-p}^2 . \tag{6.7}$$

The result process is abbreviated as GARCH(p, q). In addition to the parameter restrictions from (6.6) it is required that

$$\beta_i \geq 0 \, , \; i = 1, \ldots, p \, . \tag{6.8}$$

Jointly, these restrictions are clearly sufficient for $\sigma_t^2 > 0$ but stricter than necessary. Substantially weaker assumptions were derived by Nelson and Cao (1992).

We adopt the stationarity conditions for GARCH models from Bollerslev (1986, Theorem 1). The resulting variance will be determined in Problem 6.4. Thus, we obtain the following results.

Proposition 6.3 (Stationary GARCH)

Let $\{x_t\}$ be from (6.1) and $\{\sigma_t^2\}$ from (6.7) with (6.6) and (6.8). The process is weakly stationary if and only if

$$\sum_{j=1}^{q} \alpha_j + \sum_{j=1}^{p} \beta_j < 1 \, .$$

Then it holds for the variance that

$$Var(x_t) = \frac{\alpha_0}{1 - \sum_{j=1}^{q} \alpha_j - \sum_{j=1}^{p} \beta_j} \, .$$

It can be shown that the stationary GARCH process can be considered as an ARCH(∞) process. Under the conditions from Proposition 6.3 it holds that (see Problem 6.5):

$$\sigma_t^2 = \gamma_0 + \sum_{i=1}^{\infty} \gamma_i \, x_{t-i}^2 \quad \text{with } \gamma_i \geq 0 \text{ and } \sum_{i=1}^{\infty} |\gamma_i| < \infty \, . \tag{6.9}$$

Thus, the GARCH process allows for modeling an infinitely long dependence of the volatility on the past of the process itself with only $p + q$ parameters although this dependence decays with time (i.e. $\gamma_i \to 0$ for $i \to \infty$). The fact that GARCH can be considered as ARCH(∞) has the nice consequence that results for stationary ARCH processes also hold for GARCH models. In particular, GARCH models are again special cases of processes with volatility (6.2) and therefore examples of martingale differences, i.e. Proposition 6.1 holds true. If we assume a Gaussian distribution of $\{\varepsilon_t\}$, it follows, just as for the ARCH(q) process of finite order, that the skewness is zero and that the kurtosis exceeds the value 3.

Example 6.3 (GARCH(1,1)) Consider the GARCH(1,1) case more explicitly. It is by far the most frequently used GARCH specification in practice. Continued substitution shows under the assumption of stationarity that ($\alpha_1 + \beta_1 < 1$):

$$\sigma_t^2 = \frac{\alpha_0}{1 - \beta_1} + \alpha_1 \sum_{i=1}^{\infty} \beta_1^{i-1} \, x_{t-i}^2 \, .$$

Hence, we have an explicit ARCH(∞) representation of GARCH(1,1). Assuming that

$$1 - (\alpha_1 + \beta_1)^2 - 2\alpha_1^2 > 0,$$

the kurtosis is defined. In Problem 6.6 we show (with Gaussian distribution of $\{\varepsilon_t\}$):

$$\gamma_2 = 3 \, \frac{1 - (\alpha_1 + \beta_1)^2}{1 - (\alpha_1 + \beta_1)^2 - 2\alpha_1^2} > 3.$$

Furthermore one shows by

$$\begin{aligned} \sigma_t^2 &= \alpha_0 + \alpha_1 \, x_{t-1}^2 + \beta_1 \, \sigma_{t-1}^2 \\ &\Longleftrightarrow \\ x_t^2 &= \alpha_0 + (\alpha_1 + \beta_1) \, x_{t-1}^2 - \beta_1 \, (x_{t-1}^2 - \sigma_{t-1}^2) + x_t^2 - \sigma_t^2 \end{aligned}$$

the equation

$$x_t^2 = \alpha_0 + (\alpha_1 + \beta_1) \, x_{t-1}^2 + e_t - \beta_1 \, e_{t-1}$$

with

$$e_t = x_t^2 - \sigma_t^2\,, \quad \mathrm{E}(e_t) = 0\,.$$

The GARCH(1,1) process $\{x_t\}$ therefore corresponds to an ARMA(1,1) structure of the squares $\{x_t^2\}$. ■

In Figs. 6.3 and 6.4 the influence of the sum of the parameters $\alpha_1 + \beta_1$ is illustrated by means of simulated GARCH(1,1) observations. We therefore fix $\alpha_0 = 1$ and $\alpha_1 = 0.3$ and vary β_1 in such a way that stationarity is ensured. The larger β_1 (and therefore the sum of $\alpha_1 + \beta_1$), the more pronounced is the change from quiet periods with little or, in absolute value, moderate amplitudes to excited periods in which extreme amplitudes follow each other. Again, this pattern of volatility becomes particularly apparent with the serially correlated squares in the lower panel, respectively.

IGARCH

Considering the volatility of GARCH(1,1),

$$\sigma_t^2 = \alpha_0 + \alpha_1\, x_{t-1}^2 + \beta_1\, \sigma_{t-1}^2,$$

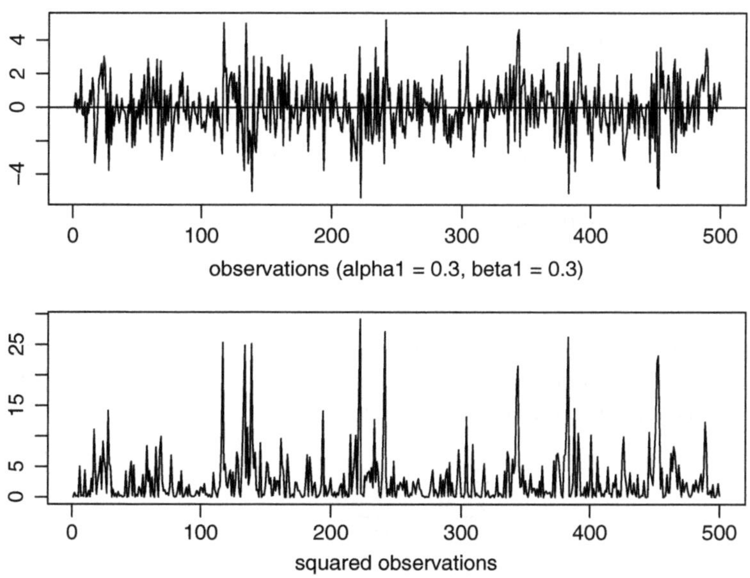

Fig. 6.3 GARCH(1,1) with $\alpha_0 = 1$, $\alpha_1 = 0.3$ and $\beta_1 = 0.3$

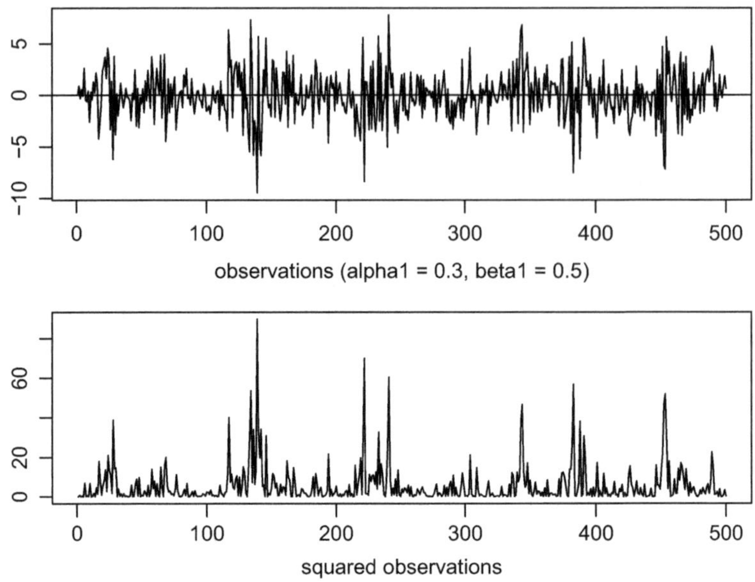

Fig. 6.4 GARCH(1,1) with $\alpha_0 = 1, \alpha_1 = 0.3$ and $\beta_1 = 0.5$

we are reminded of $s_t^2(\alpha)$ from (6.4). The difference being that $\alpha_0 = 0$, and it holds that $\alpha_1 + \beta_1 = 1$ (i.e. $\alpha_1 = 1 - \lambda$ and $\beta_1 = \lambda$, respectively). Models with such a restriction violate the stationarity condition ($\alpha_1 + \beta_1 < 1$). This can be shown when forming the expected value of σ_t^2 with $x_{t-1}^2 = \sigma_{t-1}^2 \, \varepsilon_{t-1}^2$:

$$\begin{aligned} \mathrm{E}(\sigma_t^2) &= \alpha_0 + \alpha_1 \, \mathrm{E}(\sigma_{t-1}^2) \, \mathrm{E}(\varepsilon_{t-1}^2) + \beta_1 \mathrm{E}(\sigma_{t-1}^2) \\ &= \alpha_0 + (\alpha_1 + \beta_1) \, \mathrm{E}(\sigma_{t-1}^2). \end{aligned}$$

With $\alpha_1 + \beta_1 = 1$ one obtains

$$\mathrm{E}(\sigma_t^2 - \sigma_{t-1}^2) = \alpha_0 > 0.$$

In other words: The expectations for the increments of the volatility are positive for every point in time, modeling a volatility expectation which tends to infinity with t. This idea was generalized in literature. With

$$\sum_{j=1}^{q} \alpha_j + \sum_{j=1}^{p} \beta_j = 1$$

one talks about integrated GARCH processes (IGARCH) since Engle and Bollerslev (1986). This is a naming which becomes more understandable in the chapter on integrated processes.

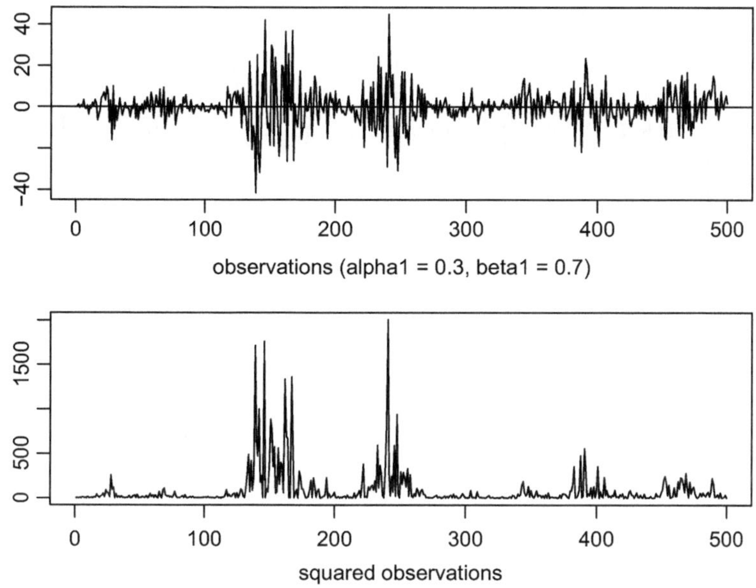

Fig. 6.5 IGARCH(1,1) with $\alpha_0 = 1$, $\alpha_1 = 0.3$ and $\beta_1 = 0.7$

In Fig. 6.5, an IGARCH(1,1) process ($\alpha_1 + \beta_1 = 1$) was simulated according to the scheme from Figs. 6.3 and 6.4. In comparison to the previous figures, in this case we find considerably more extreme volatility clusters which, however, are not exaggerated. The kind of depicted dynamics in Fig. 6.5 can be frequently observed in financial practice.

GARCH-M

We talk about "GARCH in mean"[3] (GARCH-M) if the volatility term influences the (mean) level of the process. In order to explain this with regard to contents, we think of risk premia: For a high volatility of an investment (high-risk), a higher return is expected, on average. In equation form we write this down as follows:

$$x_t = \theta\, \sigma_t + u_t\,, \tag{6.10}$$

where $\{u_t\}$ is a GARCH process:

$$u_t = \sigma_t\, \varepsilon_t\,, \quad \sigma_t^2 = \alpha_0 + \alpha_1\, u_{t-1}^2 + \cdots + \alpha_q\, u_{t-q}^2 + \beta_1\, \sigma_{t-1}^2 + \ldots + \beta_p\, \sigma_{t-p}^2\,. \tag{6.11}$$

[3]Originally, the ARCH-M model was proposed by Engle, Lilien, and Robins (1987).

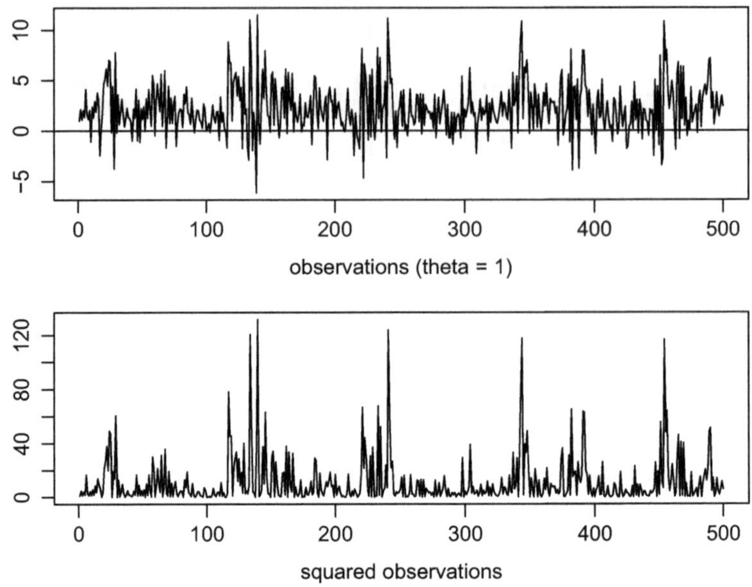

Fig. 6.6 GARCH(1,1)-M from (6.11) with $\alpha_0 = 1$, $\alpha_1 = 0.3$ and $\beta_1 = 0.5$

Therefore, in this case the mean function μ_t is set to $\theta\,\sigma_t$. In some applications it has proved successful to model the risk premium as a multiple of the variance instead of modeling it by the standard deviation ($\theta\,\sigma_t$):

$$x_t = \theta\,\sigma_t^2 + u_t\,.$$

For both mean functions the GARCH-M process $\{x_t\}$ is no longer free from serial correlation for $\theta > 0$; it is no longer a martingale difference sequence.

In Fig. 6.6 a GARCH-M series was generated as in (6.11). The volatility cluster can be well identified in the lower panel of the squares.The effect of $\theta = 1$ becomes apparent in the upper panel: In the series of x_t local, reversing trends can be spotted. Upward trends involve a high volatility, whereas quiet periods are marked by a decreasing or lower level.

EGARCH

We talk about exponential GARCH when the volatility is modeled as an exponential function of the past squares x_{t-i}^2. This suggestion originates from Nelson (1991) and was made in order to capture the asymmetries in the volatility.[4] It is observed that decreasing stock prices (negative returns) tend to involve higher volatilities than

[4]We do not exactly present Nelson's model but a slightly modified implementation which is used in the software package *EViews*.

increasing ones. This so-called **leverage effect** is not captured by ordinary GARCH models.

In *EViews* the variance for EGARCH is calculated as follows:

$$\log \sigma_t^2 = \omega + \sum_{j=1}^{p} \beta_j \log \sigma_{t-j}^2 + \sum_{j=1}^{q} \left(\alpha_j |\varepsilon_{t-j}| + \gamma_j \varepsilon_{t-j} \right),$$

where ε_{t-j} is defined as

$$\varepsilon_{t-j} = \frac{x_{t-j}}{\sigma_{t-j}}.$$

For $\gamma_j = 0$ the sign is not an issue. However, for $\gamma_j < 0$ in the negative case

$$\alpha_j |\varepsilon_{t-j}| + \gamma_j \varepsilon_{t-j} = (\alpha_j - \gamma_j)|\varepsilon_{t-j}| \quad \text{for} \quad \varepsilon_{t-j} < 0$$

has a stronger effect on $\log \sigma_t^2$ than in the positive case

$$\alpha_j |\varepsilon_{t-j}| + \gamma_j \varepsilon_{t-j} = (\alpha_j + \gamma_j)\varepsilon_{t-j} \quad \text{for} \quad \varepsilon_{t-j} > 0.$$

Note that for EGARCH the expression σ_t^2 is without parameter restrictions always positive by construction. Applying the exponential function, it results that

$$\sigma_t^2 = \exp \left[\omega + \sum_{j=1}^{p} \beta_j \log \sigma_{t-j}^2 + \sum_{j=1}^{q} \left(\alpha_j |\varepsilon_{t-j}| + \gamma_j \varepsilon_{t-j} \right) \right].$$

Example 6.4 (EGARCH (1,1)) Again, as special case we treat the situation with $p = q = 1$,

$$\log \sigma_t^2 = \omega + \beta_1 \log \sigma_{t-1}^2 + \alpha_1 |\varepsilon_{t-1}| + \gamma_1 \varepsilon_{t-1},$$

or after applying the exponential function:

$$\begin{aligned} \sigma_t^2 &= e^{\omega} \sigma_{t-1}^{2\beta_1} \cdot \exp\left(\alpha_1 |\varepsilon_{t-1}| + \gamma_1 \varepsilon_{t-1} \right) \\ &= e^{\omega} \sigma_{t-1}^{2\beta_1} \cdot \begin{cases} \exp\left(|\varepsilon_{t-1}|(\alpha_1 - \gamma_1) \right), & \varepsilon_{t-1} \le 0 \\ \exp\left(\varepsilon_{t-1}(\alpha_1 + \gamma_1) \right), & \varepsilon_{t-1} \ge 0 \end{cases}. \end{aligned}$$

In this case it is again shown that for $\gamma_1 < 0$ the leverage effect is modeled in such a way that negative observations have a larger volatility effect than positive observations of the same absolute value. ■

In Fig. 6.7 a realization of a simulated EGARCH(1,1) process is depicted. The "leverage parameter" is $\gamma_1 = -0.5$. When the graphs of the squared and the original observations are compared, it can be detected that the most extreme amplitudes are

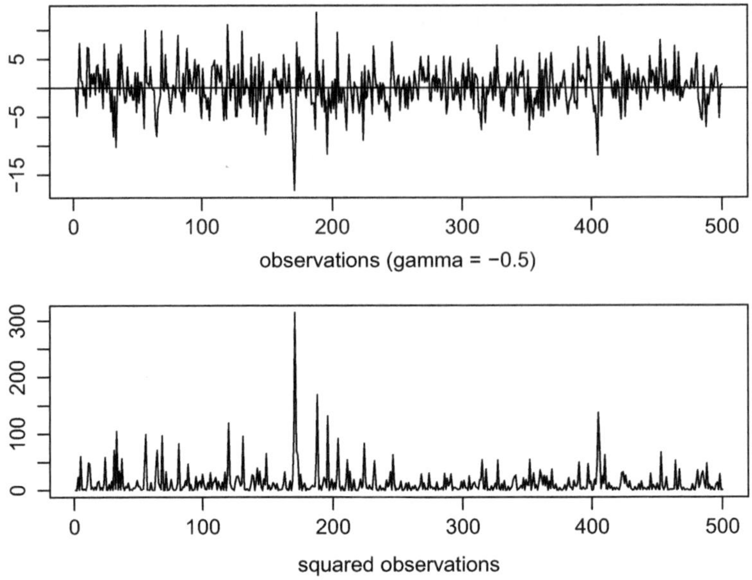

Fig. 6.7 EGARCH(1,1) with $\omega = 1$, $\alpha_1 = 0.3$, $\beta_1 = 0.5$ and $\gamma_1 = -0.5$

in fact negative. Furthermore, it can be observed that periods with predominantly negative values are characterized by a high volatility.

YAARCH

The works of Engle (1982) and Bollerslev (1986) have set the stage for a downright ARCH industry. A large number of generalizations and extensions has been published and applied in practice. Most of these versions were published under more or less appealing acronyms. When Engle (2002) balanced the books after 20 years of ARCH, he added with some irony another acronym: YAARCH standing for Yet Another ARCH. There is no end in sight for this literature.

6.5 Problems and Solutions

Problems

6.1 Prove Proposition 6.2.
Hint: According to Engle (1982, Theorem 2) the process is stationary if and only if it holds that

$$\alpha(z) = 0 \Rightarrow |z| > 1\,, \tag{6.12}$$

with $\alpha(z) := 1 - \alpha_1 z - \ldots - \alpha_q z^q$.

6.2 Show that the kurtosis γ_2 of an ARCH process exceeds the value 3. Assume a Gaussian distribution of $\{\varepsilon_t\}$ and $\mathrm{E}(\sigma_t^4) < \infty$.

6.3 Calculate the kurtosis of a stationary ARCH(1) process as given in Example 6.2 for the case that it exists. Assume a Gaussian distribution of $\{\varepsilon_t\}$.

6.4 Assume $\{x_t\}$ to be a stationary GARCH process. Determine the variance expression from Proposition 6.3.

6.5 Assume $\{x_t\}$ to be a stationary GARCH process with (6.6) and (6.8). Determine the ARCH(∞) representation from (6.9).

6.6 Calculate the kurtosis of a stationary GARCH(1,1) process as given in Example 6.3 for the case that it exists. Assume a Gaussian distribution of $\{\varepsilon_t\}$.

Solutions

6.1 We have to show the equivalence of (6.12) and the condition from Proposition 6.2, given (6.6). This condition can also be written as $\alpha(1) > 0$. Hence, we have to prove the equivalence:

$$(6.12) \iff \alpha(1) > 0.$$

We proceed in two steps.
"$\Rightarrow$": Under the condition by Engle (1982) it holds that

$$\mathrm{Var}(x_t) = \frac{\alpha_0}{1 - \alpha_1 - \ldots - \alpha_q} = \frac{\alpha_0}{\alpha(1)} \geq 0\,.$$

Due to $\alpha_0 > 0$ it immediately follows that $\alpha(1) \geq 0$. The case $\alpha(1) = 0$, however, is due to (6.12) excluded, such that $\alpha(1) > 0$ can be concluded.
"$\Leftarrow$": For a root z of $\alpha(z)$ it holds that:

$$1 = \sum_{j=1}^{q} \alpha_j z^j\,.$$

By the triangle inequality and applying (6.6) it follows that:

$$\begin{aligned} 1 \leq \sum_{j=1}^{q} \left|\alpha_j z^j\right| = \sum_{j=1}^{q} \alpha_j \left|z^j\right| &\leq \max_j \left|z^j\right| \sum_{j=1}^{q} \alpha_j \\ &= \max_j \left|z^j\right| (1 - \alpha(1)) \\ &< \max_j \left|z^j\right|, \end{aligned}$$

where the assumption was used for the last inequality. Therefore, we have shown for a root z of $\alpha(z)$ that it holds that $\max_j |z^j| > 1$ and thus $|z| > 1$.
Hence, the proof is completed.

6.2 From Example 2.4 we adopt due to the Gaussian distribution

$$\mathrm{E}(\varepsilon_t^4) = \mathrm{E}\left(\left(\frac{\varepsilon_t - 0}{1}\right)^4\right) = 3\,.$$

Therefore, in a first step it follows due to the independence of σ_t and ε_t:

$$\mathrm{E}(x_t^4) = \mathrm{E}\left(\sigma_t^4\,\varepsilon_t^4\right) = \mathrm{E}\left(\sigma_t^4\right)\,\mathrm{E}\left(\varepsilon_t^4\right) = 3\,\mathrm{E}\left(\sigma_t^4\right).$$

Hence, because of $\mathrm{E}(x_t) = 0$ and Proposition 6.1 the kurtosis of x_t results as:

$$\gamma_2 = \frac{\mathrm{E}(x_t^4)}{(\mathrm{Var}(x_t))^2} = \frac{3\,\mathrm{E}\left(\sigma_t^4\right)}{\left(\mathrm{E}\left(\sigma_t^2\right)\right)^2}\,.$$

The usual variance decomposition, see Eq. (2.1),

$$\mathrm{Var}\left(\sigma_t^2\right) = \mathrm{E}\left(\sigma_t^4\right) - \left(\mathrm{E}\left(\sigma_t^2\right)\right)^2 \geq 0\,,$$

yields

$$\frac{\mathrm{E}\left(\sigma_t^4\right)}{\left(\mathrm{E}\left(\sigma_t^2\right)\right)^2} \geq 1\,.$$

Hence, the claim is verified: $\gamma_2 \geq 3$.

6.3 As σ_t and ε_t are stochastically independent, it holds that

$$\mathrm{E}(x_t^k) = \mathrm{E}(\sigma_t^k)\mathrm{E}(\varepsilon_t^k),$$

whereby the k-th central moment is given, as $\{x_t\}$ is a martingale difference sequence with zero expectation. On the assumption of a standard normally distributed random process one obtains

$$\mathrm{E}(\varepsilon_t^2) = 1, \quad \mathrm{E}(\varepsilon_t^3) = 0 \quad \text{and} \quad \mathrm{E}(\varepsilon_t^4) = 3.$$

This implies for the ARCH(1) process that the skewness is zero due to $\mathrm{E}(x_t^3) = 0$.

In order to determine the kurtosis, we first observe that the fourth moment is constant under the condition $3\alpha_1^2 < 1$. To that end define $\mu_{4,t}$ and use $\mu_2 = \mathrm{Var}(x_t) = \frac{\alpha_0}{1-\alpha_1}$:

$$\begin{aligned}\mu_{4,t} &= \mathrm{E}(x_t^4) = \mathrm{E}(\varepsilon_t^4)\,\mathrm{E}(\sigma_t^4) = 3\,\mathrm{E}\left((\alpha_0 + \alpha_1 x_{t-1}^2)^2\right) \\ &= 3\left(\alpha_0^2 + 2\,\frac{\alpha_0^2\alpha_1}{1-\alpha_1} + \alpha_1^2\mu_{4,t-1}\right) = c + 3\,\alpha_1^2\mu_{4,t-1}\,,\end{aligned}$$

where the constant c is defined appropriately. Infinite substitution yields

$$\mu_{4,t} = c\,\left(1 + 3\alpha_1^2 + (3\alpha_1^2)^2 + \cdots\right) = \frac{c}{1-3\alpha_1^2} = \mu_4\,.$$

With a constant μ_4 (and μ_2) one obtains

$$\begin{aligned}\mu_4 &= \mathrm{E}(x_t^4) = 3\,\mathrm{E}(\sigma_t^4) = 3\,\mathrm{E}(\alpha_0^2 + 2\alpha_0\,\alpha_1\,x_{t-1}^2 + \alpha_1^2\,x_{t-1}^4) \\ &= 3[\alpha_0^2 + 2\alpha_0\,\alpha_1\,\mu_2 + \alpha_1^2\,\mu_4]\end{aligned}$$

or

$$\begin{aligned}\mu_4 &= \frac{3}{1-3\,\alpha_1^2}\left[\alpha_0^2 + \frac{2\alpha_0^2\,\alpha_1}{1-\alpha_1}\right] \\ &= \frac{3}{1-3\,\alpha_1^2}\,\frac{\alpha_0^2(1+\alpha_1)}{1-\alpha_1}.\end{aligned}$$

From this it follows that

$$\begin{aligned}\gamma_2 &= \frac{\mu_4}{(\mathrm{Var}(x_t))^2} = \frac{3}{1-3\,\alpha_1^2}(1-\alpha_1)(1+\alpha_1) \\ &= 3\,\frac{1-\alpha_1^2}{1-3\,\alpha_1^2}.\end{aligned}$$

Of course, these transformations were only possible for $1 - 3\alpha_1^2 > 0$. Hence, this is the condition for a finite, constant kurtosis.

6.4 We use the fact that σ_t from (6.7) is again independent of ε_t. This can be shown by substitution of σ_{t-j}^2 and x_{t-i}^2 according to (6.1). Thus, as in Proposition 6.1, for stationarity and for arbitrary points in time, it holds that:

$$\mathrm{Var}(x_t) = \mathrm{E}\left(\sigma_t^2\right) = \gamma(0)\,.$$

Hence, by forming the expected value we obtain from (6.7):

$$\gamma(0) = \alpha_0 + \alpha_1 \gamma(0) + \ldots + \alpha_q \gamma(0) + \beta_1 \gamma(0) + \ldots + \beta_p \gamma(0).$$

Therefore, we can solve

$$\gamma(0) = \frac{\alpha_0}{1 - \sum_{j=1}^{q} \alpha_j - \sum_{j=1}^{p} \beta_j},$$

as claimed.

6.5 We define the lag polynomial $\beta(L)$ with

$$\beta(L) = 1 - \beta_1 L - \ldots - \beta_p L^p.$$

Hence, it holds that

$$\beta(L)\,\sigma_t^2 = \alpha_0 + \alpha_1 x_{t-1}^2 + \ldots + \alpha_q x_{t-q}^2.$$

By assumption

$$\beta_j \geq 0, \quad j = 1, \ldots, p, \quad \text{and} \quad \beta(1) > 0.$$

In Problem 6.1 we have shown that this is equivalent to

$$\beta(z) = 0 \quad \Rightarrow \quad |z| > 1.$$

This is in turn the condition of invertibility known from Proposition 3.3 which guarantees a causal, absolutely summable series expansion with coefficients $\{c_j\}$:

$$\frac{1}{\beta(L)} = \sum_{j=0}^{\infty} c_j L^j, \quad \sum_{j=0}^{\infty} |c_j| < \infty.$$

By comparison of coefficients one obtains from

$$1 = \left(1 - \beta_1 L - \ldots - \beta_p L^p\right) \sum_{j=0}^{\infty} c_j L^j$$

as usual

$$\begin{aligned} c_0 &= 1 \\ c_1 &= \beta_1 c_0 = \beta_1 \geq 0 \end{aligned}$$

$$c_2 = \beta_1 c_1 + \beta_2 c_0 = \beta_1^2 + \beta_2 \geq 0$$

$$\vdots$$

$$c_j = \beta_1 c_{j-1} + \ldots + \beta_p c_{j-p} \geq 0 , \quad j \geq p .$$

Thus, the inversion of $\beta(L)$ yields:

$$\begin{aligned}
\sigma_t^2 &= \frac{\alpha_0}{\beta(1)} + \frac{\alpha_1 x_{t-1}^2 + \ldots + \alpha_q x_{t-q}^2}{\beta(L)} \\
&= \gamma_0 + \sum_{i=1}^{\infty} \gamma_i x_{t-i}^2 ,
\end{aligned}$$

where γ_i, $i > 0$, results by convolution of

$$\frac{\alpha_1 L + \ldots + \alpha_q L^q}{\beta(L)} = \sum_{k=1}^{q} \alpha_k L^k \sum_{j=1}^{\infty} c_j L^j .$$

The non-negativity and summability of $\{\alpha_k\}$ and $\{c_j\}$ is conveyed to the series $\{\gamma_i\}$. Hence, the proof is complete.

6.6 As for the ARCH(1) case it holds that

$$\mu_4 = \mathrm{E}\left(x_t^4\right) = 3\,\mathrm{E}\left(\sigma_t^4\right) .$$

Applying $\mathrm{E}\left(x_t^2\right) = \mathrm{E}\left(\sigma_t^2\right) = \frac{\alpha_0}{1-\alpha_1-\beta_1}$ yields:

$$\begin{aligned}
\mathrm{E}\left(\sigma_t^4\right) &= \mathrm{E}\left(\left[\alpha_0 + \alpha_1 x_{t-1}^2 + \beta_1 \sigma_{t-1}^2\right]^2\right) \\
&= \mathrm{E}\left(\alpha_0^2 + \alpha_1^2 x_{t-1}^4 + \beta_1^2 \sigma_{t-1}^4\right) \\
&\quad + \mathrm{E}\left(2\,\alpha_0 \alpha_1 x_{t-1}^2 + 2\,\alpha_0 \beta_1 \sigma_{t-1}^2 + 2\,\alpha_1 \beta_1 x_{t-1}^2 \sigma_{t-1}^2\right) \\
&= \alpha_0^2 + 3\,\alpha_1^2\,\mathrm{E}\left(\sigma_{t-1}^4\right) + \beta_1^2\,\mathrm{E}\left(\sigma_{t-1}^4\right) \\
&\quad + \frac{2\,\alpha_0^2 \alpha_1}{1-\alpha_1-\beta_1} + \frac{2\,\alpha_0^2 \beta_1}{1-\alpha_1-\beta_1} + 2\,\alpha_1 \beta_1\,\mathrm{E}\left(\sigma_{t-1}^4\right)\mathrm{E}\left(\varepsilon_{t-1}^2\right) .
\end{aligned}$$

As for the ARCH(1) case one has to show that $\mathrm{E}\left(\sigma_t^4\right)$ turns out to be constant under stationarity and the condition $1 - 2\,\alpha_1^2 - (\alpha_1 + \beta_1)^2 > 0$. We omit this step here and take $\mathrm{E}\left(\sigma_t^4\right) = \mathrm{E}\left(\sigma_{t-1}^4\right)$ for granted. It then holds that:

$$\mathrm{E}\left(\sigma_t^4\right) = \frac{\alpha_0^2 + \frac{2\alpha_0^2(\alpha_1+\beta_1)}{1-\alpha_1-\beta_1}}{1 - 2\,\alpha_1^2 - (\alpha_1 + \beta_1)^2} .$$

From this it follows that

$$\begin{aligned}
\gamma_2 &= 3\,\mathrm{E}\left(\sigma_t^4\right) \frac{1}{\left(\mathrm{Var}\left(x_t^2\right)\right)^2} \\
&= 3\,\frac{\alpha_0^2\,(1+\alpha_1+\beta_1)}{(1-\alpha_1-\beta_1)(1-2\,\alpha_1^2-(\alpha_1+\beta_1)^2)} \; \frac{(1-\alpha_1-\beta_1)^2}{\alpha_0^2} \\
&= 3\,\frac{(1+(\alpha_1+\beta_1))(1-(\alpha_1+\beta_1))}{1-2\,\alpha_1^2-(\alpha_1+\beta_1)^2} \\
&= 3\,\frac{1-(\alpha_1+\beta_1)^2}{1-2\,\alpha_1^2-(\alpha_1+\beta_1)^2}\,.
\end{aligned}$$

This is in accordance with the claim.

References

Bollerslev, T. (1986). Generalized autoregressive conditional heteroskedasticity. *Journal of Econometrics, 31*, 307–327.

Engle, R. F. (1982). Autoregressive conditional heteroskedasticity with estimates of the variance of U.K. inflation. *Econometrica, 50*, 987–1008.

Engle, R. F. (2002). New frontiers for ARCH models. *Journal of Applied Econometrics, 17*, 425–446.

Engle, R. F., & Bollerslev T. (1986). Modelling the persistence of conditional variances. *Econometric Reviews, 5*, 1–50.

Engle, R. F., Lilien, D. M., & Robins, R. P. (1987). Estimating time-varying risk premia in the term structure: the ARCH-M model. *Econometrica, 55*, 391–407.

Nelson, D. B. (1991). Conditional heteroskedasticity in asset returns: A new approach. *Econometrica, 59*, 347–370.

Nelson, D. B., & Cao, Ch. Q. (1992). Inequality constraints in the univariate GARCH model. *Journal of Business & Economic Statistics, 10*, 229–235.

Taylor, S.J. (1994). Modeling stochastic volatlity: A review and comparative study. *Mathematical Finance, 4*, 183–204.

Part II

Stochastic Integrals

Wiener Processes (WP) 7

7.1 Summary

The Wiener process (or the Brownian motion) is the starting point and the basis for all the following chapters.[1] This is why we will consider this process more explicitly. It is a continuous-time process having as prominent a position in stochastic calculus as the Gaussian distribution in statistics. After introducing its defining properties intuitively, we will discuss important characteristics in the third section. Examples derived from the Wiener process will conclude the exposition.

7.2 From Random Walk to Wiener Process

We consider a nonstationary special case of the AR(1) process and thereby try to arrive at the Wiener process which is the most important continuous-time process in the fields of our applications.

Random Walks

The cumulation of white noise is labeled **random walk**,

$$x_t = \sum_{j=1}^{t} \varepsilon_j , \quad t \in \{1, 2, \ldots, n\} .$$

[1]Norbert Wiener, 1894–1964, was a US-American mathematician. He succeeded in finding a mathematically solid definition and discussion of the so-called Brownian motion named after the nineteenth century British botanist Brown. With a microscope, Brown initially observed and described erratic paths of molecules.

U. Hassler, *Stochastic Processes and Calculus*, Springer Texts in Business and Economics, DOI 10.1007/978-3-319-23428-1_7

Obviously, it holds that

$$x_t = x_{t-1} + \varepsilon_t\,, \quad x_0 = 0\,.$$

In other words: The random walk results as an AR(1) process for the parameter value $a = 1$ and with the starting value zero[2]

$$x_t = a\,x_{t-1} + \varepsilon_t\,, \quad a = 1\,,\; x_0 = 0\,.$$

As the process is nonstationary,

$$\mathrm{E}(x_t) = 0\,, \quad \mathrm{Var}(x_t) = \sigma^2\,t\,,$$

it cannot have an infinitely long past, i.e. the index set is finite, $\mathbb{T} = \{1, 2, \ldots, n\}$.

In a way, the random walk models the way home of a drunk who at a point in time t turns to the left or to the right by chance and uncorrelated with his previous path. Put more formally: The random walk is a martingale. We briefly want to convince ourselves of this fact. By substitution the AR(1) process yields

$$x_{t+s} = a^s x_t + \sum_{j=0}^{s-1} a^j \varepsilon_{t+s-j}\,.$$

Therefore, for $s > 0$ it holds that

$$\mathrm{E}(x_{t+s}|\mathcal{I}_t) = a^s x_t + 0\,,$$

where $\mathcal{I}_t$ again denotes the information set of the AR(1) process. Thus, the martingale condition (2.11) for AR(1) processes is fulfilled if and only if $a = 1$. The second martingale condition, $\mathrm{E}(|x_t|) < \infty$, is given as $\sigma^2 < \infty$ and hence $\mathrm{E}(x_t^2) = t\,\sigma^2 < \infty$.

Example 7.1 (Discrete-Valued Random Walk) Let the set of outcomes contain only two elements (e.g. coin toss: heads or tails),

$$\Omega = \{\omega_0\,,\; \omega_1\}\,,$$

with probabilities $\mathrm{P}(\{\omega_1\}) = \frac{1}{2} = \mathrm{P}(\{\omega_0\})$. Let $\{\varepsilon_t\}$ be a white noise process assigning the numerical values 1 and -1 to the events,

$$\varepsilon(t;\omega_1) = 1\,, \quad \varepsilon(t;\omega_0) = -1\,, \quad t = 1, 2, \ldots, n\,.$$

[2]This special assumption for the starting value is made out of convenience; it is by no means crucial for the behavior of a random walk.

For every point in time, this induces the probabilities

$$P_\varepsilon(\varepsilon_t = 1) = P(\{\omega_1\}) = P_\varepsilon(\varepsilon_t = -1) = P(\{\omega_0\}) = \frac{1}{2}.$$

Then, for expectation and variance it is immediately obtained:

$$E(\varepsilon_t) = 0, \quad \mathrm{Var}(\varepsilon_t) = 1^2 \frac{1}{2} + (-1)^2 \frac{1}{2} = 1.$$

For $t = 1, \ldots, n$, the corresponding random walk $x_t = \sum_{j=1}^{t} \varepsilon_j$ can only take on the countably many values $\{-n, -n+1, \ldots, n-1, n\}$ and is therefore also called discrete-valued. ■

Example 7.2 (Continuous-Valued Random Walk) If $\{\varepsilon_t\}$ is a Gaussian random process,

$$\varepsilon_t \sim \mathcal{N}(0, \sigma^2),$$

then, obviously, the random walk based thereon is also Gaussian, where the variance grows linearly with time:

$$x_t = \sum_{j=1}^{t} \varepsilon_j \sim \mathcal{N}(0, \sigma^2 t).$$

In this case, $\{x_t\}$ is a continuous random variable by assumption and hence this random walk is also called continuous-valued. ■

Wiener Process

At this point, the continuous-time Wiener process will not yet be defined rigorously, but we will approach it intuitively step by step. In order to do so, we choose the index set $\mathbb{T} = [0, 1]$ with the equidistant, disjoint partition

$$[0, 1) = \bigcup_{i=1}^{n} \left[\frac{i-1}{n}, \frac{i}{n}\right).$$

Now, the random walk is multiplied by the factor $1/\sqrt{n}$ and expanded into a step function $X_n(t)$. Interval by interval, we define as continuous-time process:

$$X_n(t) = \frac{1}{\sqrt{n}} \sum_{j=1}^{i-1} \varepsilon_j \quad \text{for } t \in \left[\frac{i-1}{n}, \frac{i}{n}\right), \; i = 1, \ldots, n. \tag{7.1}$$

In addition, we assume $\varepsilon_t \in \{-1, 1\}$ and set for $t = 1$

$$X_n(1) = \frac{1}{\sqrt{n}} \sum_{j=1}^{n} \varepsilon_j \,.$$

For $t = 0$, i.e. $i = 1$ in (7.1), we follow the convention that a sum equals zero if the upper summation limit is smaller than the lower one, which is why $X_n(0) = 0$ begins in the origin. Apparently, $X_n(t)$ is a constant step function on an interval of the length $1/n$, respectively; if $X_n(t)$ was only observed at the jump discontinuities, a time-discrete random walk would be obtained. Being dependent on the choice of n (i.e. the fineness of the partition), the process $X_n(t)$ is indexed accordingly.

If ε_t is from Example 7.1, i.e. $|\varepsilon_t| = 1$, this means that each individual step of the step function has the height $1/\sqrt{n}$ in absolute value. Hence, $X_n(t)$ only takes on values from

$$\left\{ \frac{-n}{\sqrt{n}}, \frac{-n+1}{\sqrt{n}}, \ldots, \frac{n-1}{\sqrt{n}}, \frac{n}{\sqrt{n}} \right\}.$$

Therefore, $X_n(t)$ is a continuous-time but discrete-valued process.

Now, the starting point for the Wiener process is the step function $X_n(t)$ with ε_t from Example 7.1. The number of the steps obviously depends on n which indicates the fineness of the partition of the unit interval. Simultaneously, the step height of the steps with $n^{-0.5}$ becomes flatter, the finer it is partitioned. Note that, due to this fact, the range becomes finer and finer and larger and larger as n grows. Hence, with n growing, $X_n(t)$ becomes "more continuous", in the sense that the step heights $n^{-0.5}$ turn out to be smaller; simultaneously, the jump discontinuities move together more closely (the steps of the width $1/n$ get narrower) such that $X_n(t)$ can take on more and more possible values. In the limit ($n \to \infty$) a process named after Norbert Wiener is obtained which we will always denote by W in the following:

$$X_n(t) \Rightarrow W(t) \quad \text{for } n \to \infty \,,$$

where "$\Rightarrow$" denotes a mode of convergence which will be clarified in Chap. 14. Intuitively speaking, it holds that for each of the uncountably many points in time t the function $X_n(t)$ converges in distribution to $W(t)$ just at this point. The transition of discrete-time and discrete-valued step functions from (7.1) to the Wiener process (for n growing) is illustrated in Fig. 7.1.[3]

The Wiener process $W(t)$ as a limit of $X_n(t)$ is continuous-valued with range $\mathbb{R} = (-\infty, \infty)$ and of course it is continuous-time with $t \in [0, 1]$. Furthermore, the Wiener process is a Gaussian process (normally distributed) which is not that surprising. As, due to the central limit theorem for $n \to \infty$, it holds for the

[3]In order not to violate the concept of functions, strictly speaking, the vertical lines would not be allowed to occur in the graphs of the figure. We ignore this subtlety to enhance clarity.

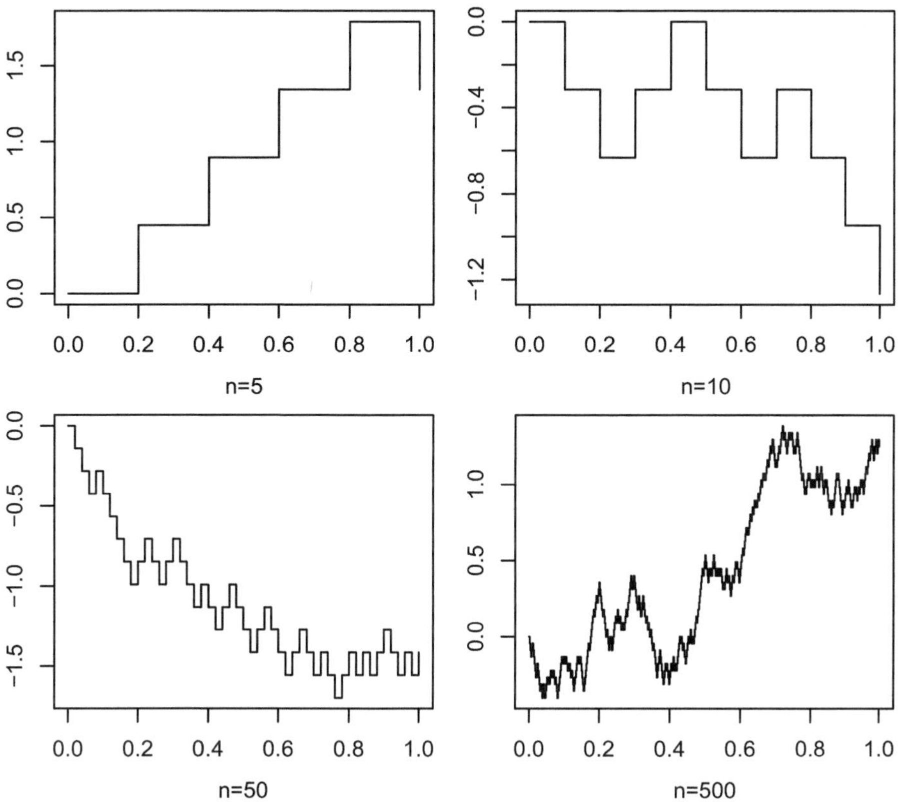

Fig. 7.1 Step function from (7.1) on the interval [0,1]

standardized sum of uncorrelated random variables $\{\varepsilon_j\}$ (whose variance is one and whose expectation is zero) that it tends to a standard normal distribution:

$$X_n(1) = \frac{1}{\sqrt{n}} \sum_{j=1}^{n} \varepsilon_j = \frac{\sum_{j=1}^{n} \varepsilon_j - \mathrm{E}\left(\sum_{j=1}^{n} \varepsilon_j\right)}{\sqrt{\mathrm{Var}\left(\sum_{j=1}^{n} \varepsilon_j\right)}} \stackrel{d}{\to} \mathcal{N}(0,1)\,. \tag{7.2}$$

Here, "$\stackrel{d}{\to}$" denotes the usual convergence in distribution; cf. Sect. 8.4. As $X_n(1)$ tends to $W(1)$ at the same time, the Wiener process has to be a standard normally distributed random variable at $t = 1$. After giving a formal definition, we will again bridge the gap from the Wiener process to $X_n(t)$.

Formal Definition

The Wiener process (WP) $W(t)$, $t \in [0, T]$, is defined by three assumptions. Put into words, they read: It is a process with starting value zero and independent, normally distributed, stationary increments. These assumptions are to be concretized and specified. Hence, the **Wiener process** is defined by:

(W1) The starting value is zero with probability one, $P(W(0) = 0) = 1$;
(W2) non-overlapping increments $W(t_1) - W(t_0)$, ..., $W(t_n) - W(t_{n-1})$, with $0 \leq t_0 \leq t_1 \leq \ldots \leq t_n$, are independent for arbitrary n;
(W3) the increments follow a Gaussian distribution with the variance equalling the difference of the arguments, $W(t) - W(s) \sim \mathcal{N}(0, t - s)$ with $0 \leq s < t$.

Note that the variance of the increments does not depend on the point in time but only on the temporal difference. Furthermore, the covariance of non-overlapping increments is zero due to the independence and the joint distribution results as the product of the marginal distributions. Hence, the joint distribution of non-overlapping increments is multivariate normal. If all increments are measured over equidistant constant time intervals, $t_i - t_{i-1} = const$, then the variances are identical. Therefore, such a series of increments is (strictly) stationary.

Although the WP is defined by its increments, they translate into properties of the level. Obviously, the first and the third property[4] imply

$$W(t) \sim \mathcal{N}(0, t), \tag{7.3}$$

i.e. the Wiener process is clearly a stochastic function being normally distributed at every point in time with linearly growing variance t. More precisely, the WP is even a Gaussian process in the sense of the definition from Chap. 2. The autocovariances being necessary for the complete characterization of the multivariate normal distribution $(W(t_1), \ldots, W(t_n))'$, are determined as follows (see Problem 7.3):

$$\text{Cov}(W(t), W(s)) = \min(s, t). \tag{7.4}$$

The Wiener process, which here will always be denoted by W, is for us a special case of the more general Brownian motion.[5] So to speak, it takes over the role of the *standard* normal distribution, and by multiplication with a constant a general **Brownian motion** is obtained as

$$B(t) = \sigma W(t), \quad \sigma > 0.$$

---

[4]To be completely accurate, this needs to read: $W(t) - W(0) \sim \mathcal{N}(0, t)$. As $W(0)$ is zero with probability one, we set $W(0)$ equal to zero here and in the following; then, the corresponding statements only hold with probability one.

[5]This convention does not hold beyond these pages. Many authors use the terms Wiener process or Brownian motion interchangeably or they apply one of them exclusively.

The assumptions (W1) to (W3) seem very natural if the WP is accepted as a limit of $X_n(t)$ from (7.1). For this process it holds by construction that

- $X_n(t) = 0$ for $t \in [0, 1/n)$,
- e.g. the increments

$$X_n\left(\frac{k+1}{n}\right) - X_n\left(\frac{k}{n}\right) = \frac{\varepsilon_{k+1}}{\sqrt{n}}$$

  and

$$X_n\left(\frac{k}{n}\right) - X_n(0) = \frac{1}{\sqrt{n}} \sum_{j=1}^{k} \varepsilon_j$$

  are uncorrelated (or even independent if $\{\varepsilon_t\}$ is a pure random process),
- $X_n(1) - X_n(0)$ is approximately normally distributed due to (7.2).

The three properties (W1) to (W3) just reflect the properties of the step function $X_n(t)$.

7.3 Properties

We have already come to know some properties of the WP, for example its autocovariance structure and the Gaussian distribution. For the understanding and handling of Wiener processes further properties are important.

Pathwise Properties

Loosely speaking, it holds that the Brownian motion is everywhere (i.e. for all t) continuous in terms of conventional calculus[6]; however, it is nowhere differentiable. These are pathwise properties, i.e. for a given ω_0, $W(t) = W(t; \omega_0)$ can be regarded as a function which is continuous in t but which is nowhere differentiable. This is a matter of properties being mathematically rather deep but which can be made

[6] Occasionally, the pathwise continuity is claimed to be the fourth defining property. This is to be understood as follows. Billingsley (1986, Theorem 37.1) proves more or less the following: If one has a WP W with (W1) to (W3) at hand, then a process W^* can be constructed which is a WP in the sense of (W1) to (W3), as well, which has the same distribution as W *and* which is pathwise continuous. As W^* and W are equal in distribution, they cannot be distinguished and therefore, w.l.o.g. it can safely be assumed that one may work with the continuous W^*.

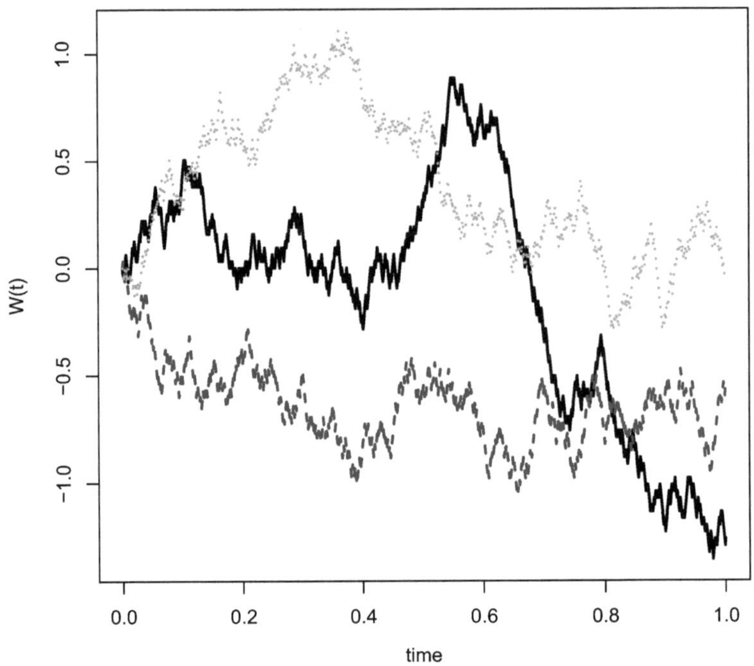

Fig. 7.2 Simulated paths of the WP on the interval [0,1]

heuristically plausible at least. Concerning this, let us consider

$$W(t+h) - W(t) \sim \mathcal{N}(0, h), \quad h > 0 .$$

For $h \to 0$ the given Gaussian distribution degenerates to zero suggesting continuity: $W(t+h) - W(t) \approx 0$ for $h \approx 0$. Analogously, we observe a difference quotient whose variance tends to infinity for $h \to 0$,

$$\frac{W(t+h) - W(t)}{h} \sim \mathcal{N}\left(0, \frac{1}{h}\right),$$

which suggests that a usual derivative does not exist. Related to contents, this means that it is not possible to add a tangent line to $W(t)$ which would allow for approximating $W(t+h)$ for an ever so small h. Three simulated paths of the WP in Fig. 7.2 illustrate these properties.

Markov and Martingale Property

In the previous section, we have learned that the random walk is a martingale. For the WP as a continuous-time counterpart, a corresponding result can be obtained (where $\mathcal{I}_t = \sigma\,(W(r),\ r \leq t)$ contains all the information about the past up to t):

$$\mathrm{E}\,(|W(t)|) < \infty\,,$$

$$\mathrm{E}\,(W(t+s)\,|\,\mathcal{I}_t) = W(t)\,.$$

The WP satisfies the Markov property (2.9) as well. In order to show this, we use the fact that the increment $W(t+s) - W(t)$ for $s > 0$ is independent of the information set $\mathcal{I}_t$ due to (W2). Hence, for $W(t) = v$ it holds that:

$$\begin{aligned}\mathrm{P}\,(W(t+s) \leq w \mid \mathcal{I}_t) &= \mathrm{P}\,(W(t+s) - W(t) \leq w - v \mid \mathcal{I}_t)\\ &= \mathrm{P}\,(W(t+s) - W(t) \leq w - v)\ .\end{aligned}$$

At the same time it holds again due to independence that:

$$\begin{aligned}\mathrm{P}\,(W(t+s) \leq w \mid W(t) = v) &= \mathrm{P}\,(W(t+s) - W(t) \leq w - v \mid W(t) = v)\\ &= \mathrm{P}\,(W(t+s) - W(t) \leq w - v)\ ,\end{aligned}$$

which just verifies the Markov property.

Scale Invariance

The Wiener process is a function being Gaussian for every point in time t with expectation zero and variance t. However, time can be measured in minutes, hours or other units. If the time scale is blown up or squeezed by the factor $\sigma > 0$, then it holds that

$$W(\sigma\, t) \ \sim\ \mathcal{N}(0, \sigma\, t)\,.$$

The same distribution is obtained for the $\sqrt{\sigma}$-fold of the Wiener process:

$$\sqrt{\sigma}\, W(t) \ \sim\ \mathcal{N}(0, \sigma\, t)\,.$$

That is why the Wiener process is called scale-invariant (or self-similar). Hence, $W(\sigma\, t)$ and $\sqrt{\sigma}\, W(t)$ are equal in distribution, which we formulate as

$$\sqrt{\sigma}\, W(t) \ \sim\ W(\sigma\, t) \tag{7.5}$$

as well. Such an equality in distribution is to be handled with care and by no means to be confused with ordinary equality. Naturally, it does not hold that, e.g. the double of $W(t)$ is equal to the value at the point in time $4t$:

$$\sqrt{\sigma}\, W(t) \ \neq\ W(\sigma\, t)\,.$$

In other words: Scale invariance is a distributional and not a pathwise property.

Up to this point, it has not been emphasized that the Wiener process is nonstationary. This can already be noted in (7.4) as for $s > 0$ the covariance $\text{Cov}(W(t), W(t+s)) = t$ does not result as dependent on the temporal distance s, but as dependent on the point in time t itself. As we have said, the increments of the WP from (W2), however, are multivariate normal with expectations and covariances of zero and variances which are not affected by a shift on the time axis. The trending behavior of the nonstationary Wiener process will now be clarified by two propositions.

Hitting Time

Let T_b be the point in time at which the WP attains (or hits) a given value $b > 0$ for the first time.[7] By variable transformation it is shown that this random variable has the distribution function (see Eq. (7.14) in Problem 7.5)

$$F_b(t) := \mathrm{P}(T_b \leq t) = \frac{2}{\sqrt{2\pi}} \int_{b/\sqrt{t}}^{\infty} e^{-y^2/2} dy.$$

Thereby statement (a) from the following proposition is proved; statement (b) is obtained by means of the corresponding density function (see Problem 7.5).

Proposition 7.1 (Hitting Time) *For the hitting time T_b, where the WP hits $b > 0$ for the first time, it holds that:*

(a) $P(T_b > t) \to 0$ for $t \to \infty$;
(b) $E(T_b)$ does not exist.

The result $T_b > t$ is tantamount to the fact that $W(s)$ has not attained the value b up to t:

$$\mathrm{P}(T_b > t) = \mathrm{P}\left(\max_{0 \leq s \leq t} W(s) < b\right).$$

Laxly formulated this proposition implies that, paradoxically, (a) sooner or later, the WP exceeds every value with certainty; (b) on average, this takes infinitely long: $E(T_b) = \infty$.

[7] The random variable T_b is a so-called "stopping time". This is a term from the theory of stochastic processes which we will not elaborate on here.

Zero Crossing

Next, let $p(t_1, t_2)$ with $0 < t_1 < t_2$ be the probability of the WP hitting the zero line at least once between t_1 and t_2 (even if not necessarily crossing it). We then talk about a zero crossing. The following proposition states how to calculate it. For a proof see e.g. Klebaner (2005, Theorem 3.25).

Proposition 7.2 (Arcus Law)
The probability of a zero crossing between t_1 and t_2, $0 < t_1 < t_2$, equals

$$p(t_1, t_2) = \frac{2}{\pi} \arctan \sqrt{\frac{t_2 - t_1}{t_1}},$$

where arctan *denotes the inverse of the tangent function* $\tan = \frac{\sin}{\cos}$.

It is interesting to fathom out the limiting cases of Proposition 7.2. From the shape of the inverse function of the tangent function it results that

$$\lim_{x\to\infty} \arctan x = \frac{\pi}{2} \quad \text{and} \quad \lim_{x\to 0} \arctan x = 0\,.$$

Hence, substantially, for $t_2 \to \infty$ it follows that the probability of attaining the zero line tends to one; for $t_2 \to t_1$, however, it naturally converges to zero.

In the literature, an equivalent formulation of the Arcus Law is found:

$$p(t_1, t_2) = \frac{2}{\pi} \arccos \sqrt{\frac{t_1}{t_2}}\,.$$

The equivalence is based on the formula

$$\arctan x = \arccos \frac{1}{\sqrt{1 + x^2}},$$

see e.g. Gradshteyn and Ryzhik (2000, 1.624-8), where "arccos" stands for the inverse of the cosine function.

7.4 Functions of Wiener Processes

When applying stochastic calculus, one is often concerned with processes derived from the Brownian motion. In this section, some of these will be covered and illustrated graphically. We simulate processes on the interval $[0, 1]$; for this purpose, the theoretically continuous processes are calculated at 1000 sampling points and plotted. The resulting graphs are based on pseudo-random variables. Details

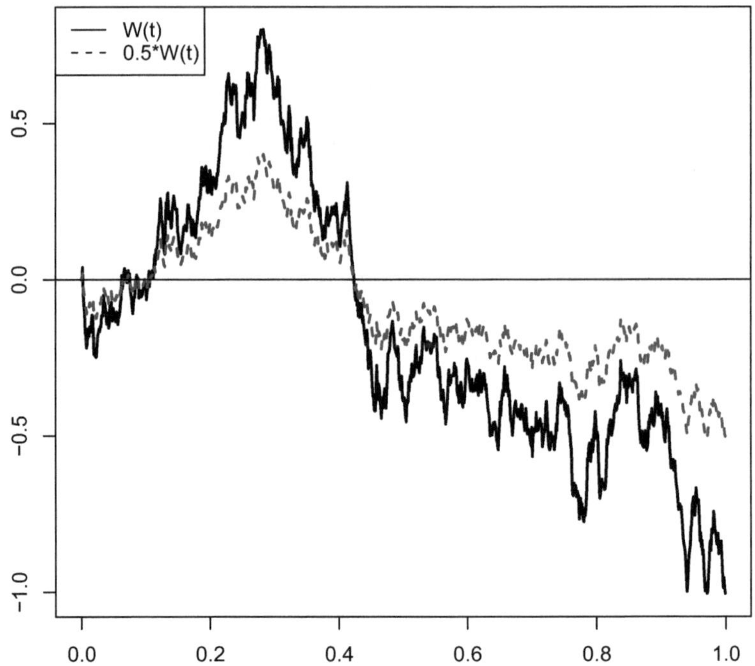

Fig. 7.3 WP and Brownian motion with $\sigma = 0.5$

regarding the simulation of stochastic processes are treated in Chap. 12 on stochastic differential equations.

Brownian Motion $B(t)$

In Fig. 7.3 a path of a WP and a Brownian motion based thereon with only half the standard deviation,

$$W(t) \text{ and } B(t) = 0.5\, W(t)\,,$$

are depicted. Obviously, the one graph is just half of the other.

Brownian Motion with Drift $X(t) = \mu\, t + \sigma\, W(t)$

Here it holds that both the expectation and the variance grow linearly with t:

$$X(t) \sim \mathcal{N}(\mu t, \sigma^2 t)\,.$$

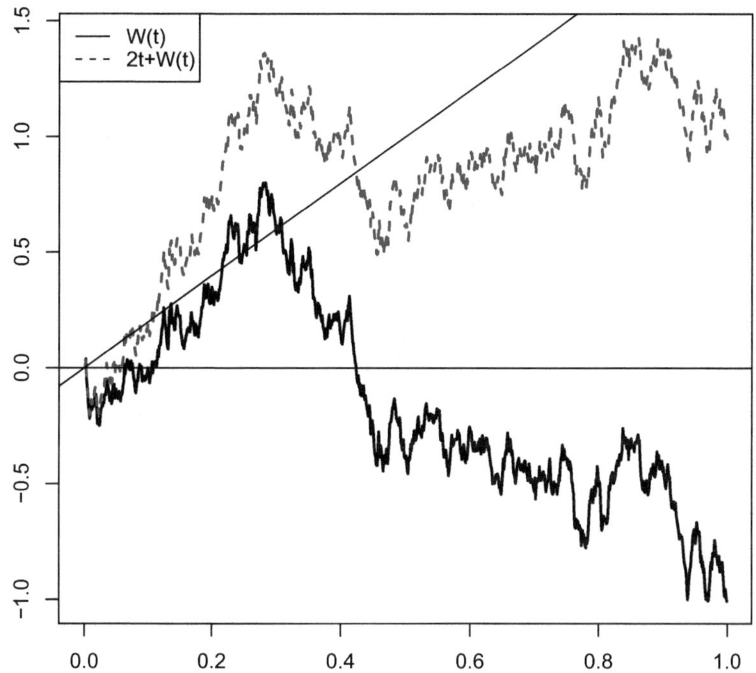

Fig. 7.4 WP and Brownian motion with drift, where $\sigma = 1$

In Fig. 7.4 the WP from Fig. 7.3 and the process based thereon with drift are depicted. The drift parameter is $\mu = 2$ for $\sigma = 1$, and the expectation function $2\,t$ is also displayed.

Brownian Bridge $X(t) = B(t) - t\,B(1)$

This process is based on the Brownian motion, $B(t) = \sigma\, W(t)$, and fundamentally, it is only defined for $t \in [0, 1]$. The name comes from the fact that the starting and the final value are equal with probability one by construction: $X(0) = X(1) = 0$. One can verify easily that (see Problem 7.6):

$$\text{Var}(X(t)) = t\,(1 - t)\,\sigma^2 < t\sigma^2\,. \tag{7.6}$$

Hence, for $t \in (0, 1]$ it holds that $\text{Var}(X(t)) < \text{Var}(B(t))$. This is intuitively clear: With being forced back to zero, the Brownian bridge has to exhibit less variability than the Brownian motion. This is also shown in Fig. 7.5 for $\sigma = 1$.

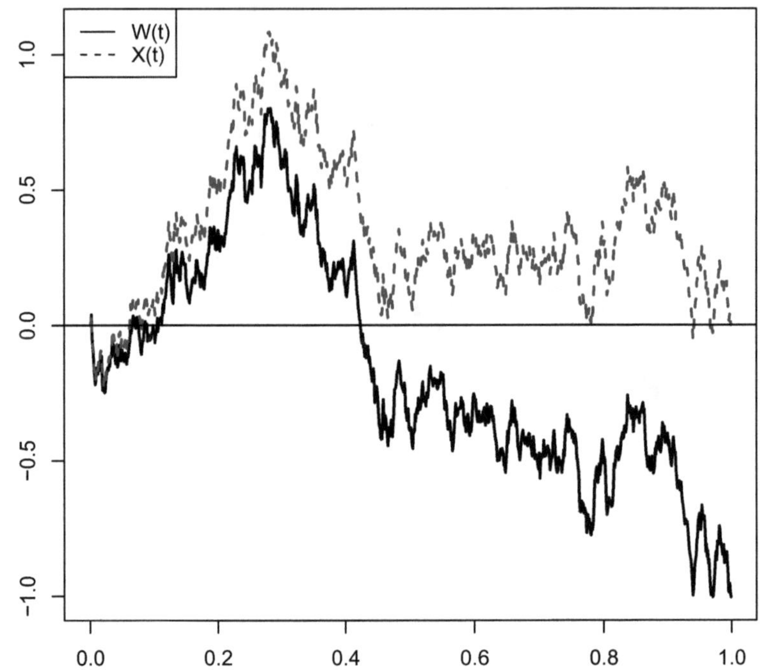

Fig. 7.5 WP and Brownian bridge ($\sigma = 1$)

Reflected Wiener Process $X(t) = |W(t)|$

Due to $W(t) \sim \mathcal{N}(0, t)$, for the distribution function it is elementary to obtain (see Problem 7.7):

$$\mathrm{P}(X(t) \le x) = \frac{2}{\sqrt{2\pi t}} \int_{-\infty}^{x} \exp\left(\frac{-y^2}{2t}\right) dy - 1 .$$

Note that one integrates over twice the density of a Gaussian random variable with expectation zero. Therefore it immediately holds that

$$\mathrm{P}(X(t) \le x) = \frac{2}{\sqrt{2\pi t}} \int_{0}^{x} \exp\left(\frac{-y^2}{2t}\right) dy . \tag{7.7}$$

Expectation and variance of the reflected Wiener process can be determined from the corresponding density function. They read (see Problem 7.7):

$$\mathrm{E}(X(t)) = \sqrt{\frac{2\,t}{\pi}}, \quad \mathrm{Var}(X(t)) = t\left(1 - \frac{2}{\pi}\right) < t = \mathrm{Var}(W(t)) . \tag{7.8}$$

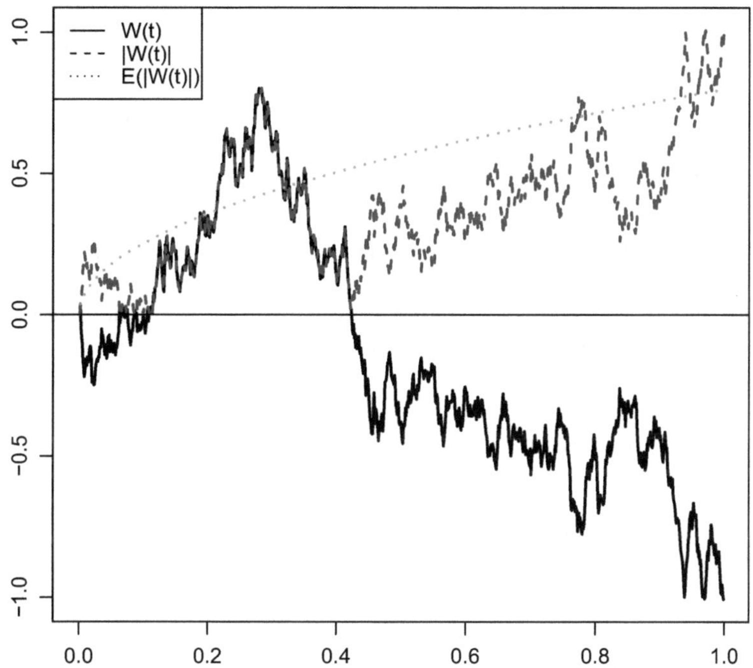

Fig. 7.6 WP and reflected WP along with expectation

As the reflected Wiener process cannot become negative, it has a positive expected value growing with t. For the same reason its variance is smaller than the one of the unrestricted Wiener process, see Fig. 7.6.

Geometric Brownian Motion $X(t) = e^{\mu t + \sigma W(t)}$

By definition, it holds in this case that the logarithm[8] of the process is a Brownian motion with drift and therefore Gaussian,

$$\log X(t) = \mu t + \sigma W(t) \sim \mathcal{N}(\mu t, \sigma^2 t) .$$

A random variable Y whose logarithm is Gaussian is called – as would seem natural – **log-normal** (logarithmically normally distributed). If it holds that

$$\log Y \sim \mathcal{N}(\mu_y, \sigma_y^2),$$

[8]By "log" we denote the natural logarithm and not the common logarithm.

then we know how the two first moments of Y look like, cf. e.g. Sydsæter, Strøm, and Berck (1999, p. 189) or Johnson, Kotz, and Balakrishnan (1994, Ch. 14):

$$\mathrm{E}(Y) = e^{\mu_y + \sigma_y^2/2}, \quad \mathrm{Var}(Y) = e^{2\mu_y + \sigma_y^2}\left(e^{\sigma_y^2} - 1\right).$$

Hence, by plugging in we obtain for the geometric Brownian motion

$$\mathrm{E}(X(t)) = e^{(\mu + \sigma^2/2)\,t} \quad \text{and} \quad \mathrm{Var}(X(t)) = e^{(2\mu + \sigma^2)\,t}\left(e^{\sigma^2 t} - 1\right). \tag{7.9}$$

While $\log X(t)$ is Gaussian with a linear trend, μt, as expectation, $X(t)$ exhibits an exponentially growing expectation function. Particularly for $\mu = 0$ and $\sigma = 1$ the results

$$\mathrm{E}(X(t)) = e^{t/2} \quad \text{and} \quad \mathrm{Var}(X(t)) = e^{t}\left(e^{t} - 1\right) \tag{7.10}$$

are obtained. The on average exponential growth in the case of $\mu > -\sigma^2/2$ is illustrated in Fig. 7.7. In Fig. 7.8 we find graphs of the WP and a geometric Brownian motion with expectation one, namely with $\mu = -0.5$ and $\sigma = 1$. Generally, for $\mu = -\sigma^2/2$ an expectation function of one is obtained. Then, one also says that the process does not exhibit a trend (or drift).

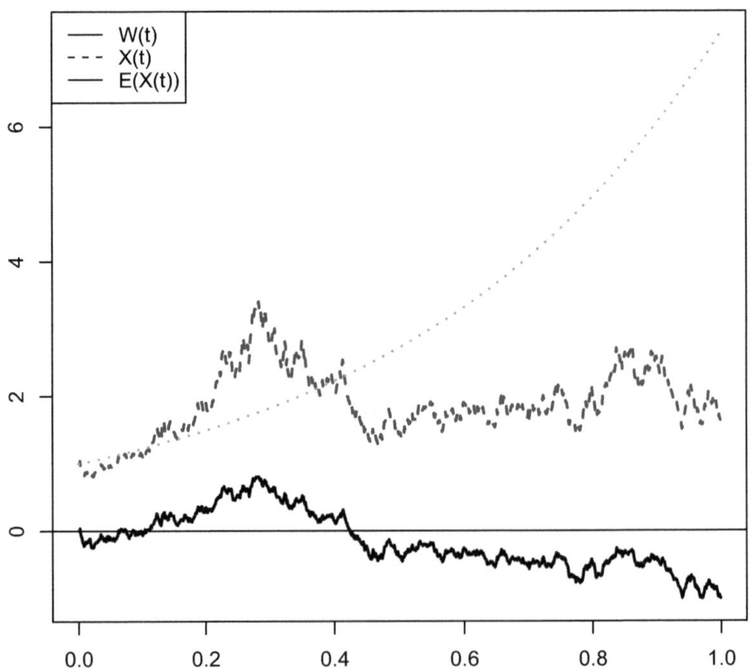

Fig. 7.7 Geometric Brownian motion with $\mu = 1.5$ and $\sigma = 1$ along with expectation

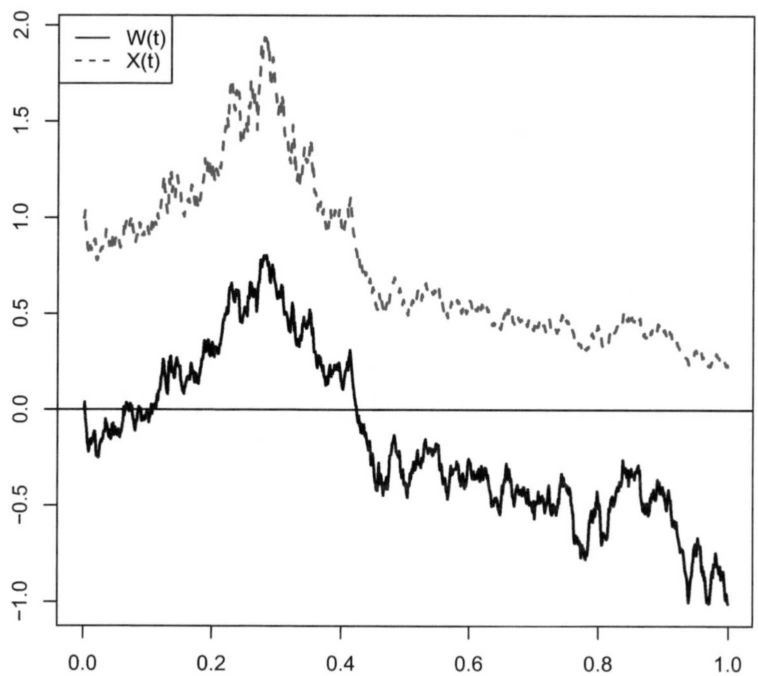

Fig. 7.8 WP and geometric Brownian motion with $\mu = -0.5$ and $\sigma = 1$

In comparison to the expectation, the median of a geometric Brownian motion does not depend on σ. Rather, it holds that (see Problem 7.8):

$$\mathrm{P}\left(e^{\mu\, t + \sigma\, W(t)} \leq e^{\mu\, t}\right) = 0.5\,,$$

such that the median results as $e^{\mu\, t}$.

Maximum of a WP $X(t) = \max_{0 \leq s \leq t} W(s)$

At t, the maximum process is assigned the maximal value which the WP has taken on up to this point in time. Therefore, in periods of a decreasing Wiener process path, $X(t)$ is constant on the historic maximum until a new relative maximum is attained. However, this process has a distribution function that we have already come to know. By applying the distribution function of the stopping time which is given above Proposition 7.1, one shows (see Problem 7.9) that the maximum process and the reflected WP are equal in distribution:

$$\mathrm{P}(X(t) \leq b) = \mathrm{P}(|W(t)| \leq b)\,.$$

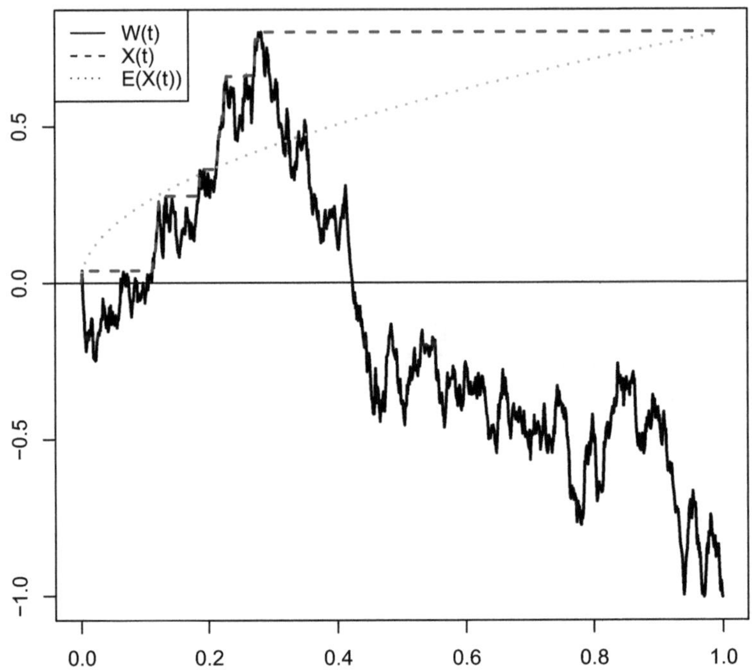

Fig. 7.9 WP and maximum process along with expectation

Therefore, expectation and variance of the maximum process of $|W(t)|$ can naturally be copied:

$$\mathrm{E}(X(t)) = \sqrt{\frac{2\,t}{\pi}}, \quad \mathrm{Var}(X(t)) = t\left(1 - \frac{2}{\pi}\right) < t = \mathrm{Var}(W(t)) . \tag{7.11}$$

The expected value is positive and grows with time as the WP again will replace a relative positive maximum by a new relative maximum. Due to the process being again and again constant over times, it is not surprising that its variance is smaller than the one of the underlying WP, cf. Fig. 7.9.

Integrated Wiener Process $X(t) = \int_0^t W(s)\,ds$

As the Brownian motion is a continuous function, the Riemann integral can be defined pathwise. Hence, e.g. the following random variable is obtained:

$$\int_0^1 B(t)\,dt = \sigma \int_0^1 W(t)\,dt .$$

Behind this random variable hides a normal distribution. The latter can be proved by using the definition of the Riemann integral or as a simple conclusion of the Proposition 8.3 below:

$$\int_0^1 W(t)\,dt \sim \mathcal{N}(0, 1/3)\,. \tag{7.12}$$

Basically, by using the integral of a WP, a new stochastic process can also be generated by making the upper limit of integration time-dependent:

$$X(t) = \int_0^t W(s)\,ds\,.$$

This idea forms the starting point for the subsequent chapter. In Fig. 7.10 the relation is shown between the WP and the integral $X(t)$ as the area beneath the graph.

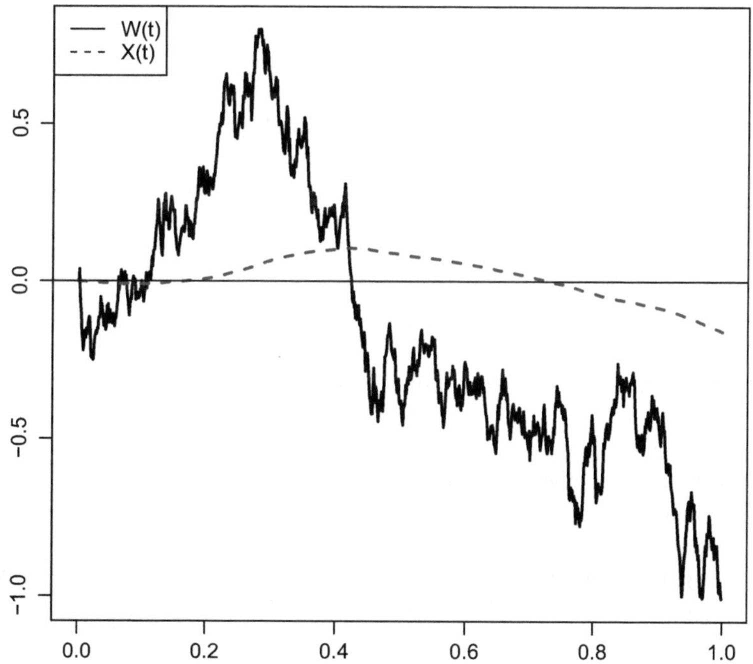

Fig. 7.10 WP and integrated WP

7.5 Problems and Solutions

Problems

7.1 Consider with $\sigma > 0$

$$X(t) = W(1) - \sigma W(1-t) \quad \text{for } 0 \le t \le 1 .$$

Determine the mean and variance of $X(t)$.

7.2 Consider

$$X(t) = t\, W(t^{-1}) \quad \text{for } t > 0 .$$

Determine the covariance of $X(t)$ and $W(t)$, $\text{Cov}(X(t), W(t))$.

7.3 Derive the autocovariance function of a WP, (7.4). Find a simple expression in t and s only for the autocorrelations

$$\rho(s,t) = \frac{\text{Cov}(W(t), W(s))}{\sqrt{\text{Var}(W(t))\,\text{Var}(W(s))}} .$$

7.4 Choose $d \in \mathbb{R}$ such that $T^{d-0.5}\, W(t)$ and $W(T\,t)$ are equal in distribution.

7.5 Prove Proposition 7.1 using the hints given in the text.

7.6 Derive the autocovariance function of a Brownian bridge, and hence show (7.6).

7.7 Determine the distribution function, (7.7), and the moments, (7.8), of a reflected Wiener process.

7.8 Show that in the general case of a geometric Brownian motion, $e^{\mu\, t + \sigma\, W(t)}$, the median is given by $e^{\mu\, t}$.

7.9 Show by means of the hints in the text that the maximum process of a WP and the corresponding reflected WP are equal in distribution:

$$\text{P}\left(\max_{0 \le s \le t} W(s) \le b\right) = \text{P}(|W(t)| \le b) .$$

Solutions

7.1 As the Wiener process is on average zero at every point in time, this obviously holds for $X(t)$ as well. Therefore, the variance is calculated as follows:

$$\begin{aligned}
\mathrm{Var}(X(t)) &= \mathrm{E}(X^2(t)) \\
&= \mathrm{E}[W^2(1) - 2\sigma W(1)W(1-t) + \sigma^2 W^2(1-t)] \\
&= \mathrm{Var}(W(1)) - 2\sigma \mathrm{Cov}(W(1), W(1-t)) + \sigma^2 \mathrm{Var}(W(1-t)) \\
&= 1 - 2\sigma \min(1,\, 1-t) + \sigma^2(1-t) \\
&= 1 - 2\sigma(1-t) + \sigma^2(1-t) \\
&= t + (1-t)(1-\sigma)^2 .
\end{aligned}$$

7.2 Due to $\mathrm{E}(W(t)) = \mathrm{E}(X(t)) = 0$ one obtains:

$$\begin{aligned}
\mathrm{Cov}(X(t), W(t)) &= \mathrm{E}(X(t)W(t)) \\
&= t\,\mathrm{E}(W(t^{-1})W(t)) \\
&= t \min(t^{-1}, t).
\end{aligned}$$

Because of

$$\min(t^{-1}, t) = \begin{cases} t, & 0 < t \leq 1 \\ t^{-1}, & t \geq 1 \end{cases}$$

it follows that

$$\mathrm{Cov}(X(t), W(t)) = \begin{cases} t^2, & 0 < t \leq 1 \\ 1, & t \geq 1 \end{cases} .$$

7.3 We simply apply the defining properties (W1), (W2) and (W3) or put differently (7.3). Due to (7.3) the WP has an expectation of zero such that

$$\mathrm{Cov}\,(W(t), W(s)) = \mathrm{E}\,(W(t)W(s)) .$$

W.l.o.g. let $s \leq t$. By using (W1) and (W2) and after adding zero, we then write:

$$\begin{aligned}
\mathrm{E}\,(W(t)W(s)) &= \mathrm{E}\,([W(s) + W(t) - W(s)]\,[W(s) - W(0)]) \\
&= \mathrm{E}\left([W(s)]^2\right) + \mathrm{E}\,([W(t) - W(s)]\,[W(s) - W(0)]) \\
&= s + 0 ,
\end{aligned}$$

where the last equality uses $\text{Var}(W(s)) = s$ and the independence of non-overlapping increments. As one could also assume $t \leq s$ w.l.o.g., (7.4) is verified.

With the autocovariance one obtains

$$\rho(s,t) = \frac{\min(s,t)}{\sqrt{ts}} = \frac{\min(s,t)}{\sqrt{\max(s,t)\,\min(s,t)}} = \sqrt{\frac{\min(s,t)}{\max(s,t)}}\,.$$

7.4 This is a problem on self-similarity or scale invariance. Due to (7.3) it obviously holds that:

$$T^{d-0.5}W(t) \sim \mathcal{N}(0, T^{2d-1}t)$$

and

$$W(Tt) \sim \mathcal{N}(0, Tt)\,.$$

Therefore, the corresponding variances are equal for $d = 1$. The corresponding result is obtained from (7.5) as well:

$$T^{0.5}W(t) \sim W(Tt)\,.$$

7.5 Proof of Proposition 7.1(a): Our proof consists of three steps. At first, we establish the equation

$$\text{P}(T_b < t) = 2\,\text{P}(W(t) > b)\,. \tag{7.13}$$

Secondly, by using this we show:

$$F_b(t) := \text{P}(T_b \leq t) = \frac{2}{\sqrt{2\pi}} \int_{b/\sqrt{t}}^{\infty} e^{-y^2/2}\,dy\,. \tag{7.14}$$

Note that in (7.14) the integrand amounts to the density function of the standard normal distribution. Hence, thirdly for $t \to \infty$ the claim immediately follows from (7.14) due to $\text{P}(T_b > t) = 1 - \text{P}(T_b \leq t)$.

In order to accept (7.13), we remember (7.3). Accordingly, for $T_b < t$ it holds that

$$W(t) - W(T_b) \sim \mathcal{N}(0, t - T_b)\,,$$

which is why from the symmetry of the Gaussian distribution with $W(T_b) = b$ it follows for the conditional probability that:

$$\text{P}\left(W(t) > b \mid T_b < t\right) = \text{P}\left(W(t) - W(T_b) > 0 \mid T_b < t\right) = \frac{1}{2}\,.$$

Hence, it results that:

$$\frac{1}{2} = \frac{\mathrm{P}(W(t) > b \text{ and } T_b < t)}{\mathrm{P}(T_b < t)} = \frac{\mathrm{P}(W(t) > b)}{\mathrm{P}(T_b < t)},$$

where the last equality is caused by the fact that $T_b < t$ is implied by $W(t) > b$. Thus, we obtain Eq. (7.13) which will now be applied for deriving (7.14).

Due to (7.3) it holds by definition that

$$\mathrm{P}(W(t) > b) = \int_b^\infty \frac{1}{\sqrt{2\pi t}} e^{-x^2/2t} dx .$$

By variable transformation, $y = \frac{x}{\sqrt{t}}$, it follows that

$$\mathrm{P}(W(t) > b) = \int_{b/\sqrt{t}}^\infty \frac{1}{\sqrt{2\pi}} e^{-y^2/2} dy ,$$

whereby (7.14) and hence claim (a) is proved due to $P(T_b < t) = P(T_b \leq t)$.

Proof of Proposition 7.1(b): With the density function $f_b(t) = F_b'(t)$ the approach for the expected value reads as follows:

$$\mathrm{E}\,(T_b) = \int_0^\infty t f_b(t)\, dt .$$

Note that the distribution function derived in (7.14) has the following structure with the antiderivative H, $H' = h$:

$$F_b(t) = \int_{g(t)}^\infty h(y)\, dy = \lim_{c\to\infty} H(c) - H(g(t)) .$$

Therefore, due to the chain rule it holds for the density that

$$F_b'(t) = -h\,(g(t))\; g'(t) = \frac{b\, e^{-\frac{b^2}{2t}} t^{-\frac{3}{2}}}{\sqrt{2\pi}} .$$

Hence, the variable transformation results in $t = b^2 u^{-2}$ with $dt = -2\, b^2\, u^{-3} du$:

$$\begin{aligned}
\mathrm{E}\,(T_b) &= \int_0^\infty t\, F_b'(t)\, dt \\
&= \frac{b}{\sqrt{2\pi}} \int_0^\infty e^{-\frac{b^2}{2t}} t^{-\frac{1}{2}} dt = \frac{-2b^2}{\sqrt{2\pi}} \int_\infty^0 e^{-\frac{u^2}{2}} u^{-2}\, du \\
&= \frac{2b^2}{\sqrt{2\pi}} \int_0^\infty e^{-\frac{u^2}{2}} u^{-2} du
\end{aligned}$$

$$\geq \frac{2b^2}{\sqrt{2\pi}} \int_0^1 e^{-\frac{u^2}{2}} u^{-2} du$$
$$\geq \frac{2b^2}{\sqrt{2\pi}} e^{-\frac{1}{2}} \int_0^1 u^{-2}\, du.$$

However, this last integral written down symbolically does not exist since the antiderivative of u^{-2} is $-u^{-1}$, and

$$\int_\varepsilon^1 u^{-2}\, du = \varepsilon^{-1} - 1$$

diverges as $\varepsilon \to 0$. This completes the proof.

7.6 First, we determine that

$$X(t) = W(t) - t\,W(1)\,, \quad t \in [0,1]\,,$$

has zero expectation:

$$\mathrm{E}(X(t)) = \mathrm{E}(W(t)) - t\,\mathrm{E}(W(1)) = 0 - 0\,.$$

By multiplying out and application of (7.4) one can show that

$$\begin{aligned}
\mathrm{Cov}(X(t), X(s)) &= \mathrm{E}(X(t)\,X(s)) \\
&= \mathrm{E}\left(W(t)W(s) - tW(1)W(s) - sW(1)W(t) + stW^2(1)\right) \\
&= \min(s,t) - t\min(s,1) - s\min(t,1) + st \\
&= \min(s,t) - st - st + st \\
&= \min(s,t) - st\,.
\end{aligned}$$

In particular, for $s = t$ the variance formula (7.6) is obtained.

7.7 At first we determine the distribution function (7.7) for $X(t) = |W(t)|$:

$$\begin{aligned}
F_x(x) &= \mathrm{P}(X(t) \leq x)\,, \quad x \geq 0\,, \\
&= \mathrm{P}(W(t) \leq x) - \mathrm{P}(W(t) \leq -x) \\
&= \mathrm{P}(W(t) \leq x) - (1 - \mathrm{P}(W(t) \leq x)) \\
&= 2\,\mathrm{P}(W(t) \leq x) - 1\,,
\end{aligned}$$

where the symmetry of the Gaussian distribution was used. With $W(t) \sim \mathcal{N}(0, t)$ we therefore have

$$F_x(x) = \frac{2}{\sqrt{2\pi t}} \int_{-\infty}^{x} e^{-\frac{y^2}{2t}} dy - 1 \,,$$

or for the density

$$f_x(x) = F_x'(x) = \frac{2}{\sqrt{2\pi t}} e^{-\frac{x^2}{2t}} \,.$$

In order to calculate expectation and variance, we determine the r-th moment in general:

$$\mathrm{E}(X^r(t)) = \int_0^{\infty} x^r f_x(x)\, dx = \frac{2}{\sqrt{2\pi t}} \int_0^{\infty} x^r e^{-\frac{x^2}{2t}} dx \,.$$

By substitution, these moments can be reduced to the Gamma function, which was introduced in Problem 5.3, see also below Eq. (5.20). For $a > 0$ and with $a\,x^2 = u$ and $du = 2ax\,dx$, we obtain:

$$\begin{aligned}
\int_0^{\infty} x^r e^{-ax^2} dx &= \int_0^{\infty} \left(\frac{u}{a}\right)^{\frac{r-1}{2}} e^{-u} \frac{du}{2a} \\
&= \frac{1}{2} a^{-\frac{r+1}{2}} \int_0^{\infty} u^{\frac{r+1}{2}-1} e^{-u} du \\
&= \frac{1}{2} a^{-\frac{r+1}{2}} \Gamma\left(\frac{r+1}{2}\right) .
\end{aligned}$$

The Gamma function possesses a number of nice properties and special values, remember in particular e.g.

$$\Gamma(1) = 1 \,, \quad \Gamma\left(\frac{1}{2}\right) = \sqrt{\pi} \,, \quad \Gamma(n+1) = n\,\Gamma(n) \,.$$

With $a = \frac{1}{2t}$, for the moments it therefore follows that:

$$\mathrm{E}(X(t)) = \sqrt{\frac{2t}{\pi}} \,, \quad \mathrm{E}(X^2(t)) = \frac{2}{\sqrt{2\pi t}} \frac{1}{2} (2t)^{\frac{3}{2}} \frac{1}{2} \Gamma\left(\frac{1}{2}\right) = t \,.$$

The variance formula is obtained by the usual variance decomposition:

$$\mathrm{Var}(X(t)) = \mathrm{E}\left(X^2(t)\right) - \left(\mathrm{E}(X(t))\right)^2 \,.$$

7.8 The random variable $\sigma W(t)$ follows for fixed t a Gaussian distribution with expectation and median equal to zero. Hence, it follows that

$$P(\sigma W(t) \leq 0) = P\left(e^{\sigma W(t)} \leq 1\right) = 0.5 .$$

By multiplying the inequality by $e^{\mu t}$, we obtain

$$P\left(e^{\mu t} e^{\sigma W(t)} \leq e^{\mu t}\right) = P\left(e^{\mu t+\sigma W(t)} \leq e^{\mu t}\right) = 0.5 .$$

Therefore, the median of $X(t) = e^{\mu t+\sigma W(t)}$ is determined independently of σ as $e^{\mu t}$ as claimed.

7.9 As $X(t) = \max_{0\leq s\leq t} W(s)$ is a continuous random variable for given t, it holds that

$$F_x(b) := P(X(t) \leq b) = P\left(\max_{0\leq s\leq t} W(s) < b\right) .$$

Remember the random variable T_b from Proposition 7.1 specifying the point in time at which $W(t)$ hits the value b for the first time. The event $\max_{0\leq s\leq t} W(s) < b$ is equivalent to the fact that the hitting time of b is larger than t. Therefore, when using the distribution function from Proposition 7.1, it holds that

$$P(X(t) \leq b) = 1 - P(T_b \leq t) = 1 - \frac{2}{\sqrt{2\pi}} \int_{b/\sqrt{t}}^{\infty} \exp\left(-\frac{z^2}{2}\right) dz .$$

Naturally, the number 1 can be written as an integral over the density of the standard normal distribution:

$$\begin{aligned} P(X(t) \leq b) &= \frac{2}{\sqrt{2\pi}} \int_{0}^{\infty} \exp\left(-\frac{z^2}{2}\right) dz - \frac{2}{\sqrt{2\pi}} \int_{b/\sqrt{t}}^{\infty} \exp\left(-\frac{z^2}{2}\right) dz \\ &= \frac{2}{\sqrt{2\pi}} \int_{0}^{b/\sqrt{t}} \exp\left(-\frac{z^2}{2}\right) dz . \end{aligned}$$

By substitution,

$$z = \frac{y}{\sqrt{t}} \quad \text{and} \quad dz = \frac{dy}{\sqrt{t}},$$

and due to (7.7) the desired result is immediately obtained:

$$\begin{aligned} P(X(t) \leq b) &= \frac{2}{\sqrt{2\pi t}} \int_{0}^{b} \exp\left(-\frac{y^2}{2t}\right) dy \\ &= P\left(|W(t)| \leq b\right) . \end{aligned}$$

References

Billingsley, P. (1986). *Probability and measure* (2nd ed.). New York: Wiley.

Gradshteyn, I. S., & Ryzhik, I. M. (2000). *Table of integrals, series, and products* (6th ed.). London/San Diego: Academic Press.

Johnson, N. L., Kotz, S., & Balakrishnan, N. (1994). *Continuous univariate distributions, Volume 1* (2nd ed.). New York: Wiley.

Klebaner, F. C. (2005). *Introduction to stochastic calculus with applications* (2nd ed.). London: Imperical College Press.

Sydsæter, K., Strøm, A., & Berck, P. (1999). *Economists' mathematical manual* (3rd ed.). Berlin/New York: Springer.

Riemann Integrals

8

8.1 Summary

In this chapter we deal with stochastic Riemann integrals, i.e. with ordinary Riemann integrals with a stochastic process as the integrand.[1] Mathematically, these constructs are relatively unsophisticated, they can be defined pathwise for continuous functions as in conventional (deterministic) calculus. However, this pathwise definition will not be possible any longer for e.g. Ito integrals in the chapter after next. Hence, at this point we propose a way of defining integrals as a limit (in mean square) which will be useful later on. If the stochastic integrand is in particular a Wiener process, then the Riemann integral follows a Gaussian distribution with zero expectation and the familiar formula for the variance. A number of examples will facilitate the understanding of this chapter.

8.2 Definition and Fubini's Theorem

As one has done in deterministic calculus, we will define the Riemann integral by an adequate partition as the limit of a sum.

Partition

In order to define an integral of a function from 0 to t, we decompose the interval into n adjacent, non-overlapping subintervals which are allowed to intersect at the

[1]Bernhard Riemann (1826–1866) studied with Gauss in Göttingen where he himself became a professor. Already before his day, integration had been used as a technique which reverses differentiation by forming an antiderivative. However, Riemann explained for the first time under which conditions a function possesses an antiderivative at all.

U. Hassler, *Stochastic Processes and Calculus*, Springer Texts in Business and Economics, DOI 10.1007/978-3-319-23428-1_8

endpoints:

$$P_n([0, t]) : \quad 0 = s_0 < s_1 < \ldots < s_n = t \,. \tag{8.1}$$

In the following, we always assume that the **partition** $P_n([0, t])$ becomes increasingly fine with n growing ("adequate partition"):

$$\max_{1 \le i \le n} (s_i - s_{i-1}) \rightarrow 0 \quad \text{for } n \rightarrow \infty \,. \tag{8.2}$$

By s_i^* we denote an arbitrary point in the i-th interval,

$$s_i^* \in [s_{i-1}, s_i] \,, \quad i = 1, \ldots, n \,.$$

Occasionally, we will sum up the lengths of the subintervals. Obviously, it holds that

$$\sum_{i=1}^{n} (s_i - s_{i-1}) = s_n - s_0 = t \,.$$

In general, for a function φ one obtains:

$$\sum_{i=1}^{n} (\varphi(s_i) - \varphi(s_{i-1})) = \varphi(t) - \varphi(0) \,. \tag{8.3}$$

Sometimes, we will operate with the example of the equidistant partition. It is given by $s_i = it/n$:

$$0 = s_0 < s_1 = \frac{t}{n} < \ldots < s_{n-1} = \frac{n-1}{n} t < s_n = t \,.$$

Due to $s_i - s_{i-1} = 1/n$ the required refinement from (8.2) for $n \rightarrow \infty$ is guaranteed.

Definition and Existence

Now, the product of a deterministic function f and a stochastic process X is to be integrated. To this end, the **Riemann sum** is defined by means of the notation introduced:

$$R_n = \sum_{i=1}^{n} f(s_i^*) \, X(s_i^*) \, (s_i - s_{i-1}). \tag{8.4}$$

Here, we have a sum of rectangular areas, each with a width of $(s_i - s_{i-1})$ and the height $f(s_i^*) \, X(s_i^*)$. With n growing, the area beneath $f(s) \, X(s)$ on $[0, t]$ is to be approximated all the better. If the limit of this sum for $n \to \infty$ exists uniquely and independently of the partition and of the choice of s_i^*, then it is defined as a (stochastic) Riemann integral. In this case, the convergence occurs in mean square ("$\xrightarrow{2}$")[2]:

$$R_n = \sum_{i=1}^{n} f(s_i^*)\, X(s_i^*)\, (s_i - s_{i-1}) \xrightarrow{2} \int_0^t f(s)\, X(s)\, ds\,.$$

One then says that the **Riemann integral** exists. For this existence, there is a sufficient and a necessary condition formulated in the following proposition. The proof is carried out with part (b) from Lemma 8.2 below, see Problem 8.1. Further elaborations on mean square convergence can be found at the end of the chapter.

Proposition 8.1 (Existence of the Riemann Integral) *The Riemann sum from Eq. (8.4) converges in mean square for $n \to \infty$ under (8.2) if and only if the double integral*

$$\int_0^t \int_0^t f(s) f(r)\, \mathrm{E}\,(X(s)\, X(r))\, drds$$

exists.

A sufficient condition for the existence of the Riemann integral is that the function f is continuous and that furthermore $\mathrm{E}\,(X(s)\,X(r))$ is continuous in both arguments. In order to find this, we define

$$\varphi(s) := f(s) \int_0^t f(r)\, \mathrm{E}\,(X(s)\, X(r))\, dr\,.$$

Now, if the function $\mathrm{E}\,(X(s)\,X(r))$ is continuous in both arguments, this implies continuity of φ for a continuous f as the integral is a continuous functional, see e.g. Trench (2013, p. 462). Therefore, the ordinary Riemann integral of φ exists,

$$\int_0^t \varphi(s)\, ds = \int_0^t \int_0^t f(s) f(r)\, \mathrm{E}\,(X(s)\, X(r))\, drds\,.$$

Hence, the Riemann sum from (8.4) converges due to Proposition 8.1.

[2] A definition and discussion of this mode of convergence can be found in the fourth section of this chapter.

Fubini's Theorem

Frequently, we are interested in the average behavior, i.e. the expected value of Riemann integrals. The expected value, however, is defined as an integral itself such that one is confronted with double integrals. For calculating these, there is a simple rule which is finally based on the fact that the order of integration does not matter for double integrals over continuous functions. In deterministic calculus, this fact is also known as "Fubini's theorem" (see Footnote 6 in Sect. 2.3). Adapted to our problem of the expected value of a Riemann integral, the corresponding circumstances are given in the following proposition, also cf. Billingsley (1986, Theorem 18.3).

Proposition 8.2 (Fubini's Theorem) *If $\int_0^t E(|X(s)|)\, ds$ exists, it holds for a continuous X that:*

$$E\left(\int_0^t X(s)\, ds\right) = \int_0^t E(X(s))\, ds\,.$$

The statement is easy to comprehend if one thinks of the integral as a finite Riemann sum. As is well known, in the discrete case, summation and expectation is interchangeable:

$$\mathrm{E}\left(\sum_{i=1}^{n} X(s_i^*)\,(s_i - s_{i-1})\right) = \sum_{i=1}^{n} \mathrm{E}(X(s_i^*))\,(s_i - s_{i-1})\,.$$

Now, Fubini's theorem just guarantees a continuation of this interchangeability for $n \to \infty$.

Example 8.1 (Expected Value of the Integrated WP) Consider the special case of the integrated WP with $X(s) = W(s)$ and $f(s) = 1$. With the WP being continuous, $|W(t)|$ is a continuous process as well. In (7.8) we have determined the following expression as the expected value:

$$\mathrm{E}(|W(t)|) = \sqrt{\frac{2t}{\pi}}.$$

Before applying Proposition 8.2, we check:

$$\begin{aligned}
\int_0^t \mathrm{E}\left(|W(s)|\right)\, ds &= \int_0^t \sqrt{\frac{2}{\pi}}\, s^{\frac{1}{2}} ds \\
&= \sqrt{\frac{2}{\pi}} \left[\frac{2}{3} s^{\frac{3}{2}}\right]_0^t \\
&= \sqrt{\frac{2}{\pi}}\, \frac{2}{3}\, t^{\frac{3}{2}}\,.
\end{aligned}$$

As this quantity is finite, the requirements of Fubini's theorem are fulfilled. Hence, it follows that

$$\mathrm{E}\left(\int_0^t W(s)\,ds\right) = \int_0^t \mathrm{E}(W(s))\,ds = 0\,. \quad \blacksquare$$

General Rules

Note that our definition of the integral seems to be unnecessarily restrictive. However, the restriction on the interval $[0, t]$ is by no means crucial. The usual rules apply and are here symbolically described for an integrand g (without proof):

$$\int_a^b g(x)\,dx = \int_a^c g(x)\,dx + \int_c^b g(x)\,dx \quad \text{for } a \leq c \leq b\,,$$

$$\int (g_1(x) + g_2(x))\,dx = \int g_1(x)\,dx + \int g_2(x)\,dx\,,$$

$$\int c\,g(x)\,dx = c \int g(x)\,dx \quad \text{for } c \in \mathbb{R}\,.$$

8.3 Riemann Integration of Wiener Processes

In this section, we concentrate on Riemann integrals where the stochastic part of the integrand is a WP: $X(t) = W(t)$.

Normal Distribution

Frequently, Gaussian random variables are hidden behind Riemann integrals. In fact, it holds that all of the integrals discussed in this section follow Gaussian distributions with zero expectation. The variances can be determined according to the following proposition (for a proof see Problem 8.3).

Proposition 8.3 (Normality of Riemann Integrals) *Let f be a continuous deterministic function on $[0, t]$. Then, it holds*

$$\int_0^t f(s)\,W(s)\,ds \sim \mathcal{N}\left(0\,, \int_0^t \int_0^t f(r)f(s)\min(r, s)drds\right)\,.$$

The normality follows from the fact that the WP is a Gaussian process. Hence, the Riemann sum R_n from (8.4) follows a Gaussian distribution for finite n. As R_n converges in mean square, it follows from Lemma 8.1 (see below) that the limit is Gaussian as well. Note that the finiteness of the variance expression from Proposition 8.3 is just sufficient and necessary for the existence of the Riemann integral (Proposition 8.1).

Example 8.2 (Variance of the Integrated WP) Consider an integrated WP with $f(s) = 1$ as in Example 8.1. We look for a closed expression for the variance of $\int_0^t W(s)ds$. Due to Proposition 8.3, the starting point is:

$$\mathrm{Var}\left(\int_0^t W(s)ds\right) = \int_0^t \int_0^t \min(r,s)\, drds\,.$$

Now, we employ a useful trick for many applications. The integral with respect to r is decomposed into the sum of two integrals with s as the integration limit such that the minimum function can be specified explicitly:

$$\begin{aligned}\int_0^t \int_0^t \min(r,s)\, drds &= \int_0^t \left[\int_0^s \min(r,s)\, dr + \int_s^t \min(r,s)\, dr\right] ds \\ &= \int_0^t \left[\int_0^s r\, dr + \int_s^t s\, dr\right] ds\,.\end{aligned}$$

Now, the integration of the power functions yields the requested variance:

$$\begin{aligned}\mathrm{Var}\left(\int_0^t W(s)ds\right) &= \int_0^t \left[\int_0^s r\, dr + \int_s^t s\, dr\right] ds \\ &= \int_0^t \left[\frac{s^2}{2} + s\,(t-s)\right] ds \\ &= \left[\frac{s^2\, t}{2} - \frac{s^3}{6}\right]_0^t \\ &= \frac{t^3}{3}\,. \quad \blacksquare\end{aligned}$$

Autocovariance Function

With the time-dependent integration limit, $\int_0^t f(s)\, W(s)\, ds$ itself is a stochastic process. Therefore, it suggests itself to not only determine the variance as in Proposition 8.3, but the covariance function as well. The general result is given in the following proposition, which will be verified in Problem 8.7.

Proposition 8.4 (Autocovariance Function) *For a continuous function f with integrable antiderivative F and $Y(t) = \int_0^t f(s)\, W(s)ds$ it holds that:*

$$E(Y(t)\, Y(t+h)) = \int_0^t f(s) \left[s\, F(s) - \int_0^s F(r)dr + s(F(t+h) - F(s)) \right] ds\,,$$

where $h \geq 0$.

Therefore, with $h = 0$ an alternative expression for the variance from Proposition 8.3 is obtained. For concrete functions f, the formula can be simplified considerably. This is to be shown by the following example.

Example 8.3 (Autocovariance of the Integrated WP) Once again, we examine the integrated WP with $f(s) = 1$ and $F(s) = s$ as in Examples 8.1 and 8.2. Then, plugging in yields:

$$\begin{aligned} \mathrm{E}\left(\int_0^t W(s)ds \int_0^{t+h} W(r)dr \right) &= \int_0^t \left[s^2 - \frac{1}{2}s^2 + s((t+h) - s) \right] ds \\ &= \int_0^t \left[s(t+h) - \frac{1}{2}s^2 \right] ds \\ &= \frac{t^2(t+h)}{2} - \frac{t^3}{6}. \end{aligned}$$

Hence, for $h = 0$ the variance of the integrated Wiener process reads:

$$\mathrm{Var}\left(\int_0^t W(s)ds \right) = \frac{t^3}{2} - \frac{t^3}{6} = \frac{t^3}{3}.$$

Of course, we already know this from Example 8.2. ■

Examples

For three special Gaussian integrals, which we will be confronted with over and over, the variances are to be calculated. We put the results in front.

Corollary 8.1 *It holds that*

$$
\begin{aligned}
&(a) \quad \int_0^1 W(s)\,ds \sim \mathcal{N}(0, 1/3)\,,\\
&(b) \; W(1) - \int_0^1 W(s)\,ds \sim \mathcal{N}(0, 1/3)\,,\\
&(c) \; \int_0^1 (s-c)\,W(s)\,ds \sim \mathcal{N}(0, \sigma_R^2)\,,
\end{aligned}
$$

where $c \in \mathbb{R}$ *and* $\sigma_R^2 = \frac{8-25c+20c^2}{60} > 0$.

The normality in (a) and (c) is clear due to Proposition 8.3. In (b) we have the sum of two Gaussian random variables which does not necessarily have to be Gaussian again unless a multivariate Gaussian distribution is present. Thus, the normality of (b) can only be proven in connection with Stieltjes integrals (see Problem 9.2).

The result from (a) is a special case of Example 8.2 with $t = 1$. We show in Problem 8.4 that the variance in (b) is just $1/3$. The proof of (c) for $c = 0$ is given in Problem 8.5; for an arbitrary c, the proof is basically similar, however, it gets computationally more involved. Note that the variance σ_R^2 cannot be zero or negative for any c (Problem 8.6).

Again, there should be a word of warning concerning equality in distribution. From (b) it follows that:

$$\int_0^1 W(s)\,ds - W(1) \sim \mathcal{N}(0, 1/3)\,.$$

Therefore, the following random variables are equal in distribution,

$$\int_0^1 W(s)\,ds - W(1) \sim \int_0^1 W(s)\,ds\,,$$

although, pathwise it obviously holds that:

$$\int_0^1 W(s)\,ds - W(1) \neq \int_0^1 W(s)\,ds\,.$$

8.4 Convergence in Mean Square

Now, we hand in some basics which are not necessary for the understanding of Riemann integrals; however, they are helpful for some technical properties. In particular, for the elaboration on the Ito integral following below, the knowledge of convergence in mean square is advantageous for a complete understanding. For a brief introduction to the basics of asymptotic theory, Pötscher and Prucha (2001) can be recommended.

Definition and Properties

Let $\{X_n\}$, $n \in \mathbb{N}$, be a sequence of real random variables with

$$\mathrm{E}(X_n^2) < \infty\,. \tag{8.5}$$

For a sequence $\{X_n\}$ and a random variable X, we define the **mean squared error** as distance or norm:

$$\mathrm{MSE}(X_n, X) := \mathrm{E}\left[(X_n - X)^2\right],$$

One says, $\{X_n\}$ converges in mean square to X for n tending to infinity if

$$\mathrm{MSE}(X_n, X) \rightarrow 0\,, \quad n \rightarrow \infty\,.$$

Abbreviating, we write for this as well

$$X_n \stackrel{2}{\rightarrow} X\,.$$

This limit is unique with probability one. Of course, it can be a random variable itself or a constant. In any case, due to (8.5) it holds that: $\mathrm{E}(X^2) < \infty$. In fact, expected value and variance of X can be determined from the moments of X_n. In particular, the limit of Gaussian random variables is again Gaussian. More precisely, the following lemma holds (see Problem 8.8 for a proof).

Lemma 8.1 (Properties of the Limit in Mean Square) *Let $\{X_n\}$ with (8.5) converge in mean square to X. Then it holds for $n \rightarrow \infty$:*

(a) $E(X_n) \rightarrow E(X)$;
(b) $E(X_n^2) \rightarrow E(X^2)$;
(c) if $\{X_n\}$ is Gaussian, then X follows a Gaussian distribution as well.

Naturally, the parameters of the Gaussian distribution X from (c) follow according to (a) and (b).

Convergence to a Constant

If the limit is a constant, then it is particularly easy to establish convergence in mean square. For this purpose, we consider the following derivation. By zero addition and the binomial formula, the following expression is obtained:

$$\begin{aligned}[X_n - X]^2 &= [(X_n - \mathrm{E}(X_n)) - (X - \mathrm{E}(X_n))]^2 \\ &= (X_n - \mathrm{E}(X_n))^2 - 2(X_n - \mathrm{E}(X_n))(X - \mathrm{E}(X_n)) + (X - \mathrm{E}(X_n))^2 .\end{aligned}$$

The expectation operator yields:

$$\begin{aligned}\mathrm{MSE}(X_n, X) = \mathrm{Var}(X_n) \\ -2\,\mathrm{E}\left[(X_n - \mathrm{E}(X_n))(X - \mathrm{E}(X_n))\right] + \mathrm{E}\left[(X - \mathrm{E}(X_n))^2\right] .\end{aligned}$$

If X is a constant (a "degenerate random variable"), $X = c$, then the second term becomes zero and the third term is the expected value of a constant. In other words, this yields:

$$\mathrm{MSE}(X_n, c) = \mathrm{Var}(X_n) + [c - \mathrm{E}(X_n)]^2 .$$

Hence, $\{X_n\}$ converges in mean square to a constant c if and only if it holds that

$$\mathrm{Var}(X_n) \to 0 \quad \text{and} \quad \mathrm{E}(X_n) \to c , \quad n \to \infty .$$

As is well known, this implies that $\{X_n\}$ converges to c in probability as well (see Lemma 8.3 below). Next, we cover criteria of convergence.

Test of Convergence

Now, we still need a convenient criterion in order to decide whether a series is convergent in mean square. In fact, we have two equivalent criteria. For the proof see Problem 8.9. The name goes back to the famous French mathematician Augustin Louis Cauchy (1789–1857).

Lemma 8.2 (Cauchy Criterion) *A series $\{X_n\}$ with (8.5) converges in mean square ...*

(a) ... if and only if it holds for arbitrary n and m that

$$E\left[(X_m - X_n)^2\right] \to 0 , \quad m, n \to \infty ;$$

(b) ... or put equivalently, if and only if it holds for arbitrary n and m that

$$E\,(X_m X_n) \to c < \infty , \quad m, n \to \infty ,$$

where $c \in \mathbb{R}$ is a constant.

Note that the convergence of the Cauchy criterion holds independently of how m and n tend to infinity. As well, the constant c results independently of the choice of m and n. As the criteria are sufficient and necessary, the proof of existence for mean square convergence can be supplied without determining the limit explicitly.

Example 8.4 (Law of Large Numbers) Let $\{\varepsilon_t\}$ be a white noise process, $\varepsilon_t \sim \text{WN}(0, \sigma^2)$. Then, it can be shown that the arithmetic mean,

$$X_n := \overline{\varepsilon}_n = \frac{1}{n}\sum_{t=1}^{n} \varepsilon_t \,,$$

converges in mean square without specifying the limit. It namely holds that

$$\begin{aligned} \mathrm{E}\left(\overline{\varepsilon}_n\, \overline{\varepsilon}_m\right) &= \frac{1}{mn}\mathrm{E}\left[\sum_{t=1}^{n} \varepsilon_t \sum_{t=1}^{m} \varepsilon_t\right] = \frac{1}{mn}\sum_{t=1}^{\min(n,m)} \mathrm{E}\left(\varepsilon_t^2\right) \\ &= \sigma^2\, \frac{\min(n,m)}{m\, n} \quad \to 0 \end{aligned}$$

for $m\,, n \to \infty$. Due to Lemma 8.2(b) we conclude that $\overline{\varepsilon}_n$ has a limit in mean square.

Let the limit of $\overline{\varepsilon}_n$ simply be called ε. Naturally, it can be determined immediately. Due to

$$\mathrm{E}\left(\overline{\varepsilon}_n\right) = 0 \quad \text{and} \quad \mathrm{Var}\left(\overline{\varepsilon}_n\right) = \frac{\sigma^2}{n}$$

it follows from Lemma 8.1(a) and (b) for the limit that

$$\mathrm{E}\left(\varepsilon\right) = 0 \quad \text{and} \quad \mathrm{Var}\left(\varepsilon\right) = 0\,.$$

Hence, the limit is equal to zero (with probability one). From this, it follows for $x_t = \mu + \varepsilon_t$ that the arithmetic mean of x_t converges in mean square to the true expected value, μ. In the literature, this fact is also known as the "law of large numbers". ■

Further Modes of Convergence

Two weaker concepts of convergence can be defined via probability statements. First, we say $\{X_n\}$ converges **in probability** to X if it holds for arbitrary $\varepsilon > 0$ that:

$$\lim_{n\to\infty} \mathrm{P}(|X_n - X| \geq \varepsilon) = 0\,.$$

Symbolically, this is denoted as[3]

$$X_n \stackrel{p}{\to} X \quad \text{for } n \to \infty\,.$$

Secondly, one talks about **convergence in distribution** of $\{X_n\}$ to X if it holds for all points $x \in \mathbb{R}$ at which the distribution function $F_n(x)$ of X_n is continuous, that $F_n(x)$ tends to the distribution function $F(x)$ of X:

$$\lim_{n\to\infty} F_n(x) = \lim_{n\to\infty} \mathrm{P}(X_n \leq x) = \mathrm{P}(X \leq x) = F(x)\,.$$

The word "distribution" suggests the symbolic notation:

$$X_n \stackrel{d}{\to} X \quad \text{for } n \to \infty\,.$$

From Grimmett and Stirzaker (2001, p. 310) or Pötscher and Prucha (2001, Theorem 5 and 9) we adopt the following results.

Lemma 8.3 (Implications of Convergence) *The following implications hold, $n \to \infty$.*

(a) Convergence in mean square implies convergence in probability:

$$(X_n \stackrel{2}{\to} X) \quad \Longrightarrow \quad (X_n \stackrel{p}{\to} X)\,.$$

(b) Convergence in probability implies convergence in distribution:

$$(X_n \stackrel{p}{\to} X) \quad \Longrightarrow \quad (X_n \stackrel{d}{\to} X)\,.$$

In general, the converse of Lemma 8.3(a) or (b) does not hold. However, if $X = c$ is a constant, then convergence in probability and convergence in distribution are equivalent, see Grimmett and Stirzaker (2001, p. 310) or Pötscher and Prucha (2001, Theorem 10).

8.5 Problems and Solutions

Problems

8.1 Prove Proposition 8.1.
Hint: Use Lemma 8.2.

[3]In particular in econometrics, one often writes alternatively $\mathrm{plim} X_n = X$ as $n \to \infty$.

8.2 Determine the expected value from Proposition 8.3.

8.3 Determine the variance from Proposition 8.3.

8.4 Calculate the variance from Corollary 8.1(b).

8.5 Calculate the variance from Corollary 8.1(c) for the special case of $c = 0$.

8.6 Show that the variance from Corollary 8.1(c) is positive.

8.7 Prove Proposition 8.4.

8.8 Prove Lemma 8.1.

8.9 Prove Lemma 8.2.

Solutions

8.1 Analogously to the partition (8.1) and the Riemann sum R_n from (8.4), we define for arbitrary m with $m \to \infty$:

$$P_m([0, t]): \quad 0 = r_0 < \ldots < r_m = t\,, \ \max_{1 \le j \le m} \left(r_j - r_{j-1}\right) \to 0\,,$$

$$R_m = \sum_{j=1}^{m} f(r_j^*)\,X(r_j^*)\,(r_j - r_{j-1})\,, \quad r_j^* \in [r_{j-1}, r_j]\,.$$

In order to apply the existence criterion from Lemma 8.2(b), we formulate the product of the two Riemann sums as follows:

$$R_n\,R_m = \sum_{i=1}^{n} \sum_{j=1}^{m} f(s_i^*) f(r_j^*)\,X(s_i^*) X(r_j^*)\,(r_j - r_{j-1})(s_i - s_{i-1})\,.$$

Hence, the Riemann integral as limit of R_n exists if and only if $\mathrm{E}(R_n\,R_m)$ converges. Further,

$$\mathrm{E}(R_n\,R_m) = \sum_{i=1}^{n} \sum_{j=1}^{m} f(s_i^*) f(r_j^*)\,\mathrm{E}\left(X(s_i^*)\,X(r_j^*)\right)\,(r_j - r_{j-1})(s_i - s_{i-1})\,,$$

converges if and only if the ordinary Riemann double integral

$$\int_0^t \int_0^t f(s) f(r)\,\mathrm{E}\left(X(s)\,X(r)\right)\,dr ds < \infty$$

exists. The Cauchy criterion from Lemma 8.2 therefore amounts to a proof of Proposition 8.1.

8.2 For the WP it holds that

$$\mathrm{E}(W(t)) = 0 \quad \text{and} \quad \mathrm{E}\left(W(s)\, W(r)\right) = \min(s, r) .$$

The minimum function is continuous in both the arguments. If f is a continuous function as well, then we know, with the considerations following Proposition 8.1, that the stochastic Riemann integral

$$\int_0^t f(s)\, W(s)\, ds$$

exists. In order to calculate the expected value, Fubini's theorem will be applied. For this purpose, we check that

$$\begin{aligned} \int_0^t \mathrm{E}\left(|f(s)\, W(s)|\right) ds &= \int_0^t |f(s)|\, \mathrm{E}\left(|W(s)|\right) ds \\ &\leq \max_{0 \leq s \leq t} |f(s)| \int_0^t \mathrm{E}\left(|W(s)|\right) ds \end{aligned}$$

is finite. The bound is based on the continuity and hence the finiteness of f. The integral

$$\int_0^t \mathrm{E}\left(|W(s)|\right) ds$$

was determined in Example 8.1 for t fixed to be finite. As the WP is continuous, Proposition 8.2 can be applied. According to this, it holds that:

$$\mathrm{E}\left(\int_0^t f(s)\, W(s)\, ds\right) = \int_0^t f(s)\, \mathrm{E}(W(s))\, ds = 0 .$$

8.3 Let us denote the Riemann integral by $Y(t)$:

$$Y(t) = \int_0^t f(s)\, W(s) ds.$$

We have already shown that $\mathrm{E}(Y(t)) = 0$. Hence, it follows that

$$\begin{aligned} \mathrm{Var}(Y(t)) &= \mathrm{E}\left[Y^2(t)\right] \\ &= \mathrm{E}\left[\int_0^t f(s)\, W(s) ds \int_0^t f(r)\, W(r) dr\right] \end{aligned}$$

$$= \mathrm{E}\left[\int_0^t \left(\int_0^t f(r)\, W(r)dr\right) f(s)\, W(s)ds\right]$$
$$= \mathrm{E}\left[\int_0^t \left(\int_0^t f(r)f(s)\, W(r)W(s)dr\right) ds\right].$$

By applying Fubini's theorem twice, we obtain:

$$\begin{aligned}
\mathrm{E}\left[Y^2(t)\right] &= \int_0^t \mathrm{E}\left(\int_0^t f(r)f(s)\, W(r)W(s)dr\right) ds \\
&= \int_0^t \int_0^t \mathrm{E}(f(r)f(s)\, W(r)W(s))dr\, ds \\
&= \int_0^t \int_0^t f(r)f(s)\, \mathrm{E}(W(r)W(s))dr\, ds \\
&= \int_0^t \int_0^t f(r)f(s) \min(r,s)dr\, ds\,,
\end{aligned}$$

which is the requested result.

8.4 Let us define

$$Y(t) = W(1) - \int_0^1 W(s)\, ds$$

with $\mathrm{E}(Y(t)) = 0$. Then, it holds that

$$\begin{aligned}
\mathrm{Var}(Y(t)) &= \mathrm{E}(Y^2(t)) \\
&= \mathrm{Var}(W(1)) + \mathrm{Var}\left[\int_0^1 W(s)\, ds\right] - 2\mathrm{E}\left[W(1)\int_0^1 W(s)ds\right] \\
&= 1 + \frac{1}{3} - 2\int_0^1 \mathrm{E}(W(1)W(s))ds,
\end{aligned}$$

where the variance from Corollary 8.1(a) and Fubini's theorem were used. On [0, 1] it holds that:

$$\mathrm{E}(W(1)\, W(s)) = \min(1,s) = s.$$

Hence, the variance results as claimed:

$$\mathrm{Var}(Y(t)) = 1 + \frac{1}{3} - 2\left[\frac{1}{2}s^2\right]_0^1 = 1 + \frac{1}{3} - 1 = \frac{1}{3}.$$

8.5 For $c = 0$ the claim reads

$$\mathrm{Var}\left(\int_0^1 s\, W(s)\, \mathrm{d}s\right) = \sigma_R^2 = \frac{2}{15}\,.$$

According to Proposition 8.3, the variance results as a double integral for $f(s) = s$,

$$\sigma_R^2 = \int_0^1 \int_0^1 rs \min(r, s)\, dr\, ds\,,$$

where the inner integral is appropriately decomposed to facilitate the calculation:

$$\begin{aligned}
\int_0^1 \int_0^1 rs \min(r, s)\, dr\, ds &= \int_0^1 s \left[\int_0^s r \min(r, s)\, dr + \int_s^1 r \min(r, s)\, dr\right] ds \\
&= \int_0^1 s \left[\int_0^s r^2\, dr + \int_s^1 rs\, dr\right] ds \\
&= \int_0^1 s \left[\frac{s^3}{3} + \frac{s}{2}\,(1 - s^2)\right] ds \\
&= \int_0^1 \left(\frac{s^4}{3} + \frac{s^2}{2} - \frac{s^4}{2}\right) ds \\
&= \int_0^1 \left(\frac{s^2}{2} - \frac{s^4}{6}\right) ds \\
&= \left[\frac{s^3}{6} - \frac{s^5}{30}\right]_0^1 \\
&= \frac{1}{6} - \frac{1}{30} = \frac{4}{30}.
\end{aligned}$$

This corresponds to the claimed result.

8.6 We consider the numerator of σ_R^2,

$$n(c) = 8 - 25c + 20c^2,$$

and show that it does not have any real zeros. Setting $n(c) = 0$ yields:

$$\begin{aligned}
c_{1,2} &= \frac{25 \pm \sqrt{25^2 - 4 \cdot 20 \cdot 8}}{2 \cdot 20} \\
&= \frac{25 \pm \sqrt{-15}}{40} \\
&= \frac{25 \pm i\sqrt{15}}{40}\,, \quad i^2 = -1\,.
\end{aligned}$$

Thus, no real zeros exist. Consequently, since $n(0) = 8 > 0$ the in c continuous function $n(c)$ cannot be zero or negative; the same holds true for σ_R^2, which proves the claim.

Of course, an alternative proof consists in determining the real extrema of $n(c)$. There exists an absolute minimum and this turns out to be positive.

8.7 If the Riemann integral is denoted by $Y(t)$, then it holds, as for the derivation of the variance, that:

$$\begin{aligned} \mathrm{E}(Y(t)\,Y(t+h)) &= \int_0^t f(s)\left[\int_0^s f(r)\min(r,s)dr + \int_s^{t+h} f(r)\min(r,s)dr\right]ds \\ &= \int_0^t f(s)\left[\int_0^s f(r)rdr + s\int_s^{t+h} f(r)dr\right]ds. \end{aligned}$$

Partial integration yields the following relation:

$$\int_0^s f(r)rdr = F(s)s - \int_0^s F(r)dr.$$

By plugging in we obtain the claim.

8.8 Proof of (a): By bounding the difference of the two expected values by means of the Cauchy-Schwarz inequality (2.5) one can immediately tell that this difference tends to zero in the case of convergence in mean square:

$$\begin{aligned} |\mathrm{E}(X_n) - \mathrm{E}(X)| &= |\mathrm{E}(X_n - X)| \\ &\leq \sqrt{\mathrm{E}\left[(X_n - X)^2\right]} = \sqrt{\mathrm{MSE}(X_n, X)} \\ &\to 0\,, \quad n \to \infty\,. \end{aligned}$$

Proof of (b): The simple trick

$$X_n^2 - X^2 = (X_n - X)^2 + 2\,(X_n - X)\,X$$

yields upon expectation:

$$\begin{aligned} \mathrm{E}(X_n^2) - \mathrm{E}(X^2) &= \mathrm{E}\left[(X_n - X)^2\right] + 2\,\mathrm{E}\left[(X_n - X)\,X\right] \\ &\leq \mathrm{E}\left[(X_n - X)^2\right] + 2\,\left|\mathrm{E}\left[(X_n - X)X\right]\right| \\ &\leq \mathrm{E}\left[(X_n - X)^2\right] + 2\,\sqrt{\mathrm{E}\left[(X_n - X)^2\right]}\,\sqrt{\mathrm{E}(X^2)} \\ &= \mathrm{MSE}(X_n, X) + 2\,\sqrt{\mathrm{MSE}(X_n, X)}\,\sqrt{\mathrm{E}(X^2)} \\ &\to 0\,, \quad n \to \infty\,, \end{aligned}$$

where again the Cauchy-Schwarz inequality (2.5) was used.

Proof of (c): As is well known, convergence in mean square implies convergence in distribution, see Lemma 8.3. The latter is equivalent to the fact that the characteristic function $\phi_n(u)$ of X_n tends to the characteristic function $\phi(u)$ of X.[4] Now we show: If $\phi_n(u)$ belongs to a Gaussian distribution, then this holds for $\phi(u)$ as well. Hence, (a) and (b) in combination with the premise of a Gaussian sequence $\{X_n\}$ imply:

$$\begin{aligned}\phi_n(u) &= \exp\left\{i\,u\mathrm{E}(X_n) - \frac{u^2\mathrm{Var}(X_n)}{2}\right\}\\ &\to \exp\left\{i\,u\mathrm{E}(X) - \frac{u^2\mathrm{Var}(X)}{2}\right\} = \phi(u)\end{aligned}$$

for $n \to \infty$. Thus, the characteristic function of X as $n \to \infty$ is that of a Gaussian distribution as well.

8.9 Proof of (a): Elementarily, it can be shown that the Cauchy criterion follows from convergence in mean square:

$$\begin{aligned}\mathrm{E}\left[(X_m - X_n)^2\right] &= \mathrm{E}\left[\left((X_m - X) + (X - X_n)\right)^2\right]\\ &= \mathrm{E}\left[(X_m - X)^2\right] + \mathrm{E}\left[(X - X_n)^2\right]\\ &\quad + 2\,\mathrm{E}\left[(X_m - X)(X - X_n)\right]\\ &\leq \mathrm{E}\left[(X_m - X)^2\right] + \mathrm{E}\left[(X - X_n)^2\right]\\ &\quad + 2\,\sqrt{\mathrm{E}\left[(X_m - X)^2\right]}\,\sqrt{\mathrm{E}\left[(X - X_n)^2\right]}\\ &= \mathrm{MSE}\,(X_m, X) + \mathrm{MSE}\,(X_n, X)\\ &\quad + 2\,\sqrt{\mathrm{MSE}\,(X_m, X)}\,\sqrt{\mathrm{MSE}\,(X_n, X)}\,,\end{aligned}$$

where the bounding is again based on the Cauchy-Schwarz inequality (2.5). It is somewhat more involved that, inversely, the condition from (a) implies convergence in mean square as well. For the proof, we refer e.g. to the exposition on Hilbert spaces in Brockwell and Davis (1991, Ch. 2).

---

[4] See e.g. sections 5.7 through 5.10 in Grimmett and Stirzaker (2001) for an introduction to the theory and application of characteristic functions. In particular, it holds for the characteristic function of a random variable with a Gaussian distribution, $Y \sim \mathcal{N}(\mu, \sigma^2)$, that:

$$\phi_y(u) = \exp\left\{i\,u\,\mu - \frac{u^2\sigma^2}{2}\right\}, \quad i^2 = -1\,,\ u \in \mathbb{R}\,.$$

Proof of (b): If we take (a) for granted, the proof is simple. Due to

$$\mathrm{E}\left[(X_m - X_n)^2\right] = \mathrm{E}(X_m^2) + \mathrm{E}(X_n^2) - 2\,\mathrm{E}(X_m X_n),$$

one can immediately tell that the condition from (a) implies:

$$\mathrm{E}(X_m X_n) \to \frac{\mathrm{E}(X^2) + \mathrm{E}(X^2)}{2} = \mathrm{E}(X^2)\,.$$

Inversely, from the condition from (b) it naturally follows that

$$\begin{aligned}\mathrm{E}\left[(X_m - X_n)^2\right] &= \mathrm{E}(X_m^2) + \mathrm{E}(X_n^2) - 2\,\mathrm{E}(X_m X_n)\\ &\to c + c - 2\,c = 0\,.\end{aligned}$$

This completes the proof.

References

Billingsley, P. (1986). *Probability and measure* (2nd ed.). New York: Wiley.

Brockwell, P. J., & Davis, R. A. (1991). *Time series: Theory and methods* (2nd ed.). New York: Springer.

Grimmett, G. R., & Stirzaker, D. R. (2001). *Probability and random processes* (3rd ed.). Oxford: Oxford University Press.

Pötscher, B. M., & Prucha, I. R. (2001). Basic elements of asymptotic theory. In B. H. Baltagi (Ed.), *A companion to theoretical econometrics* (pp. 201–229). Malden: Blackwell.

Trench, W. F. (2013). *Introduction to real analysis*. Free Hyperlinked Edition 2.04 December 2013. Downloaded on 10th May 2014 from http://digitalcommons.trinity.edu/mono/7.

Stieltjes Integrals

9

9.1 Summary

Below, we will encounter Riemann-Stieltjes integrals (or more briefly: Stieltjes integrals) as solutions of certain stochastic differential equations. They can be reduced to the sum of a Riemann integral and a multiple of the Wiener process. Stieltjes integrals are again Gaussian. As an example we consider the Ornstein-Uhlenbeck process which is defined by a Stieltjes integral and which will be dealt with in detail in the chapter on interest rate models.

9.2 Definition and Partial Integration

As a first step towards the Ito integral, we define Stieltjes[1] integrals which can be reduced to Riemann integrals by integration by parts.

Definition

The Riemann-Stieltjes integral (or Stieltjes integral), as it is considered here, integrates over a deterministic function $f(s)$. Nevertheless, the Stieltjes integral is random as it is integrated with respect to the stochastic Wiener process $W(s)$. In order to understand what is meant by this, we recall the partition (8.1):

$$P_n([0,t]): \quad 0 = s_0 < s_1 < \ldots < s_n = t\,,$$

[1]Thomas J. Stieltjes lived from 1856 to 1894. The Dutch mathematician generalized the concept of integration by Riemann.

U. Hassler, *Stochastic Processes and Calculus*, Springer Texts in Business and Economics, DOI 10.1007/978-3-319-23428-1_9

with $s_i^* \in [s_{i-1}, s_i]$. Hence, the **Riemann-Stieltjes sum** is defined as

$$RS_n = \sum_{i=1}^{n} f(s_i^*)\,(W(s_i) - W(s_{i-1}))\,. \tag{9.1}$$

If an expression well-defined in mean square follows from this for $n \to \infty$ under (8.2), then we define it as a **Stieltjes integral** with the obvious notation

$$RS_n \xrightarrow{2} \int_0^t f(s)\,dW(s)\,.$$

Note that $dW(s)$ does not stand for the derivative of the Wiener process as it does not exist. It is just a common symbolic notation.
If f is **continuously differentiable**,[2] then the existence of the Stieltjes integral is guaranteed, see Soong (1973, Theorem 4.5.2).

Integration by Parts

If f is continuously differentiable, then the Stieltjes integral can be expressed by a Riemann integral and the WP. This relation is also known as integration by parts. In Chap. 11 we will understand that it is a special case of Ito's lemma, which is why we do not have to concern ourselves with a proof of Proposition 9.1 at this point.

Proposition 9.1 (Stieltjes Integral; Integration by Parts) *For a continuously differentiable, deterministic function f we have that*

(a) the Stieltjes sum from (9.1) converges in mean square if it holds that $\max(s_i - s_{i-1}) \to 0$,
(b) and

$$\int_0^t f(s)\,dW(s) = [f(s)\,W(s)]_0^t - \int_0^t W(s)\,df(s)$$
$$= f(t)\,W(t) - \int_0^t W(s)f'(s)\,ds\,.$$

where the last equality holds with probability one.[3]

[2] We call a function continuously differentiable if it has a continuous first order derivative.

[3] Remember that we assumed $P(W(0) = 0) = 1$, which justifies the last statement. Whenever we have equalities in a stochastic setting, they are typically understood to hold with probability one for the rest of the book.

The result from (b) corresponds to the familiar rule of partial integration. As a refresher, we write this rule for two deterministic functions f and g:

$$\int_0^t f(s)\, g'(s)\, ds = [f(s)\, g(s)]_0^t - \int_0^t g(s) f'(s)\, ds\,. \tag{9.2}$$

Hence, this is the integral form of the product rule of differentiation:

$$\frac{d[f(s)\, g(s)]}{ds} = f'(s)\, g(s) + g'(s) f(s)\,,$$

or

$$d\,[f(s)\, g(s)] = g(s)\, df(s) + f(s)\, dg(s)\,,$$

or

$$[f(s)\, g(s)]_0^t = \int_0^t g(s)\, df(s) + \int_0^t f(s)\, dg(s)\,.$$

Therefore, one can make a mental note of Proposition 9.1 (b) by the well-known partial integration from (9.2).

Example 9.1 (Corollary) As an application of Proposition 9.1 we consider Riemann-Stieltjes integrals for three particularly simple functions. We will encounter these relations repeatedly. The proof amounts to a simple exercise in substitution. It holds

(a) for the identity function $f(s) = s$:

$$\int_0^t s\, dW(s) = t\, W(t) - \int_0^t W(s)\, ds\,;$$

(b) for $f(s) = 1 - s$:

$$\int_0^t (1-s)\, dW(s) = (1-t)\, W(t) + \int_0^t W(s)\, ds\,;$$

(c) for the constant function $f(s) = 1$:

$$\int_0^t dW(s) = W(t)\,.$$

In (c) we again observe a formal analogy of the WP with the random walk. Just like the latter is defined as the sum over the past of a pure random process, see (1.8), the WP is the integral of its past independent increments. ■

9.3 Gaussian Distribution and Autocovariances

The reduction of Stieltjes integrals to Riemann integrals suggests that there are Gaussian processes hiding behind them. In fact, it holds that all Stieltjes integrals follow Gaussian distributions with expectation zero.

Gaussian Distribution

The Gaussian distribution itself is obvious: The Riemann-Stieltjes sum from (9.1) is, as the sum of multivariate Gaussian random variables, Gaussian as well. Then, this also holds for the limit of the sum due to Lemma 8.1. The expected value is zero due to Propositions 9.1(b) and 8.3. The variance results as a special case of the autocovariance given in Proposition 9.3. Hence, we obtain the following proposition.

Proposition 9.2 (Normality of Stieltjes integrals) *For a continuously differentiable, deterministic function f, it holds that*

$$\int_0^t f(s)\, dW(s) \sim \mathcal{N}\left(0, \int_0^t f^2(s)\, ds\right).$$

The variance of the Stieltjes integral is well motivated as follows. For the variance of the Riemann-Stieltjes sum,

$$\text{Var}\left(\sum_{i=1}^n f(s_i^*)\ (W(s_i) - W(s_{i-1}))\right),$$

it follows for $n \to \infty$, due to the independence of the increments of the WP, that:

$$\begin{aligned}\sum_{i=1}^n f^2(s_i^*)\,\text{Var}\,(W(s_i) - W(s_{i-1})) &= \sum_{i=1}^n f^2(s_i^*)\,(s_i - s_{i-1}) \\ &\to \int_0^t f^2(s)\, ds.\end{aligned}$$

The convergence takes place as f^2 is continuous and thus Riemann-integrable. Hence, for $n \to \infty$ the expression from Proposition 9.2 is obtained.

Let us consider the integrals from Example 9.1 and calculate the variances for $t = 1$ (see Problem 9.1).

Example 9.2 (Corollary) For the functions from Example 9.1 it holds:

(a) for the identity function $f(s) = s$:

$$\int_0^1 s\,dW(s) \sim \mathcal{N}(0, 1/3)\,;$$

(b) for $f(s) = 1 - s$:

$$\int_0^1 (1-s)\,dW(s) \sim \mathcal{N}(0, 1/3)\,;$$

(c) for the constant function $f(s) = 1$:

$$W(t) = \int_0^t dW(s) \sim \mathcal{N}(0, t)\,. \quad \blacksquare$$

Autocovariance Function

As a generalization of the variance, an expression for the covariance is to be found. Hence, let us define the process $Y(t) = \int_0^t f(s)dW(s)$. The autocovariance of $Y(t)$ and $Y(t+h)$ with $h \geq 0$ can be well justified if one takes into account that the increments $dW(t)$ of the WP are stochastically independent provided they do not overlap. Therefore, one should expect $\int_0^t f(s)dW(s)$ and $\int_t^{t+h} f(r)dW(r)$ to be uncorrelated:

$$\mathrm{E}\left[\int_0^t f(s)dW(s) \int_t^{t+h} f(r)dW(r)\right] = 0.$$

If this is true, then, due to

$$\int_0^{t+h} f(r)dW(r) = \int_0^t f(r)dW(r) + \int_t^{t+h} f(r)dW(r)$$

the following result is obtained:

$$\begin{aligned}\mathrm{E}\left[\int_0^t f(s)dW(s) \int_0^{t+h} f(r)dW(r)\right] &= \mathrm{E}\left[\int_0^t f(s)dW(s) \int_0^t f(r)dW(r)\right] \\ &= \mathrm{Var}\left(\int_0^t f(s)dW(s)\right).\end{aligned}$$

Therefore, for an arbitrary $h \geq 0$ the autocovariance coincides with the variance in t. In fact, this result can be verified more rigorously (see Problem 9.5).

Proposition 9.3 (Autocovariance of Stieltjes Integrals) *For a continuously differentiable, deterministic function f it holds that*

$$E\left[\int_0^t f(s)dW(s) \int_0^{t+h} f(s)dW(s)\right] = \int_0^t f^2(s)ds$$

with $h \geq 0$.

Of course, for $h = 0$ the variance from Proposition 9.2 is obtained.

Example 9.3 (Autocovariance of the WP) As an example, let us consider $f(s) = 1$ with

$$W(t) = \int_0^t dW(s).$$

Then, it follows for $h \geq 0$:

$$\mathrm{E}(W(t)W(t+h)) = \int_0^t ds = t = \min(t, t+h).$$

Trivially, this just reproduces the autocovariance structure of the Wiener process already known from (7.4). ■

9.4 Standard Ornstein-Uhlenbeck Process

The so-called Ornstein-Uhlenbeck process has been introduced in a publication by the physicists Ornstein and Uhlenbeck in 1930.

Definition

We define the **Ornstein-Uhlenbeck process** (OUP) with starting value $X_c(0) = 0$ for an arbitrary real c as a Stieltjes integral,

$$X_c(t) := e^{ct} \int_0^t e^{-cs}\, dW(s), \quad t \geq 0,\, X_c(0) = 0\,. \tag{9.3}$$

For $c = 0$ in (9.3) the Wiener process, $X_0(t) = W(t)$, is obtained. More precisely, $X_c(t)$ from (9.3) is a standard OUP; a generalization will be offered in the chapter on interest rate dynamics. By definition, it holds that:

$$\begin{aligned} X_c(t+1) &= e^{ct}\, e^{c} \left[\int_0^t e^{-cs} dW(s) + \int_t^{t+1} e^{-cs} dW(s) \right] \\ &= e^c X_c(t) + e^{c(t+1)} \int_t^{t+1} e^{-cs} dW(s) \\ &= e^c X_c(t) + \varepsilon(t+1), \end{aligned}$$

where $\varepsilon(t+1)$ was defined implicitly. Note that the increments $dW(s)$ from t on in $\varepsilon(t+1)$ are independent of the increments up to t as they appear in $X_c(t)$. Hence, the OUP is a continuous counterpart of the AR(1) process from Chap. 3 where the autoregressive parameter is denoted by e^c. For $c < 0$ this parameter is less than one, such that in this case we expect a stable adjustment or, in a way, a quasi-stationary behavior. This will be reflected by the behavior of the variance and the covariance function which are given, among others, in the following proposition.

Properties

The proof of Proposition 9.4 will be given in an exercise problem. It comprises an application of Propositions 9.1, 9.2 and 9.3.

Proposition 9.4 (Ornstein-Uhlenbeck Process) *It holds for the Ornstein-Uhlenbeck process from (9.3) that:*

$$\begin{aligned} &(a)\ X_c(t) = W(t) + c\, e^{ct} \int_0^t e^{-cs}\, W(s)\, ds\,, \\ &(b)\ X_c(t) \sim \mathcal{N}(0,\, (e^{2ct} - 1)/2c)\,, \\ &(c)\ E\,(X_c(t)\, X_c(t+h)) = e^{ch}\, Var\,(X_c(t))\,, \end{aligned}$$

where $h \geq 0$.

Statement (a) establishes the usual relation between Stieltjes and Riemann integrals and, seen individually, it is not that thrilling. As for $c = 0$ the OUP coincides with the WP, it is interesting to examine the variance from (b) for $c \to 0$. **L'Hospital's rule** yields:

$$\lim_{c \to 0} \frac{e^{2ct} - 1}{2c} = \lim_{c \to 0} \frac{2te^{2ct}}{2} = t.$$

Hence, for $c \to 0$ the variance of the WP is embedded in (b). The covariance from (c) allows for determining the autocorrelation:

$$\begin{aligned} corr(X_c(t), X_c(t+h)) &= \frac{e^{ch}\text{Var}(X_c(t))}{\sqrt{\text{Var}(X_c(t))}\sqrt{\text{Var}(X_c(t+h))}} \\ &= e^{ch}\frac{\sqrt{\text{Var}(X_c(t))}}{\sqrt{\text{Var}(X_c(t+h))}}. \end{aligned}$$

Now, let us assume that $c < 0$. Then it holds for t growing that:

$$\lim_{t\to\infty} \text{Var}(X_c(t)) = -\frac{1}{2c} > 0.$$

Accordingly, it holds for the autocorrelation that:

$$\lim_{t\to\infty} corr(X_c(t), X_c(t+h)) = e^{ch}, \quad c < 0.$$

Thus, for $c < 0$ we obtain the "asymptotically stationary" case with asymptotically constant variance and an autocorrelation being asymptotically dependent on the lag h only. Thereby, the autocorrelation results as the h-th power of the "autoregressive parameter" $a = e^c$. With h growing, the autocovariance decays gradually. This finds its counterpart in the discrete-time AR(1) process. Just as the random walk arises from the AR(1) process with the parameter value one, the WP with $c = 0$, i.e. $a = e^0 = 1$, is the corresponding special case of the OUP. Hence, we can definitely consider the OUP as a continuous-time analog to the AR(1) process.

Simulation

The theoretical properties of the process for $c < 0$ can be illustrated graphically. In Fig. 9.1 the simulated paths of two parameter constellations are shown. It can be observed that the process oscillates about the zero line where the variance or the deviation from zero for $c = -0.1$ is much larger[4] than in the case $c = -0.9$. This is clear against the background of (b) from Proposition 9.4 in which the first moment and the variance are given: The expected value is zero and the variance decreases with the absolute value of c increasing. The positive autocorrelation (cf. Proposition 9.4(c)) is obvious as well: Positive values tend to be followed by positive values and the inverse holds for negative observations. The closer to zero c is, the stronger the autocorrelation gets. That is why the graph for $c = -0.1$ is strongly

[4] If the arithmetic mean of the 1000 observations of this time series is calculated, then by -0.72344 a notably negative number is obtained although the theoretical expected value is zero. Details on the simulation of OUP paths are to follow in Sect. 13.2.

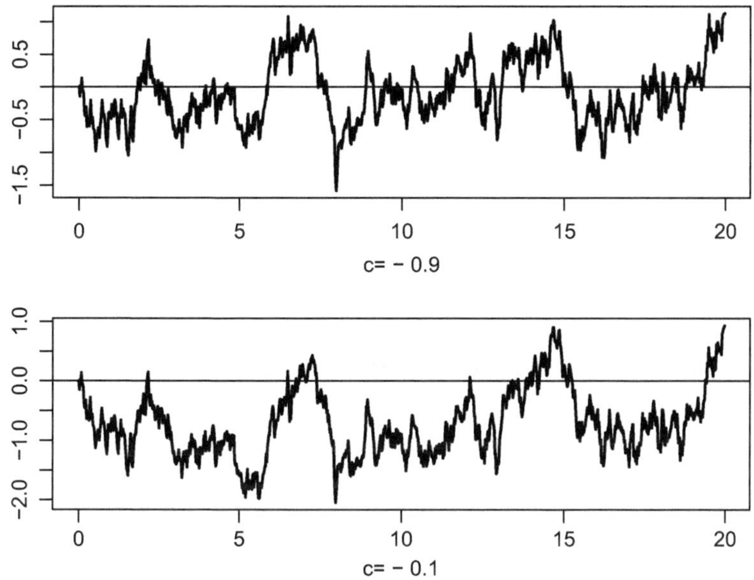

Fig. 9.1 Standard Ornstein-Uhlenbeck processes

determined by the "local" trend and does not cross the zero line for longer time spans while for $c = -0.9$ the force which pulls the observations back to the zero line is more effective such that the graph looks "more stationary" for $c = -0.9$.

9.5 Problems and Solutions

Problems

9.1 Calculate the variances from Example 9.2.

9.2 Verify the Gaussian distribution from Corollary 8.1(b).

9.3 Verify the following equality (with probability 1):

$$\int_0^t s^2 \, dW(s) = t^2 \, W(t) - 2 \int_0^t s \, W(s) \, ds \, .$$

9.4 Determine the variance of the process $X(t)$ with

$$X(t) = \int_0^t s^2 \, dW(s) \, .$$

9.5 Prove Proposition 9.3.

9.6 Prove (a) from Proposition 9.4.

9.7 Show (b) from Proposition 9.4.

9.8 Prove (c) from Proposition 9.4.

Solutions

9.1 From Proposition 9.2 it obviously follows for (a) that:

$$\int_0^1 s^2 ds = \left[\frac{s^3}{3}\right]_0^1 = \frac{1}{3}.$$

Equally, one shows (b):

$$\int_0^1 (1-s)^2 ds = \left[-\frac{(1-s)^3}{3}\right]_0^1 = \frac{1}{3}.$$

Finally, the result from (c) is known anyway.

9.2 The result follows from the examples of this chapter. From Example 9.1(a) we obtain for $t = 1$:

$$W(1) - \int_0^1 W(s)\, ds = \int_0^1 s\, dW(s)\,.$$

Due to Example 9.2(a) the claim is verified.

9.3 This is a straightforward application of Proposition 9.1. With $f(s) = s^2$ and $f'(s) = 2s$ the claim is established.

9.4 From Proposition 9.2 with $f(s) = s^2$ it follows for the variance

$$\text{Var}\left(\int_0^t s^2 dW(s)\right) = \int_0^t s^4 ds = \left[\frac{1}{5}s^5\right]_0^t = \frac{t^5}{5}.$$

9.5 With $Y(t) = \int_0^t f(s)dW(s)$ we know from Proposition 9.1 that:

$$Y(t) = f(t)\, W(t) - \int_0^t f'(s)W(s)ds.$$

Hence, the covariance results as

$$\mathrm{E}(Y(t)Y(t+h)) = A - B - C + D$$

where the expressions on the right-hand side are defined by multiplying $Y(t)$ and $Y(t+h)$. Now, we consider them one by one.

For A we obtain immediately:

$$\begin{aligned} A &= \mathrm{E}\left[f(t)f(t+h)W(t)\,W(t+h)\right] \\ &= f(t)f(t+h)\min(t,t+h) \\ &= f(t)f(t+h)t. \end{aligned}$$

By Fubini's theorem it holds for B that:

$$\begin{aligned} B &= \mathrm{E}\left[f(t+h)\int_0^t f'(s)W(s)W(t+h)ds\right] \\ &= f(t+h)\int_0^t f'(s)\min(s,t+h)ds \\ &= f(t+h)\int_0^t f'(s)sds. \end{aligned}$$

Integration by parts in the following form,

$$\int_0^t f(r)rdr = F(t)t - \int_0^t F(r)dr \quad \text{with } F' = f, \tag{9.4}$$

applied to f' yields:

$$B = f(t+h)[f(t)t - F(t) + F(0)],$$

where $F(s)$ denotes the antiderivative of $f(s)$. In the same way, we obtain

$$\begin{aligned} C &= \mathrm{E}\left[f(t)\int_0^{t+h} f'(s)W(s)W(t)ds\right] \\ &= f(t)\int_0^{t+h} f'(s)\min(s,t)ds \\ &= f(t)\left[\int_0^t f'(s)sds + \int_t^{t+h} f'(s)tds\right] \\ &= f(t)\left[f(t)t - F(t) + F(0) + t(f(t+h) - f(t))\right] \\ &= f(t)\left[F(0) - F(t) + tf(t+h)\right]. \end{aligned}$$

For the fourth expression Proposition 8.4 provides us with f' instead of f:

$$\begin{aligned}
D &= \mathrm{E}\left[\int_0^t f'(s)W(s)ds \int_0^{t+h} f'(r)W(r)dr\right] \\
&= \int_0^t f'(s)\left[f(s)s - (F(s) - F(0)) + s(f(t+h) - f(s))\right] ds \\
&= \int_0^t f'(s)dsF(0) - \int_0^t f'(s)F(s)ds + f(t+h)\int_0^t sf'(s)ds.
\end{aligned}$$

In addition to (9.4), we apply integration by parts in the form of

$$\int_0^t f'(s)F(s)ds = f(t)F(t) - f(0)F(0) - \int_0^t f^2(s)ds\,.$$

Then it holds that:

$$\begin{aligned}
D &= (f(t) - f(0))F(0) - f(t)F(t) + f(0)F(0) + \int_0^t f^2(s)ds \\
&\quad + f(t+h)\,(f(t)t - F(t) + F(0)) \\
&= \int_0^t f^2(s)ds + (F(0) - F(t))(f(t) + f(t+h)) + f(t)f(t+h)t.
\end{aligned}$$

If we assemble the terms, then we obtain the autocovariance function in the desired form:

$$\mathrm{E}\left[\int_0^t f(s)dW(s)\int_0^{t+h} f(r)dW(r)\right] = A - B - C + D = \int_0^t f^2(s)ds.$$

9.6 We use Proposition 9.1 with $f(s) = e^{-cs}$:

$$\int_0^t e^{-cs}dW(s) = e^{-ct}W(t) + c\int_0^t e^{-cs}W(s)\,ds.$$

Multiplying by e^{ct} yields

$$e^{ct}\int_0^t e^{-cs}dW(s) = W(t) + c\,e^{ct}\int_0^t e^{-cs}W(s)\,ds.$$

On the left-hand side, we have the OUP $X_c(t)$ by definition which was to be verified.

9.7 Due to Proposition 9.2 the OUP is Gaussian with expectation zero and variance

$$\begin{aligned}\mathrm{Var}(X_c(t)) &= e^{2ct}\int_0^t e^{-2cs}ds\\ &= e^{2ct}\left[\frac{e^{-2cs}}{-2c}\right]_0^t\\ &= e^{2ct}\,\frac{e^{-2ct}-1}{-2c}\\ &= \frac{1-e^{2ct}}{-2c}.\end{aligned}$$

This is equal to the claimed variance.

9.8 As for the derivation of the variance, we use

$$\int_0^t e^{-2cs}ds = \frac{1-e^{-2ct}}{2c}.$$

From Proposition 9.3 we know that this is also the expression for the autocovariance of the Stieltjes integrals ($h \geq 0$):

$$\mathrm{E}\left[\int_0^t e^{-cs}dW(s)\int_0^{t+h} e^{-cs}dW(s)\right] = \frac{1-e^{-2ct}}{2c}.$$

Hence, we obtain for the OUP:

$$\begin{aligned}\mathrm{E}(X_c(t)\,X_c(t+h)) &= e^{ct}e^{c(t+h)}\frac{1-e^{-2ct}}{2c}\\ &= e^{ch}\,\frac{e^{2ct}-1}{2c}\\ &= e^{ch}\,\mathrm{Var}(X_c(t)).\end{aligned}$$

Reference

Soong, T. T. (1973). *Random differential equations in science and engineering*. New York: Academic Press.

Ito Integrals 10

10.1 Summary

Kiyoshi Ito (1915–2008) was awarded the inaugural Gauss Prize by the International Mathematical Union in 2006. Stochastic integration in the narrow sense can be traced back to his early work published in Japanese in the forties of the last century. We precede the general definition of the Ito integral with a special case. Concluding, we discuss the (quadratic) variation of a process without which a sound understanding of Ito's lemma will not be possible.

10.2 A Special Case

We start with a special case of Ito integration, so to speak the mother of all stochastic integrals. Thereby we will understand that, besides the Ito integral, infinitely many related integrals of a similar structure exist.

Problems with the Definition

The starting point is again a partition

$$P_n([0,t]) : \quad 0 = s_0 < s_1 < \ldots < s_n = t \,,$$

that gets finer for n growing since we continue to maintain (8.2). Given this decomposition of $[0,t]$, we define analogously to the Riemann-Stieltjes sum for $s_i^* \in [s_{i-1}, s_i)$:

$$S_n(W) = \sum_{i=1}^{n} W(s_i^*)\,(W(s_i) - W(s_{i-1}))\,. \tag{10.1}$$

U. Hassler, *Stochastic Processes and Calculus*, Springer Texts in Business and Economics, DOI 10.1007/978-3-319-23428-1_10

For $n \to \infty$ we would like to denote the limit as $\int_0^t W(s)\, dW(s)$, which looks like a Stieltjes integral of a WP with respect to a WP. However, we will realize that:

1. The limit of $S_n(W)$ is not unique, but depends on the choice of s_i^*;
2. the limit of $S_n(W)$ is not defined as a Stieltjes integral.

As the Stieltjes integral has a unique limit independently of s_i^*, the second claim follows from the first one. If one chooses in particular the lower endpoint of the interval as support, $s_i^* = s_{i-1}$, then this leads to the Ito integral. Hence, this special case is called the **Ito sum**:

$$I_n(W) = \sum_{i=1}^{n} W(s_{i-1})\,(W(s_i) - W(s_{i-1}))\,. \tag{10.2}$$

The following proposition specifies the dependence on s_i^*. The special case $\gamma = 0$ leading to the Ito integral will be proved in Problem 10.1; the general result is established e.g. in Tanaka (1996, eq. (2.40)). The convergence is again in mean square.

Proposition 10.1 (Stochastic Integrals in Mean Square) *Let $s_i^* = (1-\gamma)\, s_{i-1} + \gamma\, s_i$ with $0 \leq \gamma < 1$. Then it holds for the sum from (10.1) with $n \to \infty$ under (8.2):*

$$S_n(W) \xrightarrow{2} \frac{1}{2}\left(W^2(t) - t\right) + \gamma\, t\,.$$

Before we discuss two special cases of Proposition 10.1, this striking result is to be somewhat better understood. We call it striking because it is counter-intuitive at first glance that the choice of s_i^* should matter with the intervals $[s_{i-1}, s_i)$ getting narrower and narrower for $n \to \infty$. To better understand this, we temporarily denote the limit of $S_n(W)$ as $S(\gamma)$:

$$S_n(W) \xrightarrow{2} S(\gamma)\,.$$

Then one observes immediately:

$$S(\gamma) = S(0) + \gamma\, t\,.$$

This means that the variance of all these stochastic integrals $S(\gamma)$ is identical, i.e. equal to the variance of $S(0)$. Hence, the choice of different support points s_i^* is only reflected in the expected value:

$$\begin{aligned}\mathrm{E}\left(S(\gamma)\right) &= \frac{1}{2}\left(\mathrm{E}(W^2(t)) - t\right) + \gamma\, t \\ &= \gamma\, t\,.\end{aligned}$$

However, this expected value can be well understood as for finite sums it can be shown that (see Problem 10.4):

$$\mathrm{E}\left(S_n(W)\right) = \gamma\, t\,.$$

This simply follows from the fact that $W(s_i^*)$ is not independent of $W(s_i) - W(s_{i-1})$ for $\gamma > 0$. Next, we turn towards the case $\gamma = 0$.

Ito Integral

For $\gamma = 0$, $S_n(W)$ from Proposition 10.1 merges into $I_n(W)$ from (10.2). The proposition guarantees two different things: First, that the limit of $I_n(W)$ converges in mean square. We call this limit the **Ito integral** and write instead of $S(0)$ the following integral:

$$I_n(W) \xrightarrow{2} \int_0^t W(s)\, dW(s)\,.$$

Secondly, the proposition yields an expression for this Ito integral:

$$\int_0^t W(s)\, dW(s) = \frac{1}{2}\, W^2(t) - \frac{1}{2}\, t\,. \tag{10.3}$$

By the way, (10.3) is just the "stochastified chain rule" for Wiener processes from (1.14).[1] Note the analogy and the contrast to the deterministic case (with $f(0) = 0$):

$$\int_0^t f(s)\, df(s) = \frac{1}{2} f^2(t) \quad \text{for}\, f(0) = 0\,. \tag{10.4}$$

In Eq. (10.3) we find, so to speak, the archetype of Ito calculus, i.e. of stochastic calculus using Ito's lemma. The latter will be covered in the next chapter.

[1] In particular for $t = 1$, (10.3) accomplishes the transition from (1.11) to (1.12) for the Dickey-Fuller distribution.

The moments of the Ito integral can be determined by (10.3). According to this equation it holds for the expected value:

$$\mathrm{E}\left(\int_0^t W(s)\,dW(s)\right) = 0\,.$$

The variance of the integral as well can be calculated elementarily[2]:

$$\begin{aligned}
\mathrm{Var}\left(\int_0^t W(s)\,dW(s)\right) &= \mathrm{Var}\left(\frac{W^2(t)}{2} - \frac{t}{2}\right) \\
&= \mathrm{E}\left[\left(\frac{W^2(t)}{2} - \frac{t}{2}\right)^2\right] \\
&= \frac{1}{4}\,\mathrm{E}\left(W^4(t) - 2\,t\,W^2(t) + t^2\right) \\
&= \frac{1}{4}\left(3\,t^2 - 2\,t^2 + t^2\right) \\
&= \frac{t^2}{2}\,,
\end{aligned}$$

where the kurtosis of 3 for Gaussian random variables was used. Hence, we have the first two results of the following proposition

Proposition 10.2 (Moments of $\int_0^t W(s)\,dW(s)$) *For $I(t) = \int_0^t W(s)\,dW(s)$ it holds that*

$$E(I(t)) = 0 \quad \textit{and} \quad Var\,(I(t)) = \frac{t^2}{2}\,,$$

and

$$E(I(t)\,I(t+h)) = \frac{t^2}{2} \quad \textit{for}\; h \geq 0\,.$$

[2]An alternative, interesting method uses the fact that the variance of a chi-squared distributed random variable equals twice its degrees of freedom:

$$\mathrm{Var}\left(\frac{W^2(t)}{2} - \frac{t}{2}\right) = \frac{1}{4}\,\mathrm{Var}(W^2(t)) = \frac{t^2}{4}\,\mathrm{Var}\left(\left(\frac{W(t)}{\sqrt{t}}\right)^2\right) = \frac{t^2}{4}\cdot 2 = \frac{t^2}{2}\,,$$

as it holds that $W(t)/\sqrt{t} \sim \mathcal{N}(0,1)$ and therefore

$$\left(\frac{W(t)}{\sqrt{t}}\right)^2 \sim \chi^2(1)\,.$$

As a third result it holds in Proposition 10.2, just as for the Stieltjes integral, that the autocovariance coincides with the variance, i.e.

$$\mathrm{E}(I(t)\, I(t+h)) = \mathrm{Var}\,(I(t))\ , \quad h \geq 0.$$

We will prove this in Problem 10.2.

Stratonovich Integral

For a reason that will soon become evident, sometimes a considered competitor of the Ito integral is the **Stratonovich integral**. It is defined as the limit of $S_n(W)$ from (10.1) with the midpoints of the intervals as s_i^*:

$$s_i^* = \frac{s_{i-1} + s_i}{2}\ .$$

This corresponds to the choice of $\gamma = 0.5$ in Proposition 10.1. Let the limit in mean square be denoted as follows:

$$\sum_{i=1}^{n} W\left(\frac{s_{i-1} + s_i}{2}\right) (W(s_i) - W(s_{i-1})) \overset{2}{\to} \int_0^t W(s)\, \partial W(s)\,,$$

where "∂" does not stand for the partial derivative but denotes the Stratonovich integral in contrast to the Ito integral. By the way, with $\gamma = 0.5$ Proposition 10.1 yields:

$$\int_0^t W(s)\, \partial W(s) = \frac{W^2(t)}{2}\ .$$

Hence, the Stratonovich integral stands out due to the fact that the familiar integration rule known from ordinary calculus holds true. In differential notation this rule can be formulated symbolically as follows:

$$\frac{\partial W^2(t)}{2} = W(t)\, \partial W(t)\,.$$

This just corresponds to the ordinary chain rule, cf. (10.4). Although the Ito and the Stratonovich integral are distinguished from each other only by the choice of s_i^* with intervals getting shorter and shorter, they still have drastically different properties. Obviously, it holds for the expected value

$$\mathrm{E}\left(\int_0^t W(s)\, \partial W(s)\right) = \frac{t}{2}\,,$$

while the Ito integral is zero on average. However, as aforementioned, the variances of $\int W(s)dW(s)$ and $\int W(s)\partial W(s)$ coincide, cf. Problem 10.3 as well.

Example 10.1 (Alternative Stratonovich Sum) Sometimes the Stratonovich integral is defined as the limit of the following sum, see e.g. Klebaner (2005, eq. (5.65)):

$$\sum_{i=1}^{n} \frac{W(s_{i-1}) + W(s_i)}{2} \left(W(s_i) - W(s_{i-1})\right) .$$

The intuition behind this is, due to the continuity of the WP, that

$$W(s_{i-1}) \approx W\left(\frac{s_{i-1}+s_i}{2}\right) \approx W(s_i) .$$

In fact, it can be shown more explicitly that the following difference becomes negligible in mean square:

$$\Gamma := W\left(\frac{s_{i-1}+s_i}{2}\right) - \frac{W(s_{i-1}) + W(s_i)}{2} \stackrel{2}{\to} 0 .$$

For this purpose we consider as the mean square deviation with $s_i^* = (s_{i-1} + s_i)/2$:

$$\begin{aligned} \mathrm{MSE}(\Gamma, 0) &= \mathrm{E}\left[(\Gamma - 0)^2\right] \\ &= \mathrm{E}\left[W^2(s_i^*) - W(s_i^*)\left(W(s_{i-1}) + W(s_i)\right)\right] \\ &\quad + \mathrm{E}\left[\frac{W^2(s_{i-1}) + 2\,W(s_{i-1})W(s_i) + W^2(s_i)}{4}\right] . \end{aligned}$$

Due to $s_{i-1} < s_i^* < s_i$ the familiar variance and covariance formulas yield:

$$\begin{aligned} \mathrm{MSE}(\Gamma, 0) &= s_i^* - s_{i-1} - s_i^* + \frac{s_{i-1} + 2\,s_{i-1} + s_i}{4} \\ &= \frac{s_i - s_{i-1}}{4} . \end{aligned}$$

As for $n \to \infty$ the partition gets finer and finer, $s_i - s_{i-1} \to 0$, the replacement of $W(s_i^*)$ by $\left(W(s_{i-1}) + W(s_i)\right)/2$ is asymptotically well justified. ∎

10.3 General Ito Integrals

After covering general Ito integrals, we define so-called diffusions that we will be concerned with in the following chapters.

Definition and Moments

In order to define general Ito integrals, we consider for a stochastic process X as a generalization of the sum $I_n(W)$:

$$I_n(X) = \sum_{i=1}^{n} X(s_{i-1}) \, (W(s_i) - W(s_{i-1})) \, . \qquad (10.5)$$

For the Ito integral two special things apply: First, the lower endpoint of the interval is chosen as $s_i^* = s_{i-1}$, i.e. $X(s_{i-1})$; and secondly, we integrate with respect to the WP, $(W(s_i) - W(s_{i-1}))$. If X was integrated with respect to another stochastic process, then one would obtain even more general stochastic integrals, which we are not interested in here.

If $X(t)$ is a process with finite variance where the variance varies continuously in the course of time, and if $X(t)$ only depends on the past of the WP, $W(s)$ with $s \leq t$, but not on its future, then the Ito sum converges uniquely and independently of the partition. The limit is called Ito integral and is denoted as follows:

$$\int_0^t X(s) \, dW(s) \, .$$

The assumptions about $X(t)$ are stronger than necessary, however, they guarantee the existence of the moments of an Ito integral, too. Similar assumptions can be found in Klebaner (2005, Theorem 4.3) or Øksendal (2003, Corollary 3.1.7).

Proposition 10.3 (General Ito Integral) *Let $X(s)$ be a stochastic process on $[0, t]$ with two properties:*

(i) $\mu_2(s) = E\left(X^2(s)\right) < \infty$ is a continuous function,
(ii) $X(s)$ is independent of $W(s_j) - W(s_i)$ with $s \leq s_i < s_j$.

Then it holds that

(a) the sum from (10.5) converges in mean square:

$$\sum_{i=1}^{n} X(s_{i-1}) \, (W(s_i) - W(s_{i-1})) \xrightarrow{2} \int_0^t X(s) \, dW(s) \, ;$$

(b) the moments of the Ito integral are determined as:

$$E\left(\int_0^t X(s) \, dW(s)\right) = 0 \, , \quad Var\left(\int_0^t X(s) \, dW(s)\right) = \int_0^t E\left(X^2(s)\right) ds \, .$$

Naturally, for $X(s) = W(s)$ the extensively discussed example from the previous section is obtained. In particular, in (b) the moments from Proposition 10.2 are reproduced, and it holds for the variance that:

$$\operatorname{Var}\left(\int_0^t W(s)\,dW(s)\right) = \int_0^t \operatorname{E}(W^2(s))\,ds = \int_0^t s\,ds = \frac{t^2}{2}\,.$$

Example 10.2 (Stieltjes Integral) Consider the special case where $X(s)$ is not stochastic but deterministic,

$$X(s) = f(s),$$

where $f(s)$ is continuous. Then, the conditions of existence are fulfilled which can easily be verified: The square, $\mu_2(s) = f^2(s)$, is continuous as well, and the deterministic function is independent of $W(s)$. Hence, it holds that

$$\sum_{i=1}^{n} f(s_{i-1})\,(W(s_i) - W(s_{i-1})) \xrightarrow{2} \int_0^t f(s)\,dW(s)\,.$$

In other words: For deterministic processes, $X(s) = f(s)$, the Stieltjes and the Ito integral coincide; the former is a special case of the latter. Due to $\operatorname{E}\left(f^2(s)\right) = f^2(s)$ the already familiar formulas for expectation and variance from Proposition 9.2 are embedded in the general Proposition 10.3. ■

Distribution and Further Properties

As is well known, the special case of the Stieltjes integral is Gaussian. For the Ito integral this does not hold in general. This can clearly be seen in (10.3):

$$\int_0^t W(s)\,dW(s) = \frac{W^2(t) - t}{2} \geq \frac{-t}{2}\,,$$

i.e. the support of the distribution is bounded. Then again, the integral of a WP with respect to a thereof stochastically independent WP amounts to a Gaussian distribution. The following result is by Phillips and Park (1988).

Proposition 10.4 (Ito Integral of an Independent WP) *Let $W(t)$ and $V(s)$ be stochastically independent Wiener processes. Then it holds that*

$$\left(\int_0^1 V^2(s)\,ds\right)^{-0.5} \int_0^1 V(s)\,dW(s) \sim \mathcal{N}(0,1)\,.$$

By showing the conditional distribution of the left-hand side given $V(t)$ to just follow a $\mathcal{N}(0, 1)$ distribution and therefore not to depend on this condition, one proves the result claimed in Proposition 10.4; see Phillips and Park (1988, Appendix) for details.

Note that the Ito integral again defines a stochastic process in the bounds from 0 to t whose properties could be discussed, which we will not do at this point. In the literature, however, it can be looked up that the Ito integral and the Wiener process share the continuity and the martingale properties. What is more, the integration rules outlined at the end of Sect. 8.2 hold true for Ito integrals as well, see e.g. Klebaner (2005, Thm. 4.3).

Diffusions

For economic modeling the Ito integral is an important ingredient. However, it gains its true importance only when combined with Riemann integrals. In the following chapters, the sum of both integrals constitutes so-called **diffusions**[3] (diffusion processes). Hence, we now define processes $X(t)$ (with starting value $X(0)$) as follows:

$$X(t) = X(0) + \int_0^t \mu(s)\, ds + \int_0^t \sigma(s)\, dW(s)\,.$$

Frequently, we will write this integral equation in differential form as follows:

$$dX(t) = \mu(t)\, dt + \sigma(t)\, dW(t)\,.$$

The conditions set to $\mu(s)$ and $\sigma(s)$ that guarantee the existence of such processes can be adopted from Propositions 8.1 and 10.3. In general, $\mu(s)$ and $\sigma(s)$ are stochastic; particularly, they are allowed to be dependent on $X(s)$ itself. Therefore, we write $\mu(s)$ and $\sigma(s)$ as abbreviations for functions which firstly explicitly depend on time and secondly depend on X simultaneously:

$$\mu(s) = \mu\,(s, X(s))\;, \quad \sigma(s) = \sigma\,(s, X(s))\;.$$

Processes μ and σ satisfying these conditions are used to define diffusions $X(t)$:

$$dX(t) = \mu\,(t, X(t))\; dt + \sigma\,(t, X(t))\; dW(t)\,, \quad t \in [0, T]\,. \tag{10.6}$$

[3]The name stems from molecular physics, where diffusions are used to model the change of location of a molecule due to a deterministic component (drift) and an erratic (stochastic) component. Physically, the influence of temperature on the motion hides behind the stochastics: The higher the temperature of the matter in which the particles move, the more erratic is their behavior.

Recall that this differential equation actually means the following:

$$X(t) = X(0) + \int_0^t \mu\left(s, X(s)\right)\, ds + \int_0^t \sigma\left(s, X(s)\right)\, dW(s)\,.$$

Example 10.3 (Brownian Motion with Drift) We consider the Brownian motion with drift and a starting value 0:

$$\begin{aligned} X(t) &= \mu t + \sigma W(t) \\ &= \mu \int_0^t ds + \sigma \int_0^t dW(s)\,. \end{aligned}$$

Therefore, the differential notation reads

$$dX(t) = \mu\, dt + \sigma\, dW(t)\,.$$

Hence, this is a diffusion whose drift and volatility are constant:

$$\mu\left(t, X(t)\right) = \mu \quad \text{and} \quad \sigma\left(t, X(t)\right) = \sigma\,. \quad \blacksquare$$

10.4 (Quadratic) Variation

From (10.3) we know that

$$\int_0^t W(s)\, dW(s) = \frac{W^2(t) - t}{2}\,.$$

Now, we want to understand where the expression t comes from that is subtracted from $W^2(t)$. It will be made clear that this is the so-called quadratic variation.

(Absolute) Variation

Again, the considerations are based on an adequate partition of the interval $[0, t]$,

$$P_n\left([0, t]\right) : 0 = s_0 < s_1 < \ldots < s_n = t\,.$$

For a function g the **variation** over this partition is defined as[4]:

$$V_n(g, t) = \sum_{i=1}^{n} \left| g(s_i) - g(s_{i-1}) \right|\,.$$

[4]Sometimes we speak of absolute variation in order to avoid confusion with e.g. quadratic variation.

If the limit exists independently of the decomposition for $n \to \infty$ under (8.2), then one says that g is of finite variation and writes[5]:

$$V_n(g,t) \;\to\; V(g,t)\,, \quad n \to \infty\,.$$

The finite sum $V_n(g,t)$ measures for a certain partition the absolute increments of the function g on the interval $[0,t]$. If the function evolves sufficiently smooth, then $V(g,t)$ takes on a finite value for $(n \to \infty)$. For very jagged functions, however, it may be that for an increasing refinement $(n \to \infty)$ the increments of the graph of g become larger and larger even for fixed t, such that g is not of finite variation.

Example 10.4 (Monotonic Functions) For monotonic finite functions the variation can be calculated very easily and intuitively. In this case, $V(g,t)$ is simply the absolute value of the difference of the function at endpoints of the interval, $|g(t) - g(0)|$. First, let us assume that g grows monotonically on $[0,t]$,

$$g(s_i) \geq g(s_{i-1}) \quad \text{for } s_i > s_{i-1}\,.$$

Obviously, it then holds by (8.3) that

$$V_n(g,t) = \sum_{i=1}^{n} (g(s_i) - g(s_{i-1})) = g(t) - g(0) = V(g,t)\,.$$

For a monotonically decreasing function, it results quite analogously:

$$\begin{aligned} V_n(g,t) &= \sum_{i=1}^{n} |g(s_i) - g(s_{i-1})| \\ &= -\sum_{i=1}^{n} (g(s_i) - g(s_{i-1})) \\ &= g(0) - g(t) \\ &= V(g,t)\,. \end{aligned}$$

Monotonic functions are hence of finite variation. ■

Without the requirement of monotonicity, an intuitive sufficient condition exists for the function to be smooth enough to be of finite variation where this variation then has a familiar form as well.

[5] If g is a deterministic function, then "$\to$" means the usual convergence of analysis. If we allow for $g(t)$ to be a stochastic process, then we mean the convergence in mean square: "$\xrightarrow{2}$".

Proposition 10.5 (Variation of Continuously Differentiable Functions) *Let g be a continuously differentiable function with derivative g' on $[0, t]$. Then g is of finite variation and it holds that*

$$V(g, t) = \int_0^t |g'(s)| \, \mathrm{d}s.$$

The proof is given in Problem 10.6.

Example 10.5 (Sine Wave) Let us consider a sine cycle of the frequency k on the interval $[0, 2\pi]$:

$$g_k(s) = \sin(ks), \quad k = 1, 2, \ldots .$$

The derivative reads

$$g_k'(s) = k \cos(ks).$$

Accounting for the sign one obtains as the variation:

$$\begin{aligned} V(g_1, 2\pi) &= \int_0^{2\pi} |\cos(s)| \, ds = 4 \int_0^{\pi/2} \cos(s) \, ds \\ &= 4 \left(\sin\left(\frac{\pi}{2}\right) - \sin(0) \right) \\ &= 4 \,, \end{aligned}$$

$$\begin{aligned} V(g_2, 2\pi) &= \int_0^{2\pi} 2 \, |\cos(2s)| \, ds = 8 \int_0^{\pi/4} 2\cos(2s) \, ds \\ &= 8 \left(\sin\left(\frac{\pi}{2}\right) - \sin(0) \right) \\ &= 8 \,, \end{aligned}$$

$$\begin{aligned} V(g_k, 2\pi) &= \int_0^{2\pi} k \, |\cos(ks)| \, ds = 4k \int_0^{\pi/2k} k\cos(ks) \, ds \\ &= 4k \left(\sin\left(\frac{\pi}{2}\right) - \sin(0) \right) \\ &= 4k \,. \end{aligned}$$

In Fig. 10.1 it can be observed, how the sum of (absolute) differences in amplitude grows with k growing. Accordingly, the absolute variation of $g_k(s) = \sin(ks)$ multiplies with k. For $k \to \infty$, g_k' tends to infinity such that this derivative is not

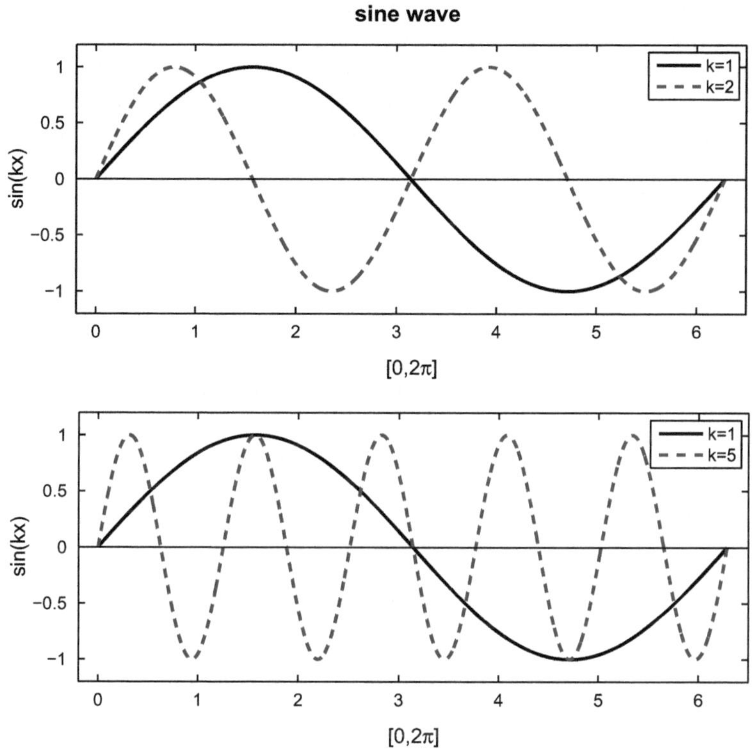

Fig. 10.1 Sine cycles of different frequencies (Example 10.5)

continuous anymore. Consequently, the absolute variation is not finite in the limiting case $k \to \infty$. ■

Quadratic Variation

In the same way as $V_n(g, t)$ a q-variation can be defined where we are only interested in the case $q = 2$, – the quadratic variation:

$$Q_n(g,t) = \sum_{i=1}^{n} |g(s_i) - g(s_{i-1})|^2 = \sum_{i=1}^{n} (g(s_i) - g(s_{i-1}))^2 .$$

As would seem natural, g is called of finite quadratic variation if it holds that

$$Q_n(g,t) \to Q(g,t), \quad n \to \infty .$$

If g is a stochastic function, i.e. a stochastic process, then $Q(g,t)$ and $V(g,t)$ are defined as limits in mean square. Between the absolute variation $V(g,t)$ and the quadratic variation $Q(g,t)$ there are connections which we want to deal with now. If a continuous function is of finite variation, then it is of finite quadratic variation as well, where the latter is in fact zero. This is the statement of the following proposition. As it seems counterintuitive at first sight that Q, as the limit of a positive sum of squares Q_n, can become zero, we start with an example.

Example 10.6 (Identity Function) Let *id* be the identity function on $[0,t]$:

$$id(s) = s.$$

As the functions increases monotonically, it is of finite variation with

$$V(id,t) = id(t) - id(0) = t\,.$$

For finite n it holds that:

$$\begin{aligned} Q_n(id,t) &= \sum_{i=1}^{n} (id(s_i) - id(s_{i-1}))^2 \\ &= \sum_{i=1}^{n} (s_i - s_{i-1})^2 \\ &> 0\,. \end{aligned}$$

Q_n consists of n terms, where the lengths $s_i - s_{i-1} > 0$ are of the magnitude $\frac{1}{n}$. Due to the squaring, the n terms are of the magnitude $\frac{1}{n^2}$. Hence, the sum converges to zero for $n \to \infty$. This intuition can be formalized as follows:

$$\begin{aligned} Q_n(id,t) &= \sum_{i=1}^{n} (s_i - s_{i-1})^2 \\ &\le \max_{1\le i\le n} (s_i - s_{i-1}) \sum_{i=1}^{n} (s_i - s_{i-1}) \\ &= \max_{1\le i\le n} (s_i - s_{i-1})\, V_n(id,t) \\ &= \max_{1\le i\le n} (s_i - s_{i-1})\, t \\ &\to 0, \end{aligned}$$

as $\max(s_i - s_{i-1}) \to 0$ for $n \to \infty$. ■

The next proposition gives a sufficient condition for the quadratic variation to vanish.

Proposition 10.6 (Absolute and Quadratic Variation) *Let g be a continuous function on $[0, t]$. It then holds under (8.2) for $n \to \infty$:*

$$V_n(g, t) \to V(g, t) < \infty$$

implies

$$Q_n(g, t) \to 0 .$$

If g is a stochastic process, then "$\to$" is to be understood as convergence in mean square.

The proof is given in Problem 10.7. From the proposition it follows by contraposition that: If we have a positive (finite) quadratic variation, then the process does not have a finite variation. Formally, we write: From

$$Q_n(g, t) \to Q(g, t) < \infty \quad \text{with} \quad Q(g, t) > 0$$

it follows that there is no finite variation:

$$V_n(g, t) \to \infty .$$

If a function g is so smooth that it has a continuous derivative, then $Q(g, t) = 0$ by Propositions 10.5 and 10.6; the other way round, values of $Q(g, t) > 0$ characterize how little smooth or jagged the function is.

Wiener Processes

As we know, the WP is nowhere differentiable, therefore it is everywhere so jagged that there is no valid tangent line approximation. Due to this extreme jaggedness the WP is of infinite variation as well, as we will show in a moment. More explicitly, we prove that the WP is of positive quadratic variation and does not have a finite absolute variation due to Proposition 10.6. We save the proof for an exercise (Problem 10.8).

Proposition 10.7 (Quadratic Variation of the WP) *For the Wiener process with $n \to \infty$ it holds under (8.2):*

$$Q_n(W,t) \overset{2}{\to} t = Q(W,t)\,.$$

The expression $Q(W,t) = t$ characterizes the level of jaggedness or irregularity of the Wiener process on the interval $[0,t]$. This non-vanishing quadratic variation causes the problems and specifics of the Ito integral. Let us recapitulate: If the Wiener process was continuously differentiable, then it would be of finite variation due to Proposition 10.5 and it would have a vanishing quadratic variation due to Proposition 10.6. However, this is just not the case.

Symbolic Notation

In finance textbooks one frequently finds a notation for time that is strange at first sight:

$$(dW(t))^2 = dt\,. \tag{10.7}$$

How is this to be understood? Formal integration yields

$$\int_0^t (dW(s))^2 = t\,.$$

As would seem natural, the "integral" on the left-hand side here stands for $Q(W,t)$:

$$Q_n(W,t) = \sum_{i=1}^{n} (W(s_i) - W(s_{i-1}))^2 \overset{2}{\to} \int_0^t (dW(s))^2 := Q(W,t)\,.$$

Therefore, the integral equation and hence (10.7) is justified by Proposition 10.7: $Q(W,t) = t$. We adopt the result into the following proposition. The expressions

$$dW(t)\,dt = 0 \quad \text{and} \quad (dt)^2 = 0 \tag{10.8}$$

are to be understood similarly, namely in the sense of Proposition 10.8.

Proposition 10.8 (Symbolic Notation) *It holds for $n \to \infty$ under (8.2):*

$$(a)\ \sum_{i=1}^{n} (W(s_i) - W(s_{i-1}))^2 \overset{2}{\to} \int_0^t (dW(s))^2 = t\,,$$

$$(b)\ \sum_{i=1}^{n} (W(s_i) - W(s_{i-1}))\ (s_i - s_{i-1}) \overset{2}{\to} \int_0^t dW(s)\, ds = 0\,,$$

$$(c)\ \sum_{i=1}^{n} (s_i - s_{i-1})^2 \ \to \int_0^t (ds)^2 = 0\,.$$

In symbols, these facts are frequently formulated as in (10.7) and (10.8).

Note that the expression in (c) in Proposition 10.8 is the quadratic variation of the identity function $id(s) = s$:

$$Q_n(id, t) \to \int_0^t (ds)^2 := Q(id, t) = 0\,.$$

Hence, the third claim is already established by Example 10.6. The expression from (b) in Proposition 10.8 is sometimes also called covariation (of $W(s)$ and $id(s) = s$). The claimed convergence to zero in mean square is shown in Problem 10.9.

10.5 Problems and Solutions

Problems

10.1 Prove Proposition 10.1 for $\gamma = 0$ (Ito integral).
Hint: Use Proposition 10.7.

10.2 Prove the autocovariance from Proposition 10.2.

10.3 Derive that the Ito integral from (10.3) and the corresponding Stratonovich integral have the same variance.

10.4 Show for $S_n(W)$ from (10.1) with s_i^* from Proposition 10.1,

$$s_i^* = (1 - \gamma)\, s_{i-1} + \gamma\, s_i\,, \quad 0 \le \gamma < 1\,,$$

that it holds:

$$\mathrm{E}\left(S_n(W)\right) = \gamma\, t\,.$$

10.5 Show: $\Gamma_n \stackrel{2}{\to} 0$ with

$$\Gamma_n = W\left((1-\gamma)\, s_{i-1} + \gamma\, s_i\right) - \left[(1-\gamma)\, W(s_{i-1}) + \gamma\, W(s_i)\right]$$

for $\gamma \in [0, 1]$ for an adequate partition, i.e. for $s_i - s_{i-1} \to 0$.

10.6 Prove Proposition 10.5.

10.7 Prove Proposition 10.6.

10.8 Determine the quadratic variation of the Wiener process, i.e. verify Proposition 10.7.

10.9 Show (b) from Proposition 10.8.

10.10 Determine the covariance of $W(s)$ and $\int_0^t W(r)\, dW(r)$ for $s \leq t$.

Solutions

10.1 In order to prove Proposition 10.1 for $\gamma = 0$, it has to be shown that $I_n(W)$ from (10.2) converges in mean square, namely to the expression given in (10.3). For this purpose, we write $I_n(W)$ as follows:

$$\begin{aligned}
I_n(W) &= \sum_{i=1}^{n} W(s_{i-1})\, \left(W(s_i) - W(s_{i-1})\right) \\
&= \frac{1}{2}\left[2\sum_{i=1}^{n} W(s_i)\, W(s_{i-1}) - 2\sum_{i=1}^{n} W^2(s_{i-1})\right] \\
&= \frac{1}{2}\left[\sum_{i=1}^{n}\left(W^2(s_i) - W^2(s_{i-1})\right)\right. \\
&\quad \left. - \sum_{i=1}^{n}\left(W^2(s_i) - 2\, W(s_i)\, W(s_{i-1}) + W^2(s_{i-1})\right)\right],
\end{aligned}$$

where for the last equation we added zero, $\left(W^2(s_i) - W^2(s_i)\right)$. Hence, it furthermore follows by means of the quadratic variation:

$$
\begin{aligned}
I_n(W) &= \frac{1}{2}\left[W^2(s_n) - W^2(s_0) - \sum_{i=1}^{n}\left(W(s_i) - W(s_{i-1})\right)^2\right] \\
&= \frac{1}{2}\left(W^2(t) - W^2(0)\right) - \frac{1}{2}Q_n(W,t)\,.
\end{aligned}
$$

The Wiener process is of finite quadratic variation and we know from Proposition 10.7: $Q_n(W,t) \stackrel{2}{\to} t$. This verifies the claim (as it holds that $W(0) = 0$ with probability 1).

10.2 Based on (10.3) we consider the process

$$
I(t) = \int_0^t W(s)dW(s) = \frac{W^2(t) - t}{2}.
$$

Due to the vanishing expected value, it holds for the autocovariance that:

$$
\begin{aligned}
\mathrm{E}(I(t)\,I(s)) &= \frac{1}{4}\mathrm{E}\left[W^2(t)W^2(s) - tW^2(s) - sW^2(t) + st\right] \\
&= \frac{1}{4}\left[\mathrm{E}\left(W^2(t)W^2(s)\right) - ts - st + st\right].
\end{aligned}
$$

By adding zero one obtains:

$$
\begin{aligned}
\mathrm{E}[W^2(t)W^2(s)] &= \mathrm{E}\left[\left(W(t) - W(s) + W(s)\right)^2 W^2(s)\right] \\
&= \mathrm{E}\left[\left(W(t) - W(s)\right)^2 W^2(s)\right] \\
&\quad +2\,\mathrm{E}\left[\left(W(t) - W(s)\right)W^3(s)\right] + \mathrm{E}\left[W^4(s)\right].
\end{aligned}
$$

If we assume w.l.o.g. that $s \leq t$, then due to the independence of non-overlapping increments of W it holds that:

$$
\begin{aligned}
\mathrm{E}\left[\left(W(t) - W(s)\right)^2 W^2(s)\right] &= \mathrm{E}\left[\left(W(t) - W(s)\right)^2\right]\mathrm{E}\left[W^2(s)\right] \\
&= \mathrm{Var}\left(W(t) - W(s)\right)\mathrm{Var}(W(s)) \\
&= (t-s)s
\end{aligned}
$$

and

$$\mathrm{E}\left[(W(t) - W(s))W^3(s)\right] = \mathrm{E}(W(t) - W(s))\mathrm{E}(W^3(s)) = 0.$$

As the kurtosis of a Gaussian random variable is 3, it follows that

$$\mathrm{E}[W^4(s)] = 3s^2.$$

Therefore, these results jointly yield the claimed outcome:

$$\begin{aligned}\mathrm{E}(I(t)\,I(s)) &= \frac{(t-s)s}{4} + \frac{3}{4}s^2 - \frac{st}{4}\\ &= \frac{s^2}{2}, \quad s \leq t.\end{aligned}$$

10.3 We know about the aforementioned Ito integral from Proposition 10.2 that its variance is $t^2/2$. Hence, it is to be shown that:

$$\mathrm{Var}\left(\int_0^t W(s)\,\partial W(s)\right) = \frac{t^2}{2}.$$

We use Proposition 10.1,

$$\int_0^t W(s)\,\partial W(s) = \frac{W^2(t)}{2},$$

from which it follows immediately that

$$\mathrm{E}\left(\int_0^t W(s)\,\partial W(s)\right) = \frac{t}{2}.$$

The usual variance decomposition, see (2.1), hence yields

$$\mathrm{Var}\left(\frac{W^2(t)}{2}\right) = \frac{1}{4}\left(\mathrm{E}\left(W^4(t)\right) - t^2\right).$$

Due to a kurtosis of 3, the fourth moment of a $\mathcal{N}(0,t)$-distribution just amounts to $3\,t^2$. Thus, one obtains

$$\mathrm{Var}\left(\frac{W^2(t)}{2}\right) = \frac{1}{4}\left(3\,t^2 - t^2\right) = \frac{t^2}{2},$$

which proves the claim.

10.4 The expectation of the sum $S_n(W)$ is equal to the sum of the expectations. Therefore, we consider an individual expectation,

$$\begin{aligned}
\mathrm{E}\left[W(s_i^*)\ (W(s_i) - W(s_{i-1})\right] &= \mathrm{E}\left[W(s_i^*)W(s_i) - W(s_i^*)W(s_{i-1})\right] \\
&= \min(s_i^*, s_i) - \min(s_i^*, s_{i-1}) \\
&= (1-\gamma)\, s_{i-1} + \gamma\, s_i - s_{i-1} \\
&= \gamma\,(s_i - s_{i-1}),
\end{aligned}$$

where simply the well-known covariance formula was used. Hence, summation yields as desired

$$\begin{aligned}
\mathrm{E}\,(S_n(W)) &= \gamma \sum_{i=1}^{n} (s_i - s_{i-1}) \\
&= \gamma(\,s_n - s_0) \\
&= \gamma\,(t-0)\,.
\end{aligned}$$

10.5 Convergence in mean square implies that the mean squared error tends to zero. The MSE with the limit zero reads $\mathrm{MSE}(\Gamma_n, 0) = \mathrm{E}(\Gamma_n^2)$. Therefore, it remains to be shown that: $\mathrm{E}(\Gamma_n^2) \to 0$.

For this purpose one considers with $s_i^* = (1-\gamma)s_{i-1} + \gamma\, s_i$:

$$\begin{aligned}
\Gamma_n^2 = W^2(s_i^*)\ &-\ 2W(s_i^*)\left[(1-\gamma)W(s_{i-1}) + \gamma\, W(s_i)\right] \\
&+ (1-\gamma)^2\, W^2(s_{i-1}) + 2\gamma\,(1-\gamma)\, W(s_{i-1})W(s_i) \\
&+ \gamma^2 W^2(s_i).
\end{aligned}$$

Forming expectation yields:

$$\begin{aligned}
\mathrm{E}(\Gamma_n^2) &= s_i^* - 2\left[(1-\gamma)s_{i-1} + \gamma\, s_i^*\right] + (1-\gamma)^2 s_{i-1} \\
&\quad +2\gamma(1-\gamma)\, s_{i-1} + \gamma^2\, s_i \\
&= (1-\gamma)s_{i-1} + \gamma\, s_i - 2(1-\gamma)\, s_{i-1} - 2\gamma(1-\gamma)s_{i-1} - 2\,\gamma^2\, s_i \\
&\quad +(1-\gamma)^2\, s_{i-1} + 2\,\gamma(1-\gamma)\, s_{i-1} + \gamma^2\, s_i \\
&= s_i(\gamma - \gamma^2) + s_{i-1}\left((1-\gamma)^2 - (1-\gamma)\right) \\
&= s_i(\gamma - \gamma^2) + s_{i-1}(\gamma^2 - \gamma) \\
&= (s_i - s_{i-1})\,\gamma\,(1-\gamma).
\end{aligned}$$

Hence, for $n \to \infty$ the required result is obtained as $s_i - s_{i-1}$ tends to zero.

10.6 For a given partition we write

$$V_n(g,t) = \sum_{i=1}^{n} |g(s_i) - g(s_{i-1})| = \sum_{i=1}^{n} \left| \int_{s_{i-1}}^{s_i} g'(s)\, ds \right| .$$

As the derivative is continuous, $|g'(s)|$ is continuous as well and hence integrable. According to the mean value theorem an $s_i^* \in [s_{i-1}, s_i]$ exists with

$$\int_{s_{i-1}}^{s_i} g'(s)\, ds = g'(s_i^*)\,(s_i - s_{i-1}) .$$

Thus it follows that

$$\begin{aligned} V_n(g,t) &= \sum_{i=1}^{n} |g'(s_i^*)|\,(s_i - s_{i-1}) \\ &\to \int_0^t |g'(s)|\, ds . \end{aligned}$$

Quod erat demonstrandum.

10.7 The claim is based on the bound

$$\begin{aligned} Q_n(g,t) &\leq \max_{1\leq i\leq n} (|g(s_i) - g(s_{i-1})|) \sum_{i=1}^{n} |g(s_i) - g(s_{i-1})| \\ &= \max_{1\leq i\leq n} (|\, g(s_i) - g(s_{i-1})\,|)\; V_n(g,t) . \end{aligned}$$

Due to continuity it holds that

$$\max_{1\leq i\leq n} (|\, g(s_i) - g(s_{i-1})\,|) \;\to\; 0 .$$

Hence, the claim immediately follows from the bound.

10.8 It is to be shown that the mean squared error,

$$\text{MSE}\,(Q_n(W,t),\ t) = \text{E}\left[(Q_n(W,t) - t)^2\right] ,$$

tends to zero. For this purpose we proceed in two steps. In the first one we show that the MSE coincides with the variance of $Q_n(W,t)$. In the second step it will be shown that the variance converges to zero.

(1) For the first step we only need to derive $\mathrm{E}(Q_n(W,t)) = t$. With

$$Q_n(W,t) = \sum_{i=1}^{n} (W(s_i) - W(s_{i-1}))^2$$

the required expectation can be easily determined:

$$\begin{aligned}
\mathrm{E}\left(Q_n(W,t)\right) &= \sum_{i=1}^{n} \mathrm{Var}\left(W(s_i) - W(s_{i-1})\right) \\
&= \sum_{i=1}^{n} (s_i - s_{i-1}) = s_n - s_0 = t - 0 \\
&= t\,.
\end{aligned}$$

(2) Due to the independence of the increments of the WP one has

$$\mathrm{Var}\left(Q_n(W,t)\right) = \sum_{i=1}^{n} \mathrm{Var}\left[(W(s_i) - W(s_{i-1}))^2\right].$$

Due to $W(s_i) - W(s_{i-1}) \sim \mathcal{N}(0, s_i - s_{i-1})$ and with a kurtosis of 3 for Gaussian random variables, it furthermore holds that:

$$\begin{aligned}
\mathrm{Var}\left[(W(s_i) - W(s_{i-1}))^2\right] &= \mathrm{E}\left[(W(s_i) - W(s_{i-1}))^4\right] - \left(\mathrm{E}\left[(W(s_i) - W(s_{i-1}))^2\right]\right)^2 \\
&= 3\ [\mathrm{Var}(W(s_i) - W(s_{i-1}))]^2 - (s_i - s_{i-1})^2 \\
&= 2\,(s_i - s_{i-1})^2\,.
\end{aligned}$$

Hence, plugging in yields

$$\begin{aligned}
\mathrm{Var}(Q_n(W,t)) &= 2 \sum_{i=1}^{n} (s_i - s_{i-1})^2 \\
&\leq 2 \max_{1\leq i\leq n} (s_i - s_{i-1}) \sum_{i=1}^{n} (s_i - s_{i-1}) \\
&= 2 \max_{1\leq i\leq n} (s_i - s_{i-1})\,(s_n - s_0) \\
&\to 0\,, \quad n \to \infty\,,
\end{aligned}$$

which completes the proof.

10.9 Let us call the aforementioned covariation CV_n:

$$CV_n = \sum_{i=1}^{n} (W(s_i) - W(s_{i-1}))\,(s_i - s_{i-1})\,.$$

The claim reads: $\text{MSE}(CV_n, 0) \to 0$. As it obviously holds that $\text{E}(CV_n) = 0$, we obtain

$$\text{MSE}(CV_n, 0) = \text{Var}(CV_n)\,.$$

Hence, it remains to be shown that this variance tends to zero: Due to the independence of the increments of the WP, one determines

$$\text{Var}(CV_n) = \sum_{i=1}^{n} \text{Var}\,(W(s_i) - W(s_{i-1}))\,(s_i - s_{i-1})^2\,,$$

and hence

$$\begin{aligned}
\text{Var}(CV_n) &= \sum_{i=1}^{n} (s_i - s_{i-1})^3 \\
&\le \max_{1\le i\le n} (s_i - s_{i-1}) \sum_{i=1}^{n} (s_i - s_{i-1})^2 \\
&= \max_{1\le i\le n} (s_i - s_{i-1})\, Q_n(id, t) \\
&\to 0\,,
\end{aligned}$$

where $Q_n(id, t)$ is the quadratic variation of the identity function, see Example 10.6. Hence, the claim is established.

10.10 We want to obtain the expected value of $Y(s, t)$ with

$$Y(s,t) := W(s) \int_0^t W(r)dW(r)\,, \quad s \le t.$$

Due to (10.3) it again holds that:

$$\begin{aligned}
\text{E}\,(Y(s,t)) &= \text{E}\left[W(s)\frac{W^2(t) - t}{2}\right] \\
&= \frac{1}{2}\text{E}\left[W(s)W^2(t)\right].
\end{aligned}$$

Therefore, we study

$$\begin{aligned}
\mathrm{E}\left[W(s)W^2(t)\right] &= \mathrm{E}\left[W(s)(W(t) - W(s) + W(s))^2\right] \\
&= \mathrm{E}\left[W(s)(W(t) - W(s))^2\right] \\
&\quad +2\mathrm{E}\left[W^2(s)(W(t) - W(s))\right] + \mathrm{E}[W^3(s)].
\end{aligned}$$

Let us consider the last three terms one by one. Due to the independence of the increments, one obtains:

$$\mathrm{E}\left[W(s)(W(t) - W(s))^2\right] = \mathrm{E}(W(s))\mathrm{E}\left((W(t) - W(s))^2\right) = 0.$$

Moreover, it is obvious that the second term is also zero. For the third term the symmetry of the Gaussian distribution yields $\mathrm{E}\left(W^3(s)\right) = 0$. Summing up, we have shown that

$$\mathrm{E}\left(W(s)W^2(t)\right) = 0\,, \quad s \leq t,$$

and hence

$$\mathrm{E}\left[W(s)\int_0^t W(r)dW(r)\right] = 0\,, \quad s \leq t.$$

References

Klebaner, F. C. (2005). *Introduction to stochastic calculus with applications* (2nd ed.). London: Imperical College Press.

Øksendal, B. (2003). *Stochastic differential equations: An introduction with applications* (6th ed.). Berlin/New York: Springer.

Phillips, P. C. B., & Park, J. Y. (1988). Asymptotic equivalence of ordinary least squares and generalized least squares in regressions with integrated regressors. *Journal of the American Statistical Association, 83*, 111–115.

Tanaka, K. (1996). *Time series analysis: Nonstationary and noninvertible distribution theory*. New York: Wiley.

Ito's Lemma

11

11.1 Summary

If a process is given as a stochastic Riemann and/or Ito integral, then one may wish to determine how a function of the process looks. This is achieved by Ito's lemma as an ingredient of stochastic calculus. In particular, stochastic integrals can be determined and stochastic differential equations can be solved with it; we will get to know stochastic variants of familiar rules of differentiation (chain and product rule). For this purpose we approach Ito's lemma step by step by first discussing it for Wiener processes, then by generalizing it for diffusion processes and finally by considering some extensions.

11.2 The Univariate Case

The WP itself is a special case of a diffusion as defined in (10.6). With

$$\mu\,(t, W(t)) = 0 \quad \text{and} \quad \sigma\,(t, W(t)) = 1$$

Eq. (10.6) becomes (with probability one)

$$W(t) = W(0) + \int_0^t dW(s) = \int_0^t dW(s)\,.$$

Thus, we consider this special case first.

U. Hassler, *Stochastic Processes and Calculus*, Springer Texts in Business and Economics, DOI 10.1007/978-3-319-23428-1_11

For Wiener Processes

As a revision, let us recall (10.3), which can be written equivalently as

$$2\int_0^t W(s)\,dW(s) = W^2(t) - t\,.$$

If $g(W) = W^2$ is defined with derivatives $g'(W) = 2W$ and $g''(W) = 2$, then this equation can also be formulated as follows:

$$\begin{aligned}\int_0^t g'\left(W(s)\right)\,dW(s) &= g\left(W(t)\right) - t \\ &= g\left(W(t)\right) - \frac{1}{2}\int_0^t g''\left(W(s)\right)\,ds\,.\end{aligned}$$

Now, this is just the form of Ito's lemma for functions g of a Wiener process. It is a corollary of the more general case (Proposition 11.1) which will be covered in the following. Throughout, we will assume that g has a continuous second derivative ("twice continuously differentiable").

Corollary 11.1 (Ito's Lemma for WP) *Let $g : \mathbb{R} \to \mathbb{R}$ be twice continuously differentiable. Then it holds that*

$$dg\left(W(t)\right) = g'\left(W(t)\right)\,dW(t) + \frac{1}{2}\,g''\left(W(t)\right)\,dt\,.$$

In integral form this corollary to Ito's lemma is to be read as follows:

$$g\left(W(t)\right) = g\left(W(0)\right) + \int_0^t g'\left(W(s)\right)\,dW(s) + \frac{1}{2}\int_0^t g''\left(W(s)\right)\,ds\,.$$

Strictly speaking, this integral equation is the statement of the corollary, which is abbreviated by the differential notation. However, in doing so it must not be forgotten that the WP is not differentiable. Sometimes one also writes even more briefly:

$$dg\left(W\right) = g'(W)\,dW + \frac{1}{2}\,g''(W)\,dt\,.$$

Example 11.1 (Powers of the WP) For $g(W) = \frac{1}{2}W^2$ this special case of Ito's lemma just proves (10.3). In general, one obtains for $m \geq 2$ from Corollary 11.1 with $g(W) = \frac{W^m}{m}$:

$$d\left(\frac{W^m(t)}{m}\right) = W^{m-1}(t)\,dW(t) + \frac{m-1}{2}\,W^{m-2}(t)\,dt\,,$$

or in integral notation

$$W^m(t) = m \int_0^t W^{m-1}(s)\, dW(s) + \frac{m(m-1)}{2} \int_0^t W^{m-2}(s)\, ds\,. \quad \blacksquare$$

Explanation and Proof

Corollary 11.1 can be considered as a stochastic chain rule and can loosely be formulated as follows: the derivative of $g(W(t))$ results as the product of the outer derivative ($g'(W)$) and the inner derivative (dW), plus an Ito-specific extra term consisting of the second derivative of g times $\frac{1}{2}$. Where this term comes from (second order Taylor series expansion) and why no further terms occur (higher order derivatives), we want to clarify now. For this purpose we prove Corollary 11.1 (almost completely) although it is, as mentioned above, a corollary to Proposition 11.1.

With $s_n = t$ and $s_0 = 0$ it holds due to (8.3) that:

$$g(W(t)) = g(W(0)) + \sum_{i=1}^{n} (g(W(s_i)) - g(W(s_{i-1})))\ .$$

Now, on the right-hand side a second order Taylor expansion of $g(W(s_i))$ about $W(s_{i-1})$ yields

$$\begin{aligned} g(W(s_i)) &= g(W(s_{i-1})) + g'(W(s_{i-1}))\ (W(s_i) - W(s_{i-1})) \\ &\quad + \frac{g''(\theta_i)}{2}\ (W(s_i) - W(s_{i-1}))^2\ , \end{aligned}$$

with θ_i between $W(s_{i-1})$ and $W(s_i)$:

$$|\ \theta_i - W(s_{i-1})| \in (0, |\ W(s_i) - W(s_{i-1})|)\ .$$

By substitution of $g(W(s_i)) - g(W(s_{i-1}))$, $g(W(t)) - g(W(0))$ can be expressed by two sums:

$$g(W(t)) - g(W(0)) = \Sigma_1 + \Sigma_2$$

with

$$\Sigma_1 = \sum_{i=1}^{n} g'(W(s_{i-1}))\,(W(s_i) - W(s_{i-1}))\,,$$

$$\Sigma_2 = \frac{1}{2} \sum_{i=1}^{n} g''(\theta_i)\,(W(s_i) - W(s_{i-1}))^2\,.$$

Now, Σ_1 just coincides with the Ito sum from (10.5) such that it holds due to Proposition 10.3 that:

$$\Sigma_1 \xrightarrow{2} \int_0^t g'(W(s))\, dW(s)\,.$$

Furthermore, we know from the section on quadratic variation (Proposition 10.8)

$$(dW(s))^2 = ds\,.$$

As the quadratic variation of the WP is not negligible (Proposition 10.7), this suggests the following approximation:

$$\begin{aligned}\Sigma_2 &\approx \frac{1}{2}\int_0^t g''(W(s))\,(dW(s))^2 \\ &= \frac{1}{2}\int_0^t g''(W(s))\,ds\,.\end{aligned}$$

A corresponding convergence in mean square can actually be established, which we will dispense with at this point. Hence, except for this technical detail, Corollary 11.1 is verified.

Additionally, we want to consider why higher order derivatives do not matter for Ito's lemma. For a third order Taylor expansion e.g. it follows

$$\begin{aligned}g(W(s_i)) - g(W(s_{i-1})) &= g'(W(s_{i-1}))(W(s_i) - W(s_{i-1})) \\ &\quad + \frac{g''(W(s_{i-1}))}{2}(W(s_i) - W(s_{i-1}))^2 \\ &\quad + \frac{g'''(\theta_i)}{6}(W(s_i) - W(s_{i-1}))^3\,.\end{aligned}$$

Thus, due to the summation, the term

$$\Sigma_3 = \sum_{i=1}^{n} g'''(\theta_i)\,(W(s_i) - W(s_{i-1}))^3$$

occurs. However, it is negligible:

$$\begin{aligned}|\Sigma_3| &\le \sum_{i=1}^{n} \left|g'''(\theta_i)\right|\,|W(s_i) - W(s_{i-1})|\,(W(s_i) - W(s_{i-1}))^2 \\ &\le \max_{1\le i\le n}\left\{\left|g'''(\theta_i)\right|\,|W(s_i) - W(s_{i-1})|\right\}\cdot Q_n(W,t) \\ &\xrightarrow{2} 0\cdot t = 0\,,\end{aligned}$$

as the quadratic variation of the WP tends to t and as it furthermore holds that

$$\text{MSE}\left[W(s_i) - W(s_{i-1}), 0\right] = \text{Var}\left(W(s_i) - W(s_{i-1})\right) = s_i - s_{i-1} \to 0 .$$

For Diffusions

Now, we turn to Ito's lemma for diffusions. In this section, we consider the univariate case of only one diffusion that depends on one WP only. The following variant of Ito's lemma is again a kind of stochastic chain rule and the idea for the proof is again based on a second order Taylor expansion.

Proposition 11.1 (Ito's Lemma with One Dependent Variable) *Let $g : \mathbb{R} \to \mathbb{R}$ be twice continuously differentiable and $X(t)$ a diffusion on $[0, T]$ with (10.6), or briefly:*

$$dX(t) = \mu(t)\, dt + \sigma(t)\, dW(t) .$$

Then it holds that

$$dg\left(X(t)\right) = g'\left(X(t)\right)\, dX(t) + \frac{1}{2}\, g''\left(X(t)\right)\, \sigma^2(t)\, dt .$$

If $X(t) = W(t)$ is a Wiener process, i.e. $\mu(t) = 0$ and $\sigma(t) = 1$, then Corollary 11.1 is obtained as a special case.

The statement in Proposition 11.1 is given somewhat succinctly It can be condensed even more by suppressing the dependence on time:

$$dg\left(X\right) = g'\left(X\right)\, dX + \frac{1}{2}\, g''\left(X\right)\, \sigma^2\, dt .$$

However, it needs to be clear that by substituting $dX(t)$ one obtains for the differential $dg\left(X(t)\right)$ the following lengthy expression:

$$\left[g'\left(X(t)\right)\, \mu(t, X(t)) + \frac{1}{2}\, g''\left(X(t)\right)\, \sigma^2(t, X(t))\right] dt + g'\left(X(t)\right)\, \sigma(t, X(t))\, dW(t) .$$

The corresponding statement in integral notation naturally looks yet more extensive.

Example 11.2 (Differential of the Exponential Function) Let a diffusion $X(t)$ be given,

$$dX(t) = \mu(t)\, dt + \sigma(t)\, dW(t) .$$

Then, how does the differential of $e^{X(t)}$ read? This example is particularly easy to calculate as it holds for $g(x) = e^x$ that:

$$g''(x) = g'(x) = g(x) = e^x.$$

Hence Ito's lemma yields:

$$\begin{aligned}
de^{X(t)} &= e^{X(t)} dX(t) + \frac{e^{X(t)}}{2} \sigma^2(t) dt \\
&= e^{X(t)} \left(\mu(t) + \frac{\sigma^2(t)}{2} \right) dt + e^{X(t)} \sigma(t) dW(t).
\end{aligned}$$

If $X(t)$ is deterministic, i.e. $\sigma(t) = 0$, then it results

$$\frac{de^{X(t)}}{dt} = e^{X(t)} \frac{dX(t)}{dt},$$

which just corresponds to the traditional chain rule (outer derivative times inner derivative). ■

On the Proof

Just as for the proof of Corollary 11.1, one obtains with θ_i, where

$$|\theta_i - X(s_{i-1})| \in (0, |X(s_i) - X(s_{i-1})|) ,$$

from the Taylor expansion:

$$\begin{aligned}
g(X(t)) - g(X(0)) &= \Sigma_1 + \Sigma_2 , \\
\Sigma_1 &= \sum_{i=1}^{n} g'(X(s_{i-1})) \, (X(s_i) - X(s_{i-1})) \\
\Sigma_2 &= \frac{1}{2} \sum_{i=1}^{n} g''(\theta_i) \, (X(s_i) - X(s_{i-1}))^2 .
\end{aligned}$$

The first sum is approximated as desired:

$$\Sigma_1 \approx \int_0^t g'(X(s)) \, dX(s) .$$

The second sum is approximated by

$$\Sigma_2 \approx \frac{1}{2} \int_0^t g''(X(s))\,(dX(s))^2 .$$

By multiplying out the square of the differential of the Ito process,

$$(dX(s))^2 = \mu^2(s)(ds)^2 + 2\,\mu(s)\,\sigma(s)\,dW(s)\,ds + \sigma^2(s)\,(dW(s))^2 ,$$

one shows due to (cf. Proposition 10.8),

$$(ds)^2 = 0\,, \quad dW(s)\,ds = 0\,, \quad (dW(s))^2 = ds,$$

for the second sum:

$$\Sigma_2 \approx \frac{1}{2} \int_0^t g''(X(s))\,\sigma^2(s)\,ds .$$

This verifies Proposition 11.1 at least heuristically.

11.3 Bivariate Diffusions with One WP

A generalization of Proposition 11.1, which is sometimes needed, is presented by the following variant of Ito's lemma. The function g be dependent on two diffusions X_1 and X_2, where both are driven by the very same Wiener process. Occasionally, we will call this case (referring to the literature on interest rate models) the one-factor case as it is the identical factor $W(t)$ driving both diffusions.

One-Factor Case

Let g be a function in two arguments, whose partial derivatives are denoted by

$$\frac{\partial g(X_1, X_2)}{\partial X_i} \quad \text{and} \quad \frac{\partial^2 g(X_1, X_2)}{\partial X_i\, \partial X_j} .$$

Then, the following proposition is a special case of Proposition 11.3.

Proposition 11.2 (Ito's Lemma with Two Dependent Variables) *Let* $g\colon \mathbb{R}\times\mathbb{R} \to \mathbb{R}$ *be twice continuously differentiable with respect to both arguments, and let* $X_i(t)$ *be diffusions on* $[0, T]$ *with the same WP:*

$$dX_i(t) = \mu_i(t)\,dt + \sigma_i(t)\,dW(t)\,, \quad i = 1, 2 .$$

Then it holds that

$$\begin{aligned}
dg\left(X_1(t),\,X_2(t)\right) &= \frac{\partial g\left(X_1(t),\,X_2(t)\right)}{\partial X_1}\,dX_1(t) + \frac{\partial g\left(X_1(t),\,X_2(t)\right)}{\partial X_2}\,dX_2(t) \\
&+ \frac{1}{2}\left[\frac{\partial^2 g\left(X_1(t),\,X_2(t)\right)}{\partial X_1^2}\,\sigma_1^2(t) + \frac{\partial^2 g\left(X_1(t),\,X_2(t)\right)}{\partial X_2^2}\,\sigma_2^2(t)\right] dt \\
&+ \frac{\partial^2 g\left(X_1(t),\,X_2(t)\right)}{\partial X_1\,\partial X_2}\,\sigma_1(t)\,\sigma_2(t)\,dt\,.
\end{aligned}$$

Note that substitution of $dX_i(t)$ in Proposition 11.2 leads again to an integral equation for the process $g\left(X_1(t),\,X_2(t)\right)$ including Riemann and Ito integrals.

Frequently, the time dependence of the processes will be suppressed in order to obtain a more economical formulation of Proposition 11.2:

$$\begin{aligned}
dg\left(X_1,\,X_2\right) &= \frac{\partial g\left(X_1,\,X_2\right)}{\partial X_1}\,dX_1 + \frac{\partial g\left(X_1,\,X_2\right)}{\partial X_2}\,dX_2 \\
&+ \frac{1}{2}\left[\frac{\partial^2 g\left(X_1,\,X_2\right)}{\partial X_1^2}\,\sigma_1^2 + \frac{\partial^2 g\left(X_1,\,X_2\right)}{\partial X_2^2}\,\sigma_2^2\right] dt \\
&+ \frac{\partial^2 g\left(X_1,\,X_2\right)}{\partial X_1\,\partial X_2}\,\sigma_1\,\sigma_2\,dt\,.
\end{aligned}$$

By this notation one recognizes, that again a second order Taylor expansion hides behind Proposition 11.2, but now of the two-dimensional function g,

$$\begin{aligned}
dg\left(X_1,\,X_2\right) &= \frac{\partial g\left(X_1,\,X_2\right)}{\partial X_1}\,dX_1 + \frac{\partial g\left(X_1,\,X_2\right)}{\partial X_2}\,dX_2 \\
&+ \frac{1}{2}\left[\frac{\partial^2 g\left(X_1,\,X_2\right)}{\partial X_1^2}\,(dX_1)^2 + \frac{\partial^2 g\left(X_1,\,X_2\right)}{\partial X_2^2}\,(dX_2)^2\right] \\
&+ \frac{1}{2}\left[\frac{\partial^2 g\left(X_1,\,X_2\right)}{\partial X_1\,\partial X_2}\,dX_1 dX_2 + \frac{\partial^2 g\left(X_1,\,X_2\right)}{\partial X_2\,\partial X_1}\,dX_2 dX_1\right],
\end{aligned}$$

because the mixed second derivatives coincide due to the continuity assumed. With (10.7) and (10.8) it can easily be shown that (we again suppress the arguments)

$$(dX_i)^2 = \mu_i^2\,(dt)^2 + 2\,\mu_i\,\sigma_i\,dtdW + \sigma_i^2\,(dW)^2 = 0 + 0 + \sigma_i^2\,dt\,,$$

and for the covariance expression as well

$$dX_1\,dX_2 = \sigma_1\,\sigma_2\,dt\,. \tag{11.1}$$

Example 11.3 (One-Factor Product Rule) Proposition 11.2 provides us with a stochastic product rule for $X_1(t)\,X_2(t)$:

$$d(X_1(t)\,X_2(t)) = X_2(t)\,dX_1(t) + X_1(t)\,dX_2(t) + \sigma_1(t)\,\sigma_2(t)\,dt\,. \tag{11.2}$$

Under $\sigma_1(t) = 0$ or $\sigma_2(t) = 0$ (no stochastics), the well-known product rule is just reproduced. The derivation of (11.2) follows for $g(x_1, x_2) = x_1\,x_2$ with

$$\frac{\partial g}{\partial x_1} = x_2, \quad \frac{\partial^2 g}{\partial x_1^2} = 0,$$
$$\frac{\partial g}{\partial x_2} = x_1, \quad \frac{\partial^2 g}{\partial x_2^2} = 0$$

and

$$\frac{\partial^2 g}{\partial x_1\,\partial x_2} = \frac{\partial^2 g}{\partial x_2\,\partial x_1} = 1\,.$$

Hence, we obtain an abbreviated form:

$$d\,(X_1\,X_2) = \frac{\partial g\,(X_1,\,X_2)}{\partial X_1}\,dX_1 + \frac{\partial g\,(X_1,\,X_2)}{\partial X_2}\,dX_2 + \sigma_1\,\sigma_2\,dt\,,$$

where the second derivatives were plugged in. If one substitutes the first derivatives, then one obtains the result from (11.2). ■

Time as a Dependent Variable

Frequently it is of interest to consider another special case of Proposition 11.2. Again, g is a function in two arguments; however, the first one is time t, and the second one is a diffusion $X(t)$:

$$\begin{aligned} g:\quad [0,T]\times\mathbb{R} &\to \mathbb{R} \\ (t,X) &\mapsto g(t,X)\,. \end{aligned}$$

We consciously suppress the fact that the diffusion is time-dependent as well. Since, when we talk about the derivative of g with respect to time, then we refer strictly formally to the partial derivative with respect to the first argument. Sometimes this is confusing for beginners. For example for

$$g(t, X(t)) = g(t, X) = t\,X(t)$$

the derivative with respect to t refers to:

$$\frac{\partial g\left(t,\, X(t)\right)}{\partial t} = X(t)\,.$$

Hence, for the partial derivatives we purposely do not consider that X itself is a function of t.

With $X_1(t) = t$ and $X(t) = X_2(t)$ from Proposition 11.2 we obtain for $\mu_1(t) = 1$ and $\sigma_1(t) = 0$ the following circumstance.

Corollary 11.2 (Ito's Lemma with Time as a Dependent Variable) *Let $g : [0, T] \times \mathbb{R} \to \mathbb{R}$ be twice continuously differentiable with respect to both arguments and let $X(t)$ be a diffusion on $[0, T]$ with (10.6), or briefly*

$$dX(t) = \mu(t)\, dt + \sigma(t)\, dW(t)\,.$$

Then it holds that

$$dg\left(t,\, X(t)\right) = \frac{\partial g\left(t,\, X(t)\right)}{\partial t}\, dt + \frac{\partial g\left(t,\, X(t)\right)}{\partial X}\, dX(t) + \frac{1}{2}\, \frac{\partial^2 g\left(t,\, X(t)\right)}{\partial X^2}\, \sigma^2(t)\, dt\,.$$

Again, suppressing time-dependence, this can be condensed to

$$dg\left(t,\, X\right) = \frac{\partial g\left(t,\, X\right)}{\partial t}\, dt + \frac{\partial g\left(t,\, X\right)}{\partial X}\, dX + \frac{1}{2}\, \frac{\partial^2 g\left(t,\, X\right)}{\partial X^2}\, \sigma^2\, dt\,.$$

Example 11.4 (OUP as a Diffusion) As an application we can just prove that the standard Ornstein-Uhlenbeck process $X_c(t)$ from (9.3) is a diffusion with

$$\mu\left(t, X_c(t)\right) = c\, X_c(t) \quad \text{and} \quad \sigma\left(t, X_c(t)\right) = 1\,.$$

For this purpose we define as an auxiliary quantity the process

$$X(t) = \int_0^t e^{-cs} dW(s)\,,$$

or

$$dX(t) = e^{-ct} dW(t)\,.$$

With this variable we define the function g such,

$$g(t, X) = e^{ct} X\,,$$

that it holds for the OUP:

$$X_c(t) = g(t, X(t)) = e^{ct} X(t).$$

With the derivatives

$$\frac{\partial g(t,X)}{\partial t} = c e^{ct} X, \quad \frac{\partial g(t,X)}{\partial X} = e^{ct}, \quad \frac{\partial^2 g(t,X)}{\partial X^2} = 0$$

it follows from Corollary 11.2:

$$\begin{aligned} dX_c(t) &= c e^{ct} X(t) dt + e^{ct} dX(t) + 0 \\ &= c X_c(t) dt + dW(t), \end{aligned}$$

where $dX(t)$ was substituted. ■

Further examples for practicing Corollary 11.2 can be found in the problem section.

K-Variate Diffusions

Concerning the contents, there is no reason why Proposition 11.2 should be written just with two processes. Let us consider as a generalization the case where g depends on K diffusions, all of them given by the identical WP:

$$g\colon \mathbb{R}^K \to \mathbb{R}, \quad \text{i.e.} \quad g = g(X_1, \ldots, X_K) \in \mathbb{R}.$$

Then it holds with $dX_k(t)$, $k = 1, \ldots, K$, due to a second order Taylor expansion that:

$$dg(X_1, \ldots, X_K) = \sum_{k=1}^{K} \frac{\partial g}{\partial X_k} dX_k + \frac{1}{2} \sum_{k=1}^{K} \sum_{j=1}^{K} \frac{\partial^2 g}{\partial X_k \, \partial X_j} dX_k \, dX_j \, .$$

As in the bivariate case, one obtains $dX_k \, dX_j = \sigma_k \, \sigma_j \, dt$, cf. (11.1). Sometimes, as in Corollary 11.2, time as a further variable is allowed for,

$$g\colon [0, T] \times \mathbb{R}^K \to \mathbb{R}, \quad \text{i.e.} \quad g = g(t, X_1, \ldots, X_K) \in \mathbb{R},$$

and

$$dg(t, X_1, \ldots, X_K) = \frac{\partial g}{\partial t} dt + \sum_{k=1}^{K} \frac{\partial g}{\partial X_k} dX_k + \frac{1}{2} \sum_{k=1}^{K} \sum_{j=1}^{K} \frac{\partial^2 g}{\partial X_k \, \partial X_j} dX_k \, dX_j \, .$$

11.4 Generalization for Independent WP

We keep to the multivariate generalization, however, allowing for several, stochastically independent Wiener processes behind the diffusions.

The General Case

Now, $W_1(t), \ldots, W_d(t)$ denote stochastically independent standard Wiener processes. We allow for d factors driving each of the K diffusions. According to this, let $\mathbf{X}(t)$ be a K-dimensional diffusion[1] $\mathbf{X}'(t) = (X_1(t), \ldots, X_K(t))$, defined by d factors $W_j(t), j = 1, \ldots, d$:

$$dX_k(t) = \mu_k(t)dt + \sum_{j=1}^{d} \sigma_{kj}(t)dW_j(t), \quad k = 1, \ldots, K.$$

In order to have a diffusion, it holds for μ_k and σ_{kj} that they may only depend on $\mathbf{X}(t)$ and t:

$$\mu_k(t) = \mu_k(t, \mathbf{X}(t)), \quad k = 1, \ldots K,$$

$$\sigma_{kj}(t) = \sigma_{kj}(t, \mathbf{X}(t)), \quad k = 1, \ldots K, \ j = 1, \ldots d.$$

For a function g, which maps $\mathbf{X}(t)$ to the real numbers, Ito's lemma reads as follows, cf. Øksendal (2003, Theorem 4.2.1).

Proposition 11.3 (Ito's Lemma (Independent WP)) *Let $g : \mathbb{R}^K \to \mathbb{R}$ be twice continuously differentiable with respect to all the arguments, and let $X_k(t)$ be diffusions on $[0, T]$ depending on d independent Wiener processes:*

$$dX_k(t) = \mu_k(t)dt + \sum_{j=1}^{d} \sigma_{kj}(t)dW_j(t), \quad k = 1, \ldots, K.$$

Then it holds for $\mathbf{X}'(t) = (X_1(t), \ldots, X_K(t))$ that

$$dg(\mathbf{X}(t)) = \sum_{k=1}^{K} \frac{\partial g(\mathbf{X}(t))}{\partial X_k} dX_k(t) + \frac{1}{2} \sum_{i=1}^{K} \sum_{k=1}^{K} \frac{\partial^2 g(\mathbf{X}(t))}{\partial X_i \partial X_k} dX_i(t) dX_k(t)$$

[1] For vectors and matrices, the superscript denotes transposition and not differentiation. Further, the dimension d of the multivariate Wiener process should not be confused with the differential operator denoted by the same symbol.

with

$$dX_i(t)dX_k(t) = \sum_{j=1}^{d} \sigma_{ij}(t)\sigma_{kj}(t)dt. \tag{11.3}$$

Heuristically, (11.3) can be well justified. For this purpose we consider vectors of length d:

$$\boldsymbol{\sigma}_k(t) = \begin{pmatrix} \sigma_{k1}(t) \\ \vdots \\ \sigma_{kd}(t) \end{pmatrix}, \; k = 1, \ldots, K, \quad \text{and } \mathbf{W}(t) = \begin{pmatrix} W_1(t) \\ \vdots \\ W_d(t) \end{pmatrix},$$

such that

$$dX_k(t) = \mu_k(t)dt + \boldsymbol{\sigma}_k'(t)d\mathbf{W}(t).$$

Neglecting the dependence on time, it follows

$$\begin{aligned} dX_i\, dX_k &= \mu_i\, \mu_k (dt)^2 + \mu_i\, \boldsymbol{\sigma}_k' d\mathbf{W}(t)dt + \mu_k\, \boldsymbol{\sigma}_i' d\mathbf{W}(t)dt \\ &\quad + \boldsymbol{\sigma}_i' d\mathbf{W}(t)\boldsymbol{\sigma}_k' d\mathbf{W}(t) \\ &= \boldsymbol{\sigma}_i' d\mathbf{W}(t)d\mathbf{W}'(t)\boldsymbol{\sigma}_k \end{aligned}$$

due to (see Proposition 10.8)

$$(dt)^2 = 0 \quad \text{and} \quad dW_j(t)dt = 0$$

and

$$\boldsymbol{\sigma}_k' d\mathbf{W}(t) = (\boldsymbol{\sigma}_k' d\mathbf{W}(t))' = d\mathbf{W}'(t)\boldsymbol{\sigma}_k.$$

Let us consider the matrix

$$d\mathbf{W}(t)\, d\mathbf{W}'(t) = \begin{pmatrix} (dW_1(t))^2 & dW_1(t)dW_2(t) & \ldots & dW_1(t)dW_d(t) \\ dW_2(t)dW_1(t) & (dW_2(t))^2 & \ldots & dW_2(t)dW_d(t) \\ \vdots & \vdots & \ddots & \vdots \\ dW_d(t)dW_1(t) & dW_d(t)dW_2(t) & \ldots & (dW_d(t))^2 \end{pmatrix}.$$

As is well known, it holds due to (10.7) that:

$$(dW_j(t))^2 = dt.$$

Furthermore, it can be shown for stochastically independent Wiener processes that

$$dW_i(t)dW_k(t) = 0, \quad i \neq k.$$

Overall, we hence obtain

$$d\mathbf{W}(t)d\mathbf{W}'(t) = I_d dt,$$

with the d-dimensional identity matrix I_d. All in all it follows

$$\begin{aligned} dX_i(t)dX_k(t) &= \boldsymbol{\sigma}_i'(t)\, I_d dt\, \boldsymbol{\sigma}_k(t) \\ &= \boldsymbol{\sigma}_i'(t)\boldsymbol{\sigma}_k(t)\, dt\,, \end{aligned}$$

which is given in (11.3).

The 2-Factor Case

Let us consider the case $K = d = 2$. Then, Proposition 11.3 becomes more clearly

$$dg(\mathbf{X}(t)) = \sum_{k=1}^{2} \frac{\partial g(\mathbf{X}(t))}{\partial X_k} dX_k(t) + \frac{1}{2} \sum_{i=1}^{2} \sum_{k=1}^{2} \frac{\partial^2 g(\mathbf{X}(t))}{\partial X_i \partial X_k} dX_i(t) dX_k(t)$$

with

$$dX_1 dX_1 = (\sigma_{11}^2 + \sigma_{12}^2)dt\,,$$

$$dX_2 dX_2 = (\sigma_{21}^2 + \sigma_{22}^2)dt\,,$$

and

$$dX_1 dX_2 = (\sigma_{11}\sigma_{21} + \sigma_{12}\sigma_{22})dt.$$

Two interesting special cases result:

1. $\sigma_{12} = \sigma_{22} = 0$ (one-factor model),
2. $\sigma_{12} = \sigma_{21} = 0$ (independent diffusions).

The first case naturally corresponds to the one from the previous section: Both diffusions only depend on the same WP. The second case is the opposite extreme where both diffusions depend only on one or other of the two stochastically independent processes:

$$dX_k(t) = \mu_k(t)dt + \sigma_{kk}(t)dW_k(t), \quad k = 1, 2.$$

We want to discuss both borderline cases based on two examples.

Example 11.5 (2-Factor Product Rule) Proposition 11.3 with $K = d = 2$ yields with the derivatives from Example 11.3 as product rule:

$$d(X_1 X_2) = X_2\, dX_1 + X_1\, dX_2 + dX_1\, dX_2\,. \tag{11.4}$$

In the borderline case of only one factor, naturally the result from Eq. (11.2) is reproduced. In the second borderline case of stochastically independent diffusions, however, it holds, as in the deterministic case, that

$$d(X_1 X_2) = X_2\, dX_1 + X_1\, dX_2\,.$$

Without restrictions (11.4) reads as follows:

$$d(X_1 X_2) = X_2\, dX_1 + X_1\, dX_2 + (\sigma_{11}\sigma_{21} + \sigma_{12}\sigma_{22})dt\,.$$

The example illustrates that a proper application of Ito's lemma needs to account for the number of factors underlying the diffusions. ■

Example 11.6 (2-Factor Quotient Rule) For $X_2(t) \neq 0$ and

$$g(X_1, X_2) = \frac{X_1}{X_2}$$

we obtain:

$$\frac{\partial g}{\partial X_1} = X_2^{-1}, \quad \frac{\partial g}{\partial X_2} = -X_1 X_2^{-2},$$

$$\frac{\partial^2 g}{\partial X_1^2} = 0, \quad \frac{\partial^2 g}{\partial X_2^2} = 2X_1 X_2^{-3}, \quad \frac{\partial^2 g}{\partial X_1 \partial X_2} = -X_2^{-2}.$$

Hence Proposition 11.3 yields with $K = d = 2$ suppressing the arguments:

$$d\left(\frac{X_1}{X_2}\right) = \frac{X_2\, dX_1 - X_1\, dX_2}{X_2^2} + \frac{X_1 X_2^{-1}(\sigma_{21}^2 + \sigma_{22}^2) - (\sigma_{11}\sigma_{21} + \sigma_{12}\sigma_{22})}{X_2^2}\, dt\,. \tag{11.5}$$

If X_2 is a deterministic function ($\sigma_{21} = \sigma_{22} = 0$), then the conventional quotient rule is reproduced. ■

11.5 Problems and Solutions

Problems

11.1 Prove part (a) from Example 9.1.
Hint: Choose $g(t, W) = tW$ in Corollary 11.2 or in (11.2).

11.2 Prove part (b) from Example 9.1.
Hint: Choose $g(t, W) = (1-t)W$ in Corollary 11.2 or in (11.2).

11.3 Prove part (b) from Proposition 9.1 (integration by parts).
Hint: Choose $g(t, W) = f(t)W$ in Corollary 11.2 or in (11.2).

11.4 Prove statement (a) from Proposition 9.4 with Ito's lemma.
Hint: Choose $g(t, W) = e^{-ct}W$.

11.5 Prove for the OUP from Proposition 9.4:

$$\int_0^t X_c(s)\, dW(s) = \frac{1}{2}\left(X_c^2(t) - t\right) - c\int_0^t X_c^2(s)\, ds$$

Note that for $c = 0$ (WP) this reproduces (10.3).
Hint: Choose $g(X_c(t)) = X_c^2(t)$ in Ito's lemma.

11.6 Determine the differential of $W(t)/e^{W(t)}$ according to the one-factor product rule (11.2).

11.7 Determine the differential of $W(t)/e^{W(t)}$ directly from Corollary 11.1.

Solutions

11.1 For the proof we use Corollary 11.2 with

$$g(t, W) = tW.$$

The derivatives needed read:

$$\frac{\partial g(t, W)}{\partial t} = W, \quad \frac{\partial g(t, W)}{\partial W} = t, \quad \frac{\partial^2 g(t, W)}{\partial W^2} = 0.$$

Hence, one determines with Corollary 11.2:

$$d(tW(t)) = W(t)dt + tdW(t) + \frac{0}{2}.$$

Due to $g(0, W(0)) = 0$, we obtain as an integral equation

$$tW(t) = \int_0^t W(s)ds + \int_0^t s dW(s),$$

which was to be shown.

11.2 As in Problem 11.1 we consider

$$g(t, W) = (1-t)W$$

with

$$\frac{\partial g(t, W)}{\partial t} = -W, \quad \frac{\partial g(t, W)}{\partial W} = (1-t), \quad \frac{\partial^2 g(t, W)}{\partial W^2} = 0.$$

Therefore, substitution into Corollary 11.2 yields

$$d((1-t)W(t)) = -W(t)dt + (1-t)dW(t) + \frac{0}{2}.$$

As $W(0) = 0$ with probability one, it follows that

$$(1-t)W(t) = -\int_0^t W(s)ds + \int_0^t (1-s)dW(s),$$

which was to be shown.

11.3 As an adequate function g we choose

$$g(t, W) = f(t)W,$$

where $f(t)$ is deterministic. Then, Corollary 11.2 is used with

$$\frac{\partial g(t, W)}{\partial t} = f'(t)W, \quad \frac{\partial g(t, W)}{\partial W} = f(t), \quad \frac{\partial^2 g(t, W)}{\partial W^2} = 0.$$

This yields for the differential:

$$dg(t, W(t)) = f'(t)W(t)dt + f(t)dW(t) + \frac{0}{2}.$$

In integral notation this reads

$$g(t, W(t)) = g(0, W(0)) + \int_0^t f'(s)W(s)ds + \int_0^t f(s)dW(s).$$

As $W(0) = 0$ with probability one, we hence obtain the desired result:

$$f(t)W(t) = \int_0^t f'(s)W(s)ds + \int_0^t f(s)dW(s).$$

11.4 If one chooses $g(t, W) = e^{-ct}W$ with

$$\frac{\partial g(t,W)}{\partial t} = -ce^{-ct}W, \quad \frac{\partial g(t,W)}{\partial W} = e^{-ct}, \quad \frac{\partial^2 g(t,W)}{\partial W^2} = 0,$$

then Corollary 11.2 allows for the following calculation:

$$d(e^{-ct}W(t)) = -ce^{-ct}W(t)dt + e^{-ct}dW(t),$$

i.e.

$$e^{-ct}W(t) = W(0) - c\int_0^t e^{-cs}W(s)ds + \int_0^t e^{-cs}dW(s)$$

or

$$\begin{aligned} W(t) &= -ce^{ct}\int_0^t e^{-cs}W(s)ds + e^{ct}\int_0^t e^{-cs}dW(s) \\ &= -ce^{ct}\int_0^t e^{-cs}W(s)ds + X_c(t). \end{aligned}$$

Rearranging terms completes the proof.

11.5 With the function $g(X_c) = X_c^2$ and its derivatives,

$$g'(X_c) = 2X_c, \quad g''(X_c) = 2,$$

Proposition 11.1 can be applied. We know that $X_c(t)$ is a diffusion with (see Example 11.4):

$$dX_c(t) = cX_c(t)dt + dW(t).$$

Plugging in into Proposition 11.1 shows:

$$\begin{aligned} dX_c^2(t) &= 2X_c(t)dX_c(t) + \frac{2}{2}dt \\ &= (2cX_c^2(t) + 1)dt + 2X_c(t)dW(t). \end{aligned}$$

With starting value $X_c(0) = 0$ this translates into the following integral equation:

$$X_c^2(t) = \int_0^t (2cX_c^2(s) + 1)ds + 2\int_0^t X_c(s)dW(s).$$

This is equivalent to

$$\int_0^t X_c(s)dW(s) = \frac{1}{2}\left(X_c^2(t) - \int_0^t ds\right) - c\int_0^t X_c^2(s)ds,$$

which amounts to the claim.

11.6 We define

$$X_1(t) = W(t), \quad X_2(t) = e^{-W(t)},$$

and we are interested in the differential of the product. In order to apply the one-factor product rule, we need the differentials of the factors. For X_1 it obviously holds that: $dX_1 = dW$. For $e^{-W(t)}$ Example 11.2 yields

$$dX_2 = de^{-W} = -e^{-W}dW + \frac{e^{-W}}{2}dt.$$

Hence, we have

$$\sigma_1(t) = 1, \quad \sigma_2(t) = -e^{-W(t)}.$$

Plugging in into the product rule (11.2) yields:

$$\begin{aligned} d\left(We^{-W}\right) &= e^{-W}dX_1 + WdX_2 - e^{-W}dt \\ &= e^{-W}dW + W\left(-e^{-W}dW + \frac{e^{-W}}{2}dt\right) - e^{-W}dt \\ &= e^{-W}\left(\frac{W}{2} - 1\right)dt + e^{-W}(1-W)dW. \end{aligned}$$

11.7 As $g(W) = \frac{W}{e^W}$ is a simple function of W, Corollary 11.1 yields a direct approach to the differential. For this purpose, we only need the derivatives (quotient rule):

$$g'(W) = \frac{e^W - W\,e^W}{e^{2W}} = \frac{1-W}{e^W},$$

$$g''(W) = \frac{-e^W - (1-W)\,e^W}{e^{2W}} = \frac{W-2}{e^W}.$$

Thus, it follows from Ito's lemma that

$$
\begin{aligned}
d\left(\frac{W}{e^W}\right) &= g'(W)\,dW + \frac{1}{2}g''(W)\,dt \\
&= \frac{1-W}{e^W}\,dW + \frac{W-2}{2e^W}\,dt \\
&= e^{-W}\left(\frac{W}{2}-1\right)dt + e^{-W}(1-W)\,dW.
\end{aligned}
$$

Of course, this result coincides with the one from the previous problem.

Reference

Øksendal, B. (2003). *Stochastic differential equations: An introduction with applications* (6th ed.). Berlin/New York: Springer.

Part III

Applications

Stochastic Differential Equations (SDE) 12

12.1 Summary

In the following section we discuss the most general stochastic differential equation considered here, whose solution is a diffusion. Then, linear differential equations (with variable coefficients) will be studied extensively. Here we obtain analytical solutions by Ito's lemma. We discuss special cases that are widespread in the literature on finance. In the fourth section we turn to numerical solutions allowing to simulate processes. The sample paths displayed in the figures of Chap. 13 are constructed that way.

12.2 Definition and Existence

After a definition and a discussion of conditions for existence, we will consider the deterministic case as well. Deterministic differential equations are embedded into the stochastic ones as special cases.

Diffusions

We defined the solution of

$$dX(t) = \mu(t)\,dt + \sigma(t)\,dW(t)$$

as a diffusion process, where $\mu(t)$ and $\sigma(t)$ are allowed to depend on t and on $X(t)$ itself. As the most general case of this chapter we consider diffusions as in (10.6):

$$dX(t) = \mu(t, X(t))\,dt + \sigma(t, X(t))\,dW(t)\,, \quad t \in [0, T]\,. \tag{12.1}$$

U. Hassler, *Stochastic Processes and Calculus*, Springer Texts in Business and Economics, DOI 10.1007/978-3-319-23428-1_12

The solutions[1] of such differential equations can also be written in integral form:

$$X(t) = X(0) + \int_0^t \mu(s, X(s))\, ds + \int_0^t \sigma(s, X(s))\, dW(s), \quad t \in [0, T]. \tag{12.2}$$

Under what conditions is such a definition possible? In other words: Which requirements have to be met by the functions $\mu(t,x)$ and $\sigma(t,x)$, such that a solution of (12.1) exists at all – and uniquely so? This mathematical aspect is not to be overly deepened at this point, however, neither is it to be completely neglected. We consider stronger but simpler sufficient conditions than necessary. For a profound discussion see e.g. Øksendal (2003). The first assumption requires that μ and σ are smooth enough in the argument x:[2]

(E1) *The partial derivatives of $\mu(t,x)$ and $\sigma(t,x)$ with respect to x exist and are continuous in x.*

Secondly, we maintain a linear restriction of the growth of the diffusion process:

(E2) *There exist constants K_1 and K_2 with*

$$|\mu(t,x)| + |\sigma(t,x)| \leq K_1 + K_2|x|.$$

And finally we need a well defined starting value, which may be stochastic:

(E3) *$X(0)$ is independent of $W(t)$ with $E(X^2(0)) < \infty$.*

Under these assumptions Øksendal (2003, Theorem 5.2.1) proves the following proposition.

Proposition 12.1 (Existence of a Unique Solution) *Under the assumptions **(E1)** to **(E3)**, Eq. (12.1) has a unique solution $X(t)$ of the form (12.2) with continuous paths and $E(X^2(t)) < \infty$.*

The assumption **(E3)** can always be met by assuming a fixed starting value. The second assumption is necessary for the existence of a (finite) solution while **(E1)**

---

[1]Strictly speaking, this is a so-called "strong solution" in contrast to a "weak solution". For a weak solution the behavior of $X(t)$ is only characterized in distribution. We will not concern ourselves with weak solutions.

[2]Normally, one demands that they satisfy a Lipschitz condition. A function f is called Lipschitz continuous if it holds for all x and y that there exists a constant K with

$$|f(x) - f(y)| \leq K\,|x - y|.$$

We can conceal this condition by requiring the stronger sufficient continuous differentiability.

guarantees the uniqueness of this solution. This is to be illustrated by means of two deterministic examples.

Example 12.1 (Violation of the Assumptions) We examine two examples known from the literature on deterministic differential equations where $\sigma(t,x) = 0$. Similar cases can be found e.g. in Øksendal (2003). In the first example we set $\mu(t, X(t)) = X^{2/3}(t)$:

$$dX(t) = X^{2/3}(t)\,dt\,, \quad X(0) = 0\,,\; t \geq 0\,.$$

We define for an arbitrary $a > 0$ infinitely many solutions:

$$X_a(t) = \begin{cases} 0 \quad , & t \leq a \\ \frac{(t-a)^3}{27}, & t > a\,. \end{cases}$$

By differentiating one can observe that any $X_a(t)$ indeed satisfies the given equation. The reason for the ambiguity of the solutions lies in the violation of **(E1)** as the partial derivative,

$$\frac{\partial \mu(t,x)}{\partial x} = \frac{2}{3} x^{-1/3}\,,$$

does not exist at $x = 0$.

The second example reads for $\mu(t, X(t)) = X^2(t)$:

$$dX(t) = X^2(t)\,dt\,, \quad X(0) = 1\,,\; t \in [0,1)\,.$$

Again, by elementary means one proves that the solution reads

$$X(t) = (1-t)^{-1}\,, \quad 0 \leq t < 1\,,$$

and hence tends to ∞ for $t \to 1$. The reason for this lies in a violation of **(E2)**: The quadratic function $\mu(t,x) = x^2$ cannot be linearly bounded. ■

Linear Coefficients

In order to be able to state analytical solutions, we frequently restrict generality and consider **linear differential equations**:

$$dX(t) = (c_1(t)\,X(t) + c_2(t))\;dt + (\sigma_1(t)\,X(t) + \sigma_2(t))\;dW(t)\,, \quad t \geq 0\,, \qquad (12.3)$$

where the variable coefficients $c_i(t)$ and $\sigma_i(t)$, $i = 1, 2$, are continuous deterministic functions of time. Here, $X(t)$ enters μ and σ just linearly. Obviously, the partial derivatives from **(E1)** are constant (in x) and thus continuous. In addition, one obtains a linear bound:

$$\begin{aligned}|\mu(t,x)| + |\sigma(t,x)| &\leq |c_1(t)|\;|x| + |c_2(t)| + |\sigma_1(t)|\;|x| + |\sigma_2(t)| \\ &= (|c_1(t)| + |\sigma_1(t)|)\;|x| + (|c_2(t)| + |\sigma_2(t)|) \\ &\leq K_2\;|x| + K_1\,.\end{aligned}$$

As $c_i(t)$ and $\sigma_i(t)$ are continuous in t and hence bounded for finite t, positive constants K_1 and K_2 can be specified such that the inequality above holds true. Therefore, **(E2)** is satisfied. Therefore, a unique solution exists for linear stochastic differential equations. What is more: Ito's lemma will allow as well for the specification of an explicit form of this analytical solution from which one can determine first and second moments as functions of time. The next section is reserved for studying equation (12.3). Before, we consider the borderline case of a deterministic linear differential equation.

Deterministic Case

By setting $\sigma_1(t) = \sigma_2(t) = 0$ in (12.3), we obtain a deterministic linear differential equation (in small letters to distinguish from the stochastic case),

$$dx(t) = (c_1(t)\,x(t) + c_2(t))\;dt\,, \quad t \geq 0\,, \tag{12.4}$$

or as well

$$x'(t) = c_1(t)\,x(t) + c_2(t)\,.$$

Frequently, one speaks of first order differential equations, as only the first derivative is involved. As is well known, the solution reads (see Problem 12.1)

$$x(t) = z(t)\left[x(0) + \int_0^t \frac{c_2(s)}{z(s)}\,ds\right] \tag{12.5}$$

with

$$z(t) = \exp\left\{\int_0^t c_1(s)\,ds\right\}\,. \tag{12.6}$$

For $c_2(t) = 0$ one obtains from (12.4) the related **homogeneous differential equation** (with starting value 1),

$$dz(t) = c_1(t)\,z(t)\,dt\,, \quad z(0) = 1\,,$$

which just has $z(t)$ from (12.6) as a solution. The following example presents the special case of constant coefficients.

Example 12.2 (Constant Coefficients) In the case of constant coefficients,

$$c_1(t) = c_1 = const\,, \quad c_2(t) = c_2 = const\,,$$

the solution from (12.5) simplifies, see Problem 12.1:

$$\begin{aligned} x(t) &= e^{c_1 t}\left[x(0) + \frac{c_2}{c_1}\left(1 - e^{-c_1 t}\right)\right] \\ &= e^{c_1 t}\left[x(0) + \frac{c_2}{c_1}\right] - \frac{c_2}{c_1}\,. \end{aligned}$$

Hence, for negative values of c_1 it holds that the equation is stable in the sense that the solution tends towards a fixed value:

$$x(t) \stackrel{t\to\infty}{\longrightarrow} -\frac{c_2}{c_1} =: \mu\,, \quad c_1 < 0\,.$$

Basically, one can already observe this from the equation itself:

$$\begin{aligned} dx(t) &= (c_1\, x(t) + c_2)\, dt \\ &= c_1\,(x(t) - \mu)\, dt\,. \end{aligned}$$

Namely, if $x(t)$ lies above the limit μ, then the expression in brackets is positive and hence the change is negative such that $x(t)$ adjusts towards the limit μ. Conversely, $x(t) < \mu$ causes a positive derivative such that $x(t)$ grows and moves towards the limit. All in all, for $c_1 < 0$ a convergence to μ is modeled. ■

In the following, we will see that the solution of the deterministic linear equation is embedded into the stochastic one for $\sigma_1(t) = \sigma_2(t) = 0$.

12.3 Linear Stochastic Differential Equations

For the solution of the equation (12.3) we expect a similar structure as in the deterministic case, (12.5), i.e. a homogeneous solution as a multiplicative factor has to be expected. Hence, we start with the solution of a homogeneous stochastic equation.

Homogeneous Solution

For $c_2(t) = \sigma_2(t) = 0$ one obtains from (12.3) the corresponding homogeneous linear equation. In doing so, we rename X and choose 1 as the starting value[3]:

$$dZ(t) = c_1(t)\, Z(t)\, dt + \sigma_1(t) Z(t)\, dW(t)\,, \quad Z(0) = 1. \tag{12.7}$$

Now, Ito's lemma (Proposition 11.1) is applied to $g(Z(t)) = \log(Z(t))$. Thus, we obtain as the solution of (12.7),

$$Z(t) = \exp\left\{\int_0^t \left(c_1(s) - \frac{1}{2}\sigma_1^2(s)\right) ds + \int_0^t \sigma_1(s)\, dW(s)\right\}\,, \tag{12.8}$$

see Problem 12.2. Hence, for $\sigma_1(t) = 0$ the deterministic solution from (12.6) is reproduced. The solution with an arbitrary starting value different from zero therefore reads

$$X(t) = X(0)\, \exp\left\{\int_0^t \left(c_1(s) - \frac{1}{2}\,\sigma_1^2(s)\right) ds + \int_0^t \sigma_1(s)\, dW(s)\right\}\,.$$

General Solution

Let us return to the solution of equation (12.3). Now, analogously to the deterministic case (12.5), let us define $Z(t)$ from (12.8) as a homogeneous solution. At the end of the section we will establish the following proposition whilst applying two versions of Ito's lemma. Two interesting, alternative proofs will be given in exercise problems.

Proposition 12.2 (Solution of Linear SDE with Variable Coefficients) *The solution of (12.3) with in t continuous deterministic coefficients is*

$$X(t) = Z(t)\left[X(0) + \int_0^t \frac{c_2(s) - \sigma_1(s)\sigma_2(s)}{Z(s)}\, ds + \int_0^t \frac{\sigma_2(s)}{Z(s)}\, dW(s)\right] \tag{12.9}$$

with the homogeneous solution

$$Z(t) = \exp\left\{\int_0^t \left(c_1(s) - \frac{1}{2}\sigma_1^2(s)\right) ds + \int_0^t \sigma_1(s)\, dW(s)\right\}\,.$$

---

[3]The renaming justifies the assumption regarding the starting value. Consider

$$dX(t) = c_1(t)\, X(t)\, dt + \sigma_1(t) X(t)\, dW(t)\,, \quad X(0) \neq 0\,,$$

with a starting value different from zero, then by division one can normalize $Z(t) = X(t)/X(0)$.

For $\sigma_1(t) = \sigma_2(t) = 0$ we again obtain the known result of a deterministic differential equation, cf. (12.5).

Expected Value and Variance

The process defined by (12.3) reads in integral notation

$$X(t) = X(0) + \int_0^t (c_1(s)X(s) + c_2(s))\, ds + \int_0^t (\sigma_1(s)X(s) + \sigma_2(s))\, dW(s) .$$

Let us define the expectation function as

$$\mu_1(t) := \mathrm{E}(X(t)) ,$$

then it holds due to Propositions 8.2 (Fubini) and 10.3 that:

$$\begin{aligned}\mu_1(t) &= \mathrm{E}\left(X(0)\right) + \int_0^t (c_1(s)\,\mathrm{E}\left(X(s)\right) + c_2(s))\, ds + 0 \\ &= \mu_1(0) + \int_0^t (c_1(s)\,\mu_1(s) + c_2(s))\, ds .\end{aligned}$$

This corresponds exactly with the deterministic equation (12.4). Hence, the solution is known from (12.5) and one obtains the form given in Proposition 12.3. The derivation of an expression for the second moment is somewhat more complex,

$$\mu_2(t) := \mathrm{E}(X^2(t)) ,$$

see Problem 12.3.

Proposition 12.3 (Moments of the Solution of a Linear SDE) *Under the assumptions of Proposition 12.2 it holds that*

$$\mu_1(t) = z(t)\left[\mu_1(0) + \int_0^t \frac{c_2(s)}{z(s)}\, ds\right], \quad z(t) = \exp\left\{\int_0^t c_1(s)ds\right\} \tag{12.10}$$

and

$$\mu_2(t) = \zeta(t)\left[\mu_2(0) + \int_0^t \frac{\gamma_2(s)}{\zeta(s)}\, ds\right], \quad \zeta(t) = \exp\left\{\int_0^t \gamma_1(s)ds\right\} , \tag{12.11}$$

where

$$\gamma_1(t) = 2\,c_1(t) + \sigma_1^2(t) , \quad \gamma_2(t) = 2\,\left[c_2(t) + \sigma_1(t)\,\sigma_2(t)\right]\,\mu_1(t) + \sigma_2^2(t) .$$

Example 12.3 (Homogeneous Linear SDE (Constant Coefficients)) Since the works by Black and Scholes (1973) and Merton (1973) one assumes for the stock price $X(t)$ the model of a homogeneous linear SDE with constant coefficients (and starting value $X(0)$, cf. (1.3)):

$$dX(t) = c_1\, X(t)\, dt + \sigma_1\, X(t)\, dW(t)\,.$$

The solution resulting from (12.9) or rather from Proposition 12.2 is a geometric Brownian motion,

$$X(t) = X(0)\, \exp\left\{\left(c_1 - \frac{1}{2}\,\sigma_1^2\right)\, t + \sigma_1\, W(t)\right\}\,.$$

This process has already been discussed in Chap. 7. With the generally derived formulas we can now recheck the moment functions from (7.9). Proposition 12.3 yields (see Problem 12.4)

$$\mu_1(t) = \mu_1(0)\, \exp\left(c_1\, t\right)\,,$$

$$\mu_2(t) = \mu_2(0)\, \exp\left\{\left(2\, c_1 + \sigma_1^2\right)\, t\right\}\,.$$

Now, assume a fixed starting value $X(0)$. Then, it holds that

$$\mu_1(0) = X(0) \text{ and } \mu_2(0) = X^2(0)\,,$$

and hence

$$\begin{aligned}\operatorname{Var}(X(t)) &= \mu_2(t) - \mu_1^2(t)\\ &= X^2(0)\, \exp\left(2\, c_1\, t\right)\, \left(\exp\left(\sigma_1^2\, t\right) - 1\right)\,.\end{aligned}$$

With $X(0) = 1$, $\mu = c_1 - \frac{1}{2}\sigma_1^2$ and $\sigma = \sigma_1$ this corresponds to the notation from Chap. 7. The moments from (7.9) are indeed reproduced. ■

Inhomogeneous Linear SDE with Additive Noise

For $c_2(t) \neq 0$ the linear SDE is inhomogeneous. However, at the same time the increments of the Wiener process ("noise") enter into (12.3) additively, i.e. $\sigma_1(t) = 0$:

$$dX(t) = \left(c_1(t)\, X(t) + c_2(t)\right)\, dt + \sigma_2(t)\, dW(t)\,. \tag{12.12}$$

The solution results from (12.9) in Proposition 12.2 as

$$X(t) = z(t)\left[X(0) + \int_0^t \frac{c_2(s)}{z(s)}\,ds + \int_0^t \frac{\sigma_2(s)}{z(s)}\,dW(s)\right], \tag{12.13}$$

where $z(t)$ is a deterministic function:

$$z(t) = \exp\left\{\int_0^t c_1(s)\,ds\right\}.$$

Note that $X(t)$, as a Stieltjes integral, is a Gaussian process due to Proposition 9.2. Its moments result correspondingly (for a fixed starting value $X(0)$). We collect these results in a corollary.

Corollary 12.1 (Additive Noise) *The solution of (12.12) with in t continuous deterministic coefficients is given by (12.13). The starting value $X(0)$ be deterministic. Then, the process is Gaussian with:*

$$\mu_1(t) = z(t)\left[X(0) + \int_0^t \frac{c_2(s)}{z(s)}\,ds\right], \quad z(t) = \exp\left\{\int_0^t c_1(s)\,ds\right\}, \tag{12.14}$$

$$Var(X(t)) = z^2(t)\int_0^t \left(\frac{\sigma_2(s)}{z(s)}\right)^2 ds. \tag{12.15}$$

We illustrate the corollary with the following example.

Example 12.4 (Convergence to Zero) As a concrete example, let us consider the process given by the following equation with starting value 0:

$$dX(t) = -X(t)\,dt + \frac{dW(t)}{\sqrt{1+t}}, \quad t \geq 0,\ X(0) = 0.$$

This equation is a special case of additive noise as it holds that $\sigma_1(t) = 0$. The remaining coefficient restrictions read:

$$c_1(t) = -1, \quad c_2(t) = 0, \quad \sigma_2(t) = \frac{1}{\sqrt{1+t}}.$$

What behavior is to be expected intuitively for $X(t)$? The volatility term, $\sigma_2(t)$, tends to zero with t growing; does this also hold true for the variance of the process? And $c_1(t) = -1$ implies that positive values influence the change negatively and vice versa; does the process hence fluctuate around the expectation of zero? In fact, we can show that the process with vanishing variance varies around zero and therefore converges to zero.[4] For this we need the first two moments. These can be obtained from (12.14) and (12.15):

$$\mathrm{E}(X(t)) = 0\,,$$

$$\mathrm{Var}(X(t)) = e^{-2t}\int_0^t \frac{e^{2s}}{1+s}ds.$$

What can be learned from this about the variance for t increasing? In Problem 12.7 we show

$$\int_0^t \frac{e^{2s}}{1+s}\,ds \le \frac{e^{2t}}{1+t} - 1\,.$$

Then, this proves $\mathrm{Var}(X(t)) \to 0$ for $t \to \infty$. Hence, it is obvious that $X(t)$ indeed tends to zero in mean square. ■

Proof of Proposition 12.2

With the homogeneous solution

$$Z(t) := \exp\left\{\int_0^t \left(c_1(s) - \frac{1}{2}\sigma_1^2(s)\right)\,ds + \int_0^t \sigma_1(s)\,dW(s)\right\}\,,$$

of

$$dZ(t) = c_1(t)\,Z(t)\,dt + \sigma_1\,(t)Z(t)\,dW(t)$$

we define the two auxiliary quantities

$$X_1(t) := Z^{-1}(t)\,,\quad X_2(t) := X(t)\,.$$

Note that $X(t)$ is the process defined by (12.3) such that the differential of $X_2(t)$ is shown in (12.3). The proof suggested here uses the product rule for $d\,(X_1(t)\,X_2(t))$. However, for a valid application the derivation of $dX_1(t)$ is necessary as well.

As a first step we use Ito's lemma in the form of Proposition 11.1 in order to determine the differential for $X_1(t)$ with

$$g(Z) = Z^{-1}\,,\quad g'(Z) = -Z^{-2}\,,\quad g''(Z) = 2\,Z^{-3}\,.$$

[4]For this purpose we do not need an explicit expression for the process which, however, can be easily obtained from (12.13) with $X(0) = 0$:

$$X(t) = e^{-t}\int_0^t \frac{e^s}{\sqrt{1+s}}dW(s).$$

The differential becomes

$$\begin{aligned}
dX_1(t) &= g'(Z(t))\,dZ(t) + \frac{1}{2}\,g''(Z(t))\,\sigma_1^2(t)\,Z^2(t)\,dt \\
&= -\frac{c_1(t)\,Z(t)\,dt + \sigma_1(t)Z(t)\,dW(t)}{Z^2(t)} + \frac{2}{2}\,\frac{\sigma_1^2(t)\,Z^2(t)}{Z^3(t)}\,dt \\
&= \frac{\sigma_1^2(t) - c_1(t)}{Z(t)}\,dt - \frac{\sigma_1(t)}{Z(t)}\,dW(t) \\
&= \left(\sigma_1^2(t) - c_1(t)\right)\,X_1(t)\,dt - \sigma_1(t)\,X_1(t)\,dW(t)\,.
\end{aligned}$$

In a second step, we can now apply the stochastic product rule (see Eq. (11.2)) as an implication of Proposition 11.2 to the auxiliary quantities[5]:

$$\begin{aligned}
d\,(X_1(t)\,X_2(t)) = {} & X_1(t)\,dX_2(t) + X_2(t)\,dX_1(t) \\
& - (\sigma_1(t)\,X_2(t) + \sigma_2(t))\,\sigma_1(t)\,X_1(t)\,dt\,.
\end{aligned}$$

If the differentials $dX_1(t)$ and $dX_2(t)$ are plugged in, then some terms cancel each other such that it just remains:

$$\begin{aligned}
d\,(X_1(t)\,X_2(t)) &= X_1(t)\,\left(c_2(t)\,dt + \sigma_2(t)\,dW(t)\right) - \sigma_1(t)\,\sigma_2(t)X_1(t)\,dt \\
&= \frac{c_2(t) - \sigma_1(t)\sigma_2(t)}{Z(t)}\,dt + \frac{\sigma_2(t)}{Z(t)}\,dW(t)\,.
\end{aligned}$$

Due to

$$X_1(t)\,X_2(t) = \frac{X(t)}{Z(t)}\,,$$

it follows by integrating in a third step:

$$\frac{X(t)}{Z(t)} = \frac{X(0)}{Z(0)} + \int_0^t \frac{c_2(s) - \sigma_1(s)\sigma_2(s)}{Z(s)}\,ds + \int_0^t \frac{\sigma_2(s)}{Z(s)}\,dW(s)\,.$$

As $Z(0) = 1$, we have established (12.9) and hence completed the proof. Two alternative proofs, which are again based on Ito's lemma (or implications thereof), are covered as exercise problems.

[5]There is the risk of confusing the symbols σ_i, $i = 1, 2$, from Eq. (12.3) with the ones from Eq. (11.1). Note that the volatility of X_1 (i.e. "σ_1") is given by $-\sigma_1 X_1$ while the volatility term "σ_2" of X_2 just reads $\sigma_1 X_2 + \sigma_2$!

12.4 Numerical Solutions

Even if an analytical expression for the solution of a SDE is known, numerical solutions in the sense of simulated approximations to paths of a process are of interest. Such a simulation of a solution is, on the one hand, desired for reasons of a graphic illustration; on the other hand, in practice a whole family of numerical solutions is simulated in order to obtain a whole scenario of possible trajectories.

Euler Approximation

The interval $[0, T]$ from (12.1) is divided w.l.o.g. in n equidistant intervals of the length $\frac{T}{n}$. The corresponding partition reads:

$$0 = t_0 < t_1 = \frac{T}{n} < \ldots < t_i = \frac{iT}{n} < \ldots < t_n = T\,.$$

The theoretical solution from (12.2) of an arbitrary diffusion is now considered on the subinterval $[t_{i-1}, t_i]$, $i = 1, \ldots, n$:

$$X(t_i) = X(t_{i-1}) + \int_{t_{i-1}}^{t_i} \mu(s, X(s))\, ds + \int_{t_{i-1}}^{t_i} \sigma(s, X(s))\, dW(s)\,.$$

This allows for the following approximation[6] as it is discussed e.g. in Mikosch (1998):

$$\begin{aligned} X(t_i) &\approx X(t_{i-1}) \\ &\quad + \int_{t_{i-1}}^{t_i} \mu(t_{i-1}, X(t_{i-1}))\, ds + \int_{t_{i-1}}^{t_i} \sigma(t_{i-1}, X(t_{i-1}))\, dW(s)\,, \end{aligned}$$

which can also be written as:

$$\begin{aligned} X(t_i) &\approx X(t_{i-1}) \\ &\quad + \mu(t_{i-1}, X(t_{i-1}))\, \frac{T}{n} + \sigma(t_{i-1}, X(t_{i-1}))\, (W(t_i) - W(t_{i-1}))\,. \end{aligned}$$

For this purpose

$$\int_{t_{i-1}}^{t_i} ds = t_i - t_{i-1} = \frac{T}{n} \quad \text{and} \quad \int_{t_{i-1}}^{t_i} dW(s) = W(t_i) - W(t_{i-1})$$

[6]In the literature, one speaks of an Euler approximation. An improvement is known under the keyword Milstein approximation. In order to explain what is meant by "improve" in this case, one would have to become more involved in numerics.

was used. Hence, we have a recursive scheme. Given $X_0 = X(0)$ one calculates for $i = 1$:

$$X_1 = X_0 + \mu(0, X_0)\,\frac{T}{n} + \sigma(0, X_0)\left(W\left(\frac{T}{n}\right) - W(0)\right),$$

and in general, for $i = 1, \ldots, n$:

$$X_i = X_{i-1} + \mu\left(t_{i-1}, X_{i-1}\right)\frac{T}{n} + \sigma\left(t_{i-1}, X_{i-1}\right)\left(W\left(t_i\right) - W\left(t_{i-1}\right)\right). \qquad (12.16)$$

Thus we obtain n observations X_i (i.e. $n + 1$ observations including the starting value), with which a path of the continuous-time process $X(t)$ on $[0, T]$ is simulated. However, this simulation requires Gaussian pseudo random numbers in (12.16),

$$W\left(t_i\right) - W\left(t_{i-1}\right) = W\left(\frac{iT}{n}\right) - W\left(\frac{(i-1)\,T}{n}\right) \sim \text{ii}\mathcal{N}\left(0, \frac{T}{n}\right).$$

For this purpose, a series of stochastically independent $\mathcal{N}\left(0, \frac{T}{n}\right)$-distributed random variables ε_i need to be simulated instead of $W(t_i) - W(t_{i-1})$, in order to obtain a numerical solution X_i, $i = 1, \ldots, n$ for the diffusion $X(t)$ from (12.2) according to (12.16). Naturally, with n growing the approximation of a numerical solution improves.

12.5 Problems and Solutions

Problems

12.1 Show that the function given in (12.5) solves the deterministic differential equation (12.4). How does it look like in the case of constant coefficients?

12.2 Show that $Z(t)$ from (12.8) solves the homogeneous SDE (12.7) with $Z(0) = 1$.

Hint: See the text.

12.3 Prove (12.11) from Proposition 12.3.

Hint: Determine for $g\left(X(t)\right) = X^2(t)$ an expression with Ito's lemma.

12.4 Derive the expectation and the variance of the geometric Brownian motion with Proposition 12.3,

$$X(t) = X(0)\,\exp\left\{\left(c_1 - \frac{1}{2}\sigma_1^2\right)t + \sigma_1\,W(t)\right\}.$$

12.5 Determine the process $X(t)$ for which it holds that:

$$dX(t) = X(t)\, dW(t)\,, \quad X(0) = 1\,.$$

Hint: Proposition 12.2.

12.6 Find the solution of

$$dX(t) = \frac{-X(t)}{1+t}\, dt + \frac{dW(t)}{1+t}\,, \quad t \geq 0\,,$$

for $X(0) = 0$. Show that it tends to zero in mean square.
Hint: Proposition 12.2.

12.7 Show for the Example 12.4:

$$\int_0^t \frac{e^{2s}}{1+s}\, ds \;\leq\; \frac{e^{2t}}{1+t} - 1\,.$$

12.8 Determine the solution of

$$dX(t) = -\frac{X(t)}{1-t}\, dt + dW(t)\,, \quad 0 \leq t < 1\,,$$

with $X(0) = 0$. Show that $\mathrm{Var}(X(t)) = (1-t)\,t$ and hence that $X(t)$ tends to zero in mean square for $t \to 1$. (This reminds us of the Brownian bridge, see (7.6). In fact, the above SDE defines a Brownian bridge, cf. Grimmett & Stirzaker, 2001, p. 535.)

12.9 Prove Proposition 12.2 by directly applying Proposition 11.2.
Hint: Choose $g(X, Z) = X/Z$.

12.10 Prove Proposition 12.2 with the quotient rule from (11.5).
Hint: First derive the quotient rule for the one-factor case ($d = 1$) as a special case of (11.5).

Solutions

12.1 The solution from (12.5) reads

$$x(t) = z(t) \left[x(0) + \int_0^t \frac{c_2(s)}{z(s)} ds \right]$$

with

$$z(t) = \exp\left\{\int_0^t c_1(s)ds\right\}.$$

Let us define the square bracket as $b(t)$:

$$b(t) = \left[x(0) + \int_0^t \frac{c_2(s)}{z(s)}ds\right] = \frac{x(t)}{z(t)}$$

and

$$b'(t) = \frac{c_2(t)}{z(t)}.$$

The derivative of $z(t)$ is

$$z'(t) = c_1(t)\exp\left\{\int_0^t c_1(s)ds\right\} = c_1(t)z(t).$$

Hence, the product rule yields:

$$\begin{aligned} x'(t) &= z'(t)\,b(t) + z(t)\,b'(t) \\ &= c_1(t)\,z(t)\,\frac{x(t)}{z(t)} + z(t)\,\frac{c_2(t)}{z(t)} \\ &= c_1(t)\,x(t) + c_2(t)\,, \end{aligned}$$

which just corresponds to the claim.

In the case of constant coefficients, $x(t)$ from (12.5) with $z(t) = e^{c_1 t}$ becomes:

$$\begin{aligned} x(t) &= e^{c_1 t}[x(0) + c_2 \int_0^t e^{-c_1 s}ds] \\ &= e^{c_1 t}[x(0) - \frac{c_2}{c_1}(e^{-c_1 t} - 1)] \\ &= e^{c_1 t}\left(x(0) + \frac{c_2}{c_1}\right) - \frac{c_2}{c_1}. \end{aligned}$$

12.2 For $Z(t)$ from (12.7) it holds that

$$dZ(t) = \mu(t, Z(t))dt + \sigma(t, Z(t))dW(t)$$

with

$$\mu(t, Z(t)) = c_1(t)Z(t), \quad \sigma(t, Z(t)) = \sigma_1(t)Z(t).$$

Therefore, Proposition 11.1 yields:

$$dg(Z(t)) = g'(Z(t))dZ(t) + \frac{1}{2}g''(Z(t))\sigma^2(t, Z(t))dt.$$

With

$$g(x) = \log(x), \quad g'(x) = \frac{1}{x}, \quad g''(x) = -\frac{1}{x^2},$$

we hence obtain

$$\begin{aligned} d\log(Z(t)) &= \frac{\mu(t, Z(t))dt + \sigma(t, Z(t))dW(t)}{Z(t)} - \frac{1}{2}\frac{\sigma^2(t, Z(t))}{Z^2(t)}dt \\ &= c_1(t)dt + \sigma_1(t)dW(t) - \frac{\sigma_1^2(t)}{2}dt\,. \end{aligned}$$

Integration yields

$$\log(Z(t)) = \log(Z(0)) + \int_0^t \left(c_1(s) - \frac{\sigma_1^2(s)}{2}\right) ds + \int_0^t \sigma_1(s)dW(s).$$

Because of $Z(0) = 1$, the exponential function yields as desired:

$$Z(t) = \exp\left\{\int_0^t \left(c_1(s) - \frac{\sigma_1^2(s)}{2}\right) ds + \int_0^t \sigma_1(s)dW(s)\right\}.$$

12.3 Proposition 11.1 with

$$g(x) = x^2, \quad g'(x) = 2x, \quad g''(x) = 2$$

is applied to X^2 where the differential $dX(t)$ is given by Eq. (12.3). This leads to

$$\begin{aligned} dX^2(t) &= 2X(t)dX(t) + (\sigma_1(t)X(t) + \sigma_2(t))^2dt \\ &= [2X(t)\,(c_1(t)X(t) + c_2(t)) + (\sigma_1(t)X(t) + \sigma_2(t))^2]dt \\ &\quad +2X(t)(\sigma_1(t)X(t) + \sigma_2(t))dW(t). \end{aligned}$$

As an integral equation this reads as follows:

$$\begin{aligned} X^2(t) = X^2(0) &+ \int_0^t \left[2X(s)\,(c_1(s)X(s) + c_2(s)) + (\sigma_1(s)X(s) + \sigma_2(s))^2\right] ds \\ &+2\int_0^t X(s)(\sigma_1(s)X(s) + \sigma_2(s))dW(s). \end{aligned}$$

The expectation of the second integral is zero due to Proposition 10.3. The expectation of the first integral results due to Fubini's theorem as

$$\int_0^t \mathrm{E}\left[2c_1(s)X^2(s) + 2c_2(s)X(s) + \sigma_1^2(s)X^2(s) + 2\sigma_1(s)\sigma_2(s)X(s) + \sigma_2^2(s)\right] ds.$$

With the definition of $\mu_1(s)$ and $\mu_2(s)$ it thus follows that

$$\begin{aligned}\mu_2(t) = \mu_2(0) &+ \int_0^t \left[\left(2\,c_1(s) + \sigma_1^2(s)\right)\,\mu_2(s)\right] ds \\ &+ 2\int_0^t \left[\left(c_2(s) + \sigma_1(s)\,\sigma_2(s)\right)\,\mu_1(s) + \sigma_2^2(s)\right] ds\,.\end{aligned}$$

In differential notation this equation reads

$$d\mu_2(t) = \left(\gamma_1(t)\,\mu_2(t) + \gamma_2(t)\right)\,dt\,,$$

where the functions $\gamma_1(t)$ and $\gamma_2(t)$ in the proposition following (12.11) were adequately defined. Therefore, the second moment results as the solution of a deterministic differential equation of the form (12.4). Its solution can be found in (12.5). Hence, the proposition is verified.

12.4 The geometric Brownian motion solves the homogeneous linear equation with constant coefficients:

$$c_2(t) = \sigma_2(t) = 0, \quad c_1(t) = c_1 = \text{const.}, \quad \sigma_1(t) = \sigma_1 = \text{const.}$$

For a stochastic starting value it holds that:

$$\mu_1(0) = \mathrm{E}(X(0)), \quad \mu_2(0) = \mathrm{E}(X^2(0)).$$

By plugging in, Proposition 12.3 yields

$$\mu_1(t) = e^{c_1 t}[\mu_1(0) + 0] = \mu_1(0)\,\exp(c_1 t)\,.$$

With the definitions from Proposition 12.3 one determines

$$\gamma_1(t) = 2\,c_1 + \sigma_1^2 =: \gamma_1 \quad \text{and} \quad \gamma_2(t) = 0\,.$$

Hence, substitution yields

$$\mu_2(t) = e^{\gamma_1 t}[\mu_2(0) + 0] = \mu_2(0)\,\exp(\gamma_1 t)\,.$$

Thus, the variance is calculated as

$$\begin{aligned}\mathrm{Var}(X(t)) &= \mu_2(t) - \mu_1^2(t)\\ &= \mu_2(0)\,\exp\{(2c_1 + \sigma_1^2)t\} - \mu_1^2(0)\,\exp\{2c_1 t\}.\end{aligned}$$

12.5 The equation at hand is linear with

$$c_1(t) = c_2(t) = 0, \quad \sigma_2(t) = 0.$$

Furthermore, it holds that

$$\sigma_1(t) = 1.$$

Therefore, the solution deduced from Proposition 12.2 reads:

$$X(t) = Z(t)[X(0) + 0]$$

with

$$Z(t) = \exp\left\{-\frac{t}{2} + W(t)\right\}.$$

In particular for $X(0) = 1$ (analogously to $e^0 = 1$) it hence holds that:

$$X(t) = \exp\left\{W(t) - \frac{t}{2}\right\}.$$

Due to the analogy to $de^t = e^t\,dt$ with $e^0 = 1$ this process $X(t)$ is sometimes called "Ito exponential". It is noteworthy that the Ito exponential is not given by $\exp\{W(t)\}$.

12.6 The equation is linear, see Eq. (12.3), and corresponds to the special case of additive noise, cf. (12.12), i.e. $\sigma_1(t) = 0$. The remaining coefficients read:

$$c_1(t) = -\frac{1}{1+t}, \quad c_2(t) = 0, \quad \sigma_2(t) = \frac{1}{1+t}.$$

Hence, the expression for the solution from (12.13) yields with $X(0) = 0$:

$$X(t) = z(t)\int_0^t \frac{1}{(1+s)z(s)}dW(s),$$

where

$$\begin{aligned}z(t) &= \exp\left\{-\int_0^t (1+s)^{-1} ds\right\}\\ &= \exp\{-[\log(1+s)]_0^t\}\end{aligned}$$

$$= \exp\{-\log(1+t)+0\}$$

$$= \exp\left\{\log\left(\frac{1}{1+t}\right)\right\} = \frac{1}{1+t}.$$

Since $\sigma_2(t)/z(t) = 1$, the solution simplifies radically:

$$X(t) = \frac{1}{1+t}\int_0^t dW(s) = \frac{W(t)}{1+t}.$$

For this solution it obviously holds that:

$$\mathrm{E}(X(t)) = 0$$

$$\mathrm{Var}(X(t)) = \frac{t}{(1+t)^2} \to 0, \quad t \to \infty.$$

Thus, for $t \to \infty$ we have established

$$\mathrm{MSE}\,(X(t), 0) = \mathrm{E}[(X(t)-0)^2] \to 0$$

which just corresponds to the required convergence in mean square.

12.7 In order to prove the inequality claimed, we define the function

$$g(s) = \frac{1}{2}\frac{e^{2s}}{1+s}$$

with the derivative (quotient rule)

$$g'(s) = \frac{e^{2s}}{1+s} - \frac{1}{2}\frac{e^{2s}}{(1+s)^2}.$$

Let us call the integral of interest I,

$$I = \int_0^t \frac{e^{2s}}{1+s}\,ds.$$

Then it follows

$$I = \int_0^t g'(s)ds + \frac{1}{2}\int_0^t \frac{e^{2s}}{(1+s)^2}ds$$

$$= g(t) - g(0) + \frac{1}{2}\int_0^t \frac{e^{2s}}{(1+s)^2}ds$$

$$\leq g(t) - g(0) + \frac{1}{2}I,$$

where the bound follows from $(1+s) \leq (1+s)^2$. By rearranging terms it results that

$$I \leq 2(g(t) - g(0)).$$

With the definition of g it follows

$$I \leq \frac{e^{2t}}{(1+t)} - 1\,,$$

which was to be shown.

12.8 This is again an inhomogeneous linear equation with additive noise:

$$c_1(t) = -\frac{1}{1-t}, \quad c_2(t) = 0, \quad \sigma_1(t) = 0, \quad \sigma_2(t) = 1.$$

With $X(0) = 0$, $X(t)$ from (12.13) turns out to be:

$$X(t) = z(t) \int_0^t (z(s))^{-1}\, dW(s)$$

with

$$\begin{aligned} z(t) &= \exp\left\{-\int_0^t \frac{1}{1-s} ds\right\} \\ &= \exp\left\{[\log(1-s)]_0^t\right\} \\ &= 1 - t, \end{aligned}$$

i.e.

$$X(t) = (1-t) \int_0^t \frac{1}{1-s} dW(s).$$

Due to $c_2(t) = X(0) = 0$, (12.14) yields:

$$\mathrm{E}(X(t)) = 0.$$

Due to (12.15), the variance is:

$$\begin{aligned} \mathrm{Var}(X(t)) &= (1-t)^2 \int_0^t \frac{1}{(1-s)^2} ds \\ &= (1-t)^2 \left[\frac{1}{(1-s)}\right]_0^t \end{aligned}$$

$$= (1-t)^2 \left(\frac{1}{1-t} - 1 \right) =$$

$$= (1-t)^2 \frac{t}{1-t} = (1-t)t.$$

For $t \to 1$ the variance shrinks to zero such that it holds that $X(t)$ tends to 0 in mean square:

$$\mathrm{MSE}(X(t), 0) = \mathrm{Var}(X(t)) = \mathrm{E}[(X(t) - 0)^2] \to 0.$$

12.9 The key problem with this exercise is not to confuse the different meanings of $\sigma_i(t)$, $i = 1, 2$, in Proposition 11.2 and Eq. (12.3). Hence, firstly we adapt Proposition 11.2 for the processes $X(t)$ from (12.3) and $Z(t)$ from (12.7):

$$dX(t) = \mu_x(t)\, dt + \sigma_x(t)\, dW(t)\,,$$

$$\mu_x(t) = c_1(t)\, X(t) + c_2(t)\,, \quad \sigma_x(t) = \sigma_1(t)\, X(t) + \sigma_2(t)\,,$$

$$dZ(t) = \mu_z(t)\, dt + \sigma_z(t)\, dW(t)\,,$$

$$\mu_z(t) = c_1(t)\, Z(t)\,, \quad \sigma_z(t) = \sigma_1(t)\, Z(t)\,.$$

Following the hint, we consider

$$g(X, Z) = \frac{X}{Z} = X\, Z^{-1}$$

with

$$\frac{\partial g}{\partial X} = Z^{-1}\,, \quad \frac{\partial^2 g}{\partial X^2} = 0$$

$$\frac{\partial g}{\partial Z} = -XZ^{-2}\,, \quad \frac{\partial^2 g}{\partial Z^2} = 2XZ^{-3}\,, \quad \frac{\partial^2 g}{\partial X\, \partial Z} = -Z^{-2}\,.$$

Hence, Ito's lemma (Proposition 11.2) yields:

$$\begin{aligned}
d\left(\frac{X}{Z}\right) &= Z^{-1}\, dX - XZ^{-2}\, dZ + \frac{1}{2}\left[0 + 2XZ^{-3}\, \sigma_z^2\right] dt - Z^{-2} \sigma_x\, \sigma_z\, dt \\
&= Z^{-1}\,(c_1 X + c_2)\, dt + Z^{-1}\,(\sigma_1 X + \sigma_2)\, dW - XZ^{-2} c_1 Z\, dt \\
&\quad - XZ^{-2} \sigma_1 Z\, dW + XZ^{-3} \sigma_1^2 Z^2\, dt - Z^{-2}(\sigma_1 X + \sigma_2)\, \sigma_1 Z\, dt \\
&= \left(Z^{-1} c_2 - Z^{-1} \sigma_1 \sigma_2\right) dt + Z^{-1} \sigma_2\, dW.
\end{aligned}$$

Integration yields:

$$\frac{X(t)}{Z(t)} = \frac{X(0)}{Z(0)} + \int_0^t \frac{c_2(s) - \sigma_1(s)\sigma_2(s)}{Z(s)}\, ds + \int_0^t \frac{\sigma_2(s)}{Z(s)}\, dW(s)\,.$$

If this equation is multiplied by $Z(t)$, then, due to $Z(0) = 1$, one obtains the desired result.

12.10 We apply (11.5) with

$$X_1 = X \quad \text{and} \quad X_2 = Z\,,$$

where X and Z are driven by the same Wiener process, say $W_1 = W$. Then the one-factor quotient rule is obtained by the following restrictions:

$$\sigma_{11} = \sigma_x \quad \text{and} \quad \sigma_{12} = 0\,,$$

$$\sigma_{21} = \sigma_z \quad \text{and} \quad \sigma_{22} = 0\,.$$

For this purpose, σ_x and σ_z were defined in the previous problem. Then, the one-factor quotient rule yields:

$$\begin{aligned}
d\left(\frac{X}{Z}\right) &= \frac{ZdX - XdZ}{Z^2} + \frac{XZ^{-1}\,\sigma_z^2 - \sigma_x\sigma_z}{Z^2}\, dt \\
&= \frac{Z(c_1X + c_2)\, dt + Z(\sigma_1 X + \sigma_2)\, dW - Xc_1Z\, dt - X\sigma_1 Z\, dW}{Z^2} \\
&\quad + \frac{XZ^{-1}\sigma_1^2 Z^2 - (\sigma_1 X + \sigma_2)\sigma_1 Z}{Z^2}\, dt \\
&= \frac{Z(c_2 - \sigma_1\sigma_2)}{Z^2}\, dt + \frac{Z\sigma_2}{Z^2}\, dW \\
&= \frac{(c_2 - \sigma_1\sigma_2)}{Z}\, dt + \frac{\sigma_2}{Z}\, dW\,.
\end{aligned}$$

As before, we obtain the desired result by integration and multiplication by $Z(t)$.

References

Black, F., & Scholes, M. (1973). The pricing of options and corporate liabilities. *The Journal of Political Economy, 81*, 637–654.

Grimmett, G. R., & Stirzaker, D. R. (2001). *Probability and random processes* (3rd ed.). Oxford: Oxford University Press.

Merton, R. C. (1973). Theory of rational option pricing. *The Bell Journal of Economics and Management Science, 4*, 141–183.

Mikosch, Th. (1998). *Elementary stochastic calculus with finance in view*. Singapore: World Scientific Publishing.

Øksendal, B. (2003). *Stochastic differential equations: An introduction with applications* (6th ed.). Berlin/New York: Springer.

Interest Rate Models 13

13.1 Summary

The results from the previous chapter will be applied to stochastic differential equations that were suggested in the literature for modeling interest rate dynamics. However, we do not model yield curves with various maturities, but consider the model for one interest rate only driven by one Wiener process (one-factor model). The next section starts with the general Ornstein-Uhlenbeck process which has the drawback of allowing for negative values. Subsequently, we discuss linear models for which negativity is ruled out. Finally, a class of nonlinear models will be considered.

13.2 Ornstein-Uhlenbeck Process (OUP)

We have already encountered the standard OUP in the chapter on Stieltjes integrals. Now, we discuss the general case, which has served as an interest rate model in the literature on finance.

Vasicek

We now assume constant coefficients for the inhomogeneous linear SDE with additive noise in (12.12):

$$c_1(t) = c_1 = const,\ c_2(t) = c_2 = const,\ \sigma_2(t) = \sigma_2 = const,\ \sigma_1(t) = 0\,.$$

This defines the general **Ornstein-Uhlenbeck process**,

$$dX(t) = (c_1 X(t) + c_2)\, dt + \sigma_2\, dW(t)\,. \tag{13.1}$$

U. Hassler, *Stochastic Processes and Calculus*, Springer Texts in Business and Economics, DOI 10.1007/978-3-319-23428-1_13

According to Corollary 12.1 the solution is

$$X(t) = e^{c_1 t}\left[X(0) - \frac{c_2}{c_1}\left(e^{-c_1 t} - 1\right) + \int_0^t \sigma_2\, e^{-c_1 s}\, dW(s)\right]. \tag{13.2}$$

In particular for $c_2 = 0$ and $\sigma_2 = 1$ we obtain the standard OUP with $X(0) = 0$ from (9.3) in Sect. 9.4. The following equation sheds additional light on the OUP:

$$dX(t) = c_1\,(X(t) - \mu)\,dt + \sigma_2\,dW(t)\,, \quad \text{with } \mu := -\frac{c_2}{c_1}. \tag{13.3}$$

In this manner, Vasicek (1977) modeled the interest rate dynamics, cf. (1.7). Due to (13.2), the solution of this interest rate equation reads:

$$X(t) = e^{c_1 t}\left[X(0) + \mu\left(e^{-c_1 t} - 1\right) + \int_0^t \sigma_2\, e^{-c_1 s}\, dW(s)\right].$$

Setting the starting value to μ, $X(0) = \mu$, then one obtains the form immediately corresponding to the standard OUP (9.3),

$$X(t) = \mu + e^{c_1 t}\int_0^t \sigma_2\, e^{-c_1 s}\, dW(s)\,,$$

with expectation μ. From (12.14) and (12.15) we obtain for an arbitrary fixed starting value $X(0)$:

$$\mu_1(t) = \mathrm{E}(X(t)) = e^{c_1 t} X(0) + \mu\left(1 - e^{c_1 t}\right), \tag{13.4}$$

$$\mathrm{Var}(X(t)) = \frac{\sigma_2^2}{-2\,c_1}\left(1 - e^{2\,c_1 t}\right). \tag{13.5}$$

Particularly the mean value function $\mu_1(t)$ results as a convex combination of the long-term mean value μ and the starting value $X(0)$. For $c_1 < 0$, these moments tend to a fixed value and the process can be understood as asymptotically stationary:

$$\mu_1(t) \to \mu \ \text{ for } c_1 < 0\,,$$

$$\mathrm{Var}(X(t)) \to \frac{\sigma_2^2}{-2\,c_1} \ \text{ for } c_1 < 0\,,$$

where the limits are taken as $t \to \infty$. Processes with this property are also called "mean-reverting". The adjustment parameter $c_1 < 0$ measures the "speed of mean-reversion", i.e. the strength of adjustment: The smaller (more negative) c_1, the more strongly $dX(t)$ reacts as a function of the deviation of $X(t)$ from μ. If $X(t) > \mu$, then $c_1 < 0$ causes $c_1(X(t) - \mu)$ to have a negative impact on $X(t)$ such that the process tends to decrease and therefore approaches μ; conversely, it holds that for $X(t) < \mu$ the process experiences a positive impulse. Furthermore, one observes the distinct influence of the parameter c_1 on the (asymptotic) variance: The smaller the negative c_1, the smaller is the asymptotic expression; however, for a negative c_1 near zero the variance becomes large and the OUP loses the property of "mean reversion" for $c_1 = 0$.

Determining the autocovariance function for $c_1 < 0$ is also useful. We denote it by $\gamma(t, t+h)$ at lag h:

$$
\begin{aligned}
\gamma(t, t+h) &= \mathrm{E}\left[(X(t) - \mu_1(t))(X(t+h) - \mu_1(t+h))\right], \quad h \geq 0, \\
&= \mathrm{E}\left[e^{c_1 t}\sigma_2 \int_0^t e^{-c_1 s}\, dW(s)\, e^{c_1(t+h)}\sigma_2 \int_0^{t+h} e^{-c_1 s}\, dW(s)\right] \\
&= \sigma_2^2\, \mathrm{E}\left[X_{c_1}(t)\, X_{c_1}(t+h)\right],
\end{aligned}
$$

where $X_{c_1}(t)$ is the standard OUP with zero expectation. From Proposition 9.4 we hence adopt

$$
\begin{aligned}
\gamma(t, t+h) &= \sigma_2^2\, e^{c_1 h}\, \frac{e^{2c_1 t} - 1}{2c_1} \\
&\to -\frac{\sigma_2^2 e^{c_1 h}}{2c_1}, \quad t \to \infty.
\end{aligned}
$$

Thus, for a large t there results an autocovariance function only depending on the temporal distance h. All in all, this is why the OUP with $c_1 < 0$ can be labeled as asymptotically ($t \to \infty$) weakly stationary.

Simulations

In the following, processes with $T = 20$ and $n = 1000$ are simulated. For reasons of graphical comparability, the same WP is always assumed, i.e. the 1000 random variables filtered by the recursion (12.16) are always the same ones.

First, we examine the impact of the adjustment parameter c_1 on the behavior of the OUP. In Fig. 13.1, $\sigma_2 = 0.01$. As we want to think about interest rates when looking at this figure, expectation and starting value are chosen to be $\mu = 5\,(\%)$. Here, it is obvious that the solid line deviates less strongly from the expected value for $c_1 = -0.9$ and is thus "more stationary" than the dashed graph for $c_1 = -0.1$. This is evident as the smaller (i.e. the more negative) c_1, the smaller is the variance $\sigma_2^2/(-2c_1)$ for t growing.

In the second figure, we have an OUP with the same parameter set-up for $c_1 = -0.9$, however, with a starting value different from $\mu = 5$, $X(0) = 5.1$. Furthermore, the expected value function $\mu_1(t)$ is given and one can observe how rather rapidly it approaches the value $\mu = 5$ (Fig. 13.2).

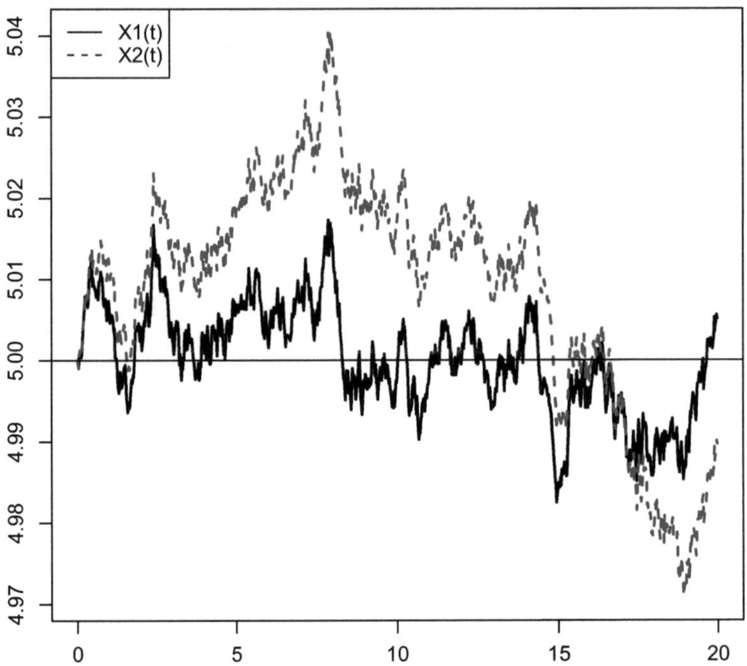

Fig. 13.1 OUP for $c_1 = -0.9$ (X_1) and $c_1 = -0.1$ (X_2) ($X(0) = \mu = 5$, $\sigma_2 = 0.01$)

Despite the convenient property of mean reversion, the OUP is only partly suitable for interest rate modeling: Note that the process takes on negative values with a positive probability. This is due to the fact that the OUP, as a Stieltjes integral, is Gaussian:

$$X(t) \sim \mathcal{N}\left(\mu_1(t), \operatorname{Var}(X(t))\right) .$$

Subsequent to Vasicek (1977), interest rate models without this drawback have been discussed.

13.3 Positive Linear Interest Rate Models

For now we stay in the class of linear SDEs, however, we restrict the discussion to the case in which positivity (more precisely: nonnegativity) is guaranteed.

Sufficient Condition

A sufficient condition for a positive evolution of the solution of a linear SDE is easy to be specified. For this purpose we naturally consider the general solution from

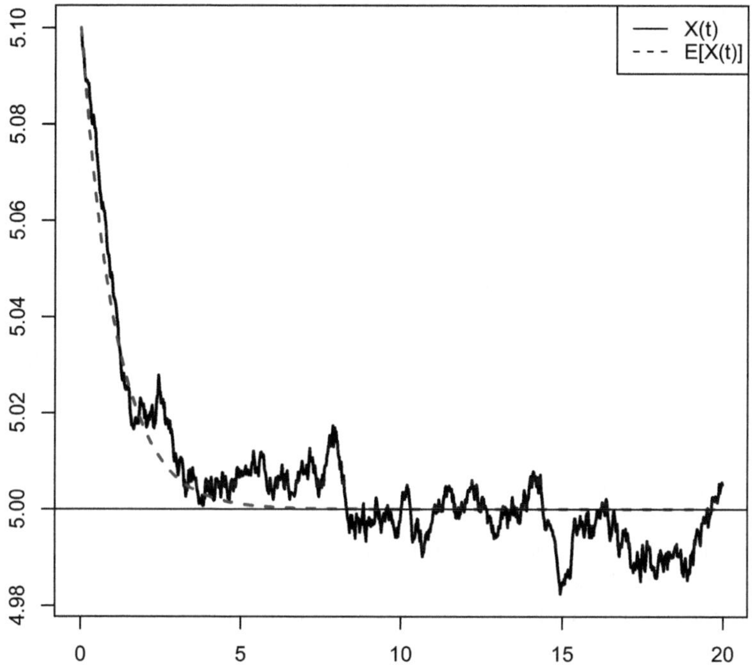

Fig. 13.2 OUP for $c_1 = -0.9$ and Starting Value $X(0) = 5.1$ including Expected Value Function ($\mu = 5, \sigma_2 = 0.01$)

Proposition 12.2. Note that $Z(t)$, as an exponential function, is always positive. With the restriction $\sigma_2(t) = 0$ the following diffusion is obtained:

$$X(t) = Z(t)\left[X(0) + \int_0^t \frac{c_2(s)}{Z(s)} ds\right]. \tag{13.6}$$

With a positive starting value and $c_2(t) \geq 0$, a positive evolution of $X(t)$ is ensured. The models in this section are of the form (13.6).

Dothan

Let us consider a special case more extensively. Dothan (1978) suggested for the interest rate dynamics a special case of the geometric Brownian motion:

$$dX(t) = \sigma_1 X(t)\, dW(t), \quad X(0) > 0.$$

With $c_2(t) = \sigma_2(t) = 0$ it holds in this case that

$$X(t) = X(0) \exp\left\{\left(-\frac{1}{2}\sigma_1^2\right) t + \sigma_1 W(t)\right\},$$

and hence, the interest rate $X(t)$ can in fact not get negative. Not $X(t)$ follows a Gaussian distribution but $\log(X(t))$ does. Furthermore, in Example 12.3 we have determined the moments (for a fixed starting value):

$$\mu_1(t) = X(0) \text{ and } \operatorname{Var}(X(t)) = X^2(0)\left(\exp(\sigma_1^2 t) - 1\right).$$

Thus, the variance of the process increases exponentially which is why the model may not be satisfactory for interest rates.

Brennan-Schwartz

Brennan and Schwartz (1980) suggested another attractive variant. It consists of a combination of Vasicek (1977) and Dothan (1978); we choose the drift component just as for the Ornstein-Uhlenbeck process and the volatility just as for the geometric Brownian motion:

$$dX(t) = c_1 (X(t) - \mu)\, dt + \sigma_1 X(t)\, dW(t), \quad X(0) = \mu > 0, \tag{13.7}$$

where, for simplicity, the starting value is set equal to μ. For $c_1 < 0$ it holds that $c_2 = -c_1\mu > 0$ such that we have indeed a positive interest rate dynamics. For this model one can show (see Problem 13.4) that the expected value results just as for Dothan (1978),

$$\mu_1(t) = \mu = X(0),$$

while it holds for the variance:

$$\operatorname{Var}(X(t)) = \frac{\mu^2 \sigma_1^2}{2c_1 + \sigma_1^2}\left(\exp\left((2c_1 + \sigma_1^2)\, t\right) - 1\right).$$

If $c_1 < -\sigma_1^2/2$ (i.e. $2c_1 + \sigma_1^2 < 0$), then it holds that the variance tends to a fixed positive value ($t \to \infty$):

$$\operatorname{Var}(X(t)) \to -\frac{\mu^2 \sigma_1^2}{2c_1 + \sigma_1^2} \quad \text{for } c_1 < -\frac{\sigma_1^2}{2}.$$

If the volatility parameter σ_1 is relatively small compared to the absolute value of the negative adjustment parameter c_1, then the model (13.7) provides a process with a fixed expected value and an asymptotically constant variance. Again, one speaks of "mean reversion". Interestingly, the variance is not only influenced by σ_1 and c_1 in an obvious manner: The greater σ_1, the greater $\text{Var}(X(t))$, and the greater c_1 in absolute value, the more strongly or the faster the adjustment happens (and the smaller is the variance). The parameter $\mu > 0$ from the drift function as well has a positive effect on the variance. Intuitively, this is obvious: The smaller μ (i.e. the closer to zero), the lesser $X(t)$ can spread as the process does not get negative; conversely, it holds that the scope for the variance between the zero line and μ increases with μ growing.

Simulations

Again, processes with $T = 20$ and $T = 1000$ were simulated. For reasons of graphical comparability, the same WP as in the previous section is assumed, i.e. the 1000 random variables being filtered by the recursion (12.16) are identical.

In Fig. 13.3 it is obvious how the variance of the geometric Brownian motion suggested by Dothan (1978) increases with the parameter σ_1. Although the expected value is constant and equal to the starting value, long periods are possible and probable in which the process does not cross the expected value. A more plausible interest rate dynamics can be observed in Fig. 13.4 for two values of the adjustment

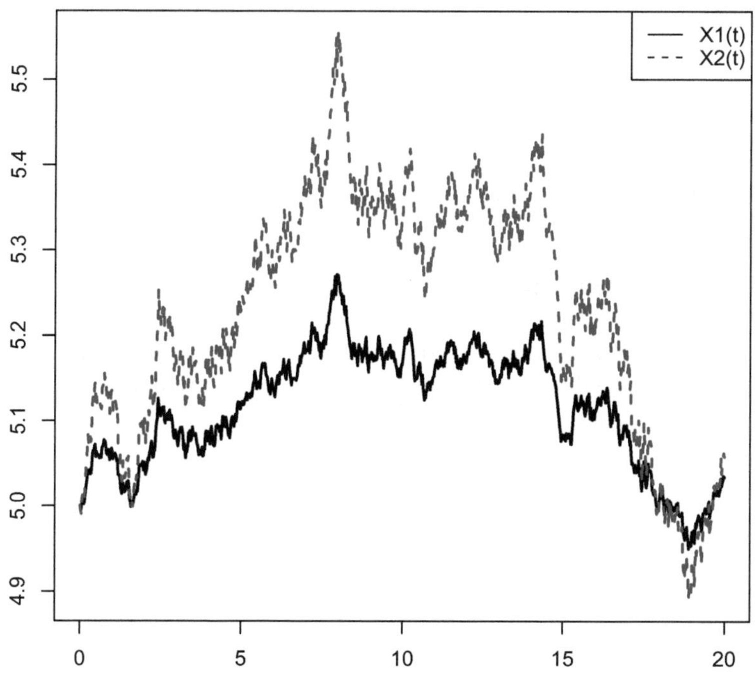

Fig. 13.3 Dothan for $\sigma_1 = 0.01$ (X_1) and $\sigma_1 = 0.02$ (X_2) ($X(0) = \mu = 5$)

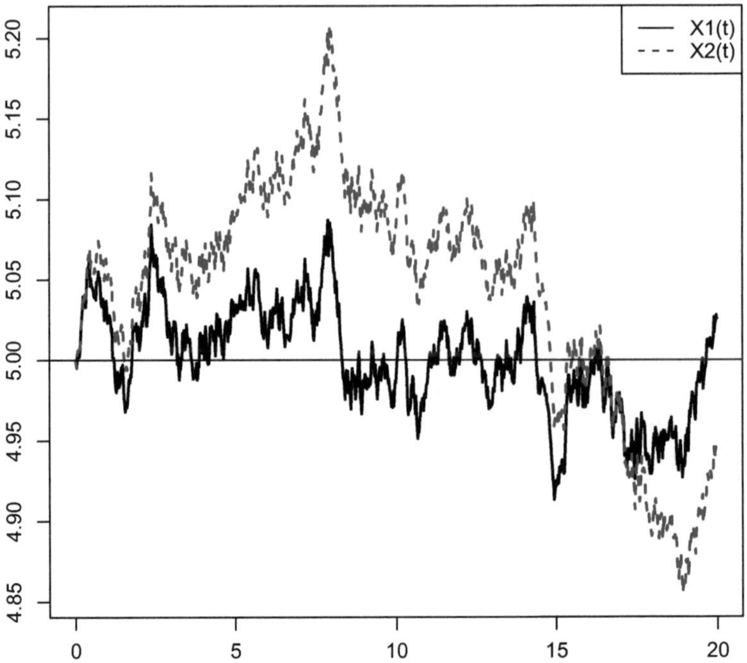

Fig. 13.4 Brennan-Schwartz for $c_1 = -0.9$ (X_1) and $c_1 = -0.1$ (X_2) ($X(0) = \mu = 5$, $\sigma_1 = 0.01$)

parameter c_1. The values are chosen small enough (relative to σ_1), such that the variance remains bounded and converges to a fixed value. It is obvious: The greater c_1 in absolute value, the smaller the variance.

13.4 Nonlinear Models

Chan, Karolyi, Longstaff, and Sanders (1992) [in short: CKLS] considered the following class of nonlinear equations for modeling short-term interest rates which is covered in this section:

$$dX(t) = c_1\,(X(t) - \mu)\,dt + \sigma\,X^{\gamma}(t)\,dW(t)\,, \quad \mu > 0\,,\ 0 \le \gamma \le 1\,. \tag{13.8}$$

Thus, the modeling of the drift component always corresponds to the one by Vasicek (1977). The OUP from (13.1) just results for $\gamma = 0$ while $\gamma = 1$ leads to the just discussed process from (13.7). Noninteger values of γ in between provide a nonlinear interest rate dynamics. The process from (13.8) is sometimes also called model with constant elasticity as it holds for the elasticity with the derivative of the volatility $\sigma X^\gamma(t)$ with respect to X that:

$$\frac{d(\sigma X^\gamma)}{dX}\frac{X}{\sigma X^\gamma} = \gamma\,.$$

In order to show the interpretation of γ as an elasticity, we consider a discretization of the CKLS process as for the computer simulation. For this purpose, we define for discrete steps of the length 1, $t = 1, 2, \ldots, T$:

$$x_t := X(t)\,, \quad \varepsilon_t := \Delta W(t) = W(t) - W(t-1)\,.$$

The discrete-time version of (13.8) hence reads

$$\Delta x_t = c_1\,(x_{t-1} - \mu) + \sigma x_{t-1}^\gamma \varepsilon_t\,, \quad \varepsilon_t \sim \text{ii}\mathcal{N}(0,1)\,,$$

or

$$x_t = x_{t-1} + c_1\,(x_{t-1} - \mu) + \sigma x_{t-1}^\gamma \varepsilon_t\,.$$

For the conditional variance it holds:

$$\text{Var}(x_t|x_{t-1}) = \sigma^2 x_{t-1}^{2\gamma}\,.$$

Correspondingly, it holds for the conditional standard deviation that e.g. a doubling of x_{t-1} leads to a multiplication by the factor 2^γ:

$$\begin{aligned}\sqrt{\text{Var}\,(x_t|\tilde{x}_{t-1})} &= \sigma\tilde{x}_{t-1}^\gamma \quad \text{for} \quad \tilde{x}_{t-1} = 2\,x_{t-1}\,,\\ &= \sigma\,2^\gamma x_{t-1}^\gamma\\ &= 2^\gamma\sqrt{\text{Var}\,(x_t|x_{t-1})}\,.\end{aligned}$$

Two simulated paths of the CKLS model are depicted in Fig. 13.5. They only differ in the elasticity γ. It is not surprising that the deviations from μ get greater with a greater γ.

Cox, Ingersoll & Ross [CIR]

A particularly prominent representative of (13.8) is obtained for $\gamma = 0.5$. This model is often used following Cox, Ingersoll, and Ross (1985):

$$dX(t) = c_1\,(X(t) - \mu)\,dt + \sigma\sqrt{X(t)}\,dW(t)\,, \quad \mu > 0\,, c_1 < 0\,. \tag{13.9}$$

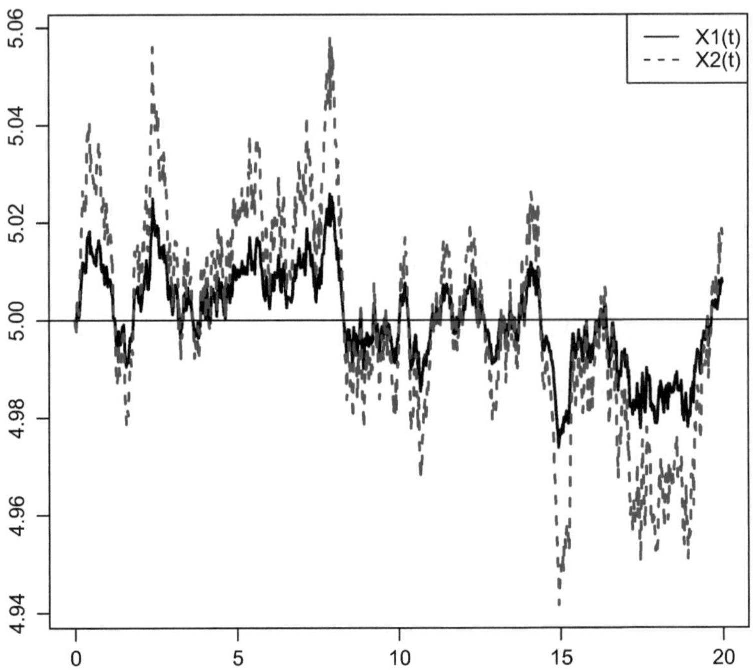

Fig. 13.5 CKLS with $\gamma = 0.25$ (X_1) and $\gamma = 0.75$ (X_2) for $c_1 = -0.9$ ($X(0) = \mu = 5$, $\sigma = 0.01$)

The conditional standard deviation is modeled as a square root which is why one also speaks of (13.9) as a "square root process". Consequently, the conditional variance of the increments is proportional to the level of the process.

For this nonlinear SDE it can be formally shown, which is also intuitive: If $X(t)$ (starting from a positive starting value $X(0) > 0$) takes on the value zero, then the variance is zero as well, but the change $dX(t)$ gets a positive impulse of the strength $-c_1\,\mu$ such that the process is reflected on the zero line for $\mu > 0$. Insofar the square root process overcomes the deficiency of the OUP as an interest rate model. However, an analytical representation of the solution of (13.9) is not known.

Already for the ordinary square root process from (13.9) with $\sigma(t,x) = \sigma\sqrt{x}$ the condition of existence **(E1)** from Proposition 12.1 is not fulfilled anymore as the derivative at zero does not exist. Fortunately, there are weaker conditions ensuring the existence of a solution of (13.9) – however, they do not guarantee the finiteness of the first two moments anymore. In order to show that finite moments exist up to the second order, we would need more fundamental arguments. Instead, we start with calculating the moments (finiteness assumed).

For reasons of simplicity, assume a fixed starting value equal to μ in the following: $X(0) = \mu$. Then we obtain (see Problem 13.5), as for the OUP, on average

$$\mu_1(t) = \mathrm{E}(X(t)) = e^{c_1 t}X(0) + \mu\left(1 - e^{c_1 t}\right) = \mu\,.$$

Under the assumption on the starting value $X(0) = \mu$, for the second moment we obtain (cf. Problem 13.6)

$$\mu_2(t) = \mu^2 - \frac{\sigma^2 \mu}{2 c_1} (1 - e^{2 c_1 t}),$$

from which it immediately follows for the variance

$$\text{Var}(X(t)) = \frac{\sigma^2 \mu}{-2 c_1} (1 - e^{2 c_1 t}) \rightarrow \frac{\sigma^2 \mu}{-2 c_1}, \quad t \rightarrow \infty.$$

The asymptotic variance for $t \rightarrow \infty$ hence coincides with the one of the OUP if $\mu = 1$; for $\mu < 1$ it turns out to be smaller (as the process is reflected on the zero line and therefore varies in a narrow band) while it is obviously greater for $\mu > 1$. The border case $\mu = 0$ makes sense as well: Here, the asymptotic variance is zero as, sooner or later, the process is absorbed by the zero line.

For Fig. 13.6 an OUP with $c_1 = -0.9$ and $\sigma_2 = 0.01$ was simulated but the expected value of 5 % is now written as 0.05. In the example it becomes clear that the OUP can definitely become negative. In comparison, we observe a numerical solution of the corresponding square root process from (13.9) with the

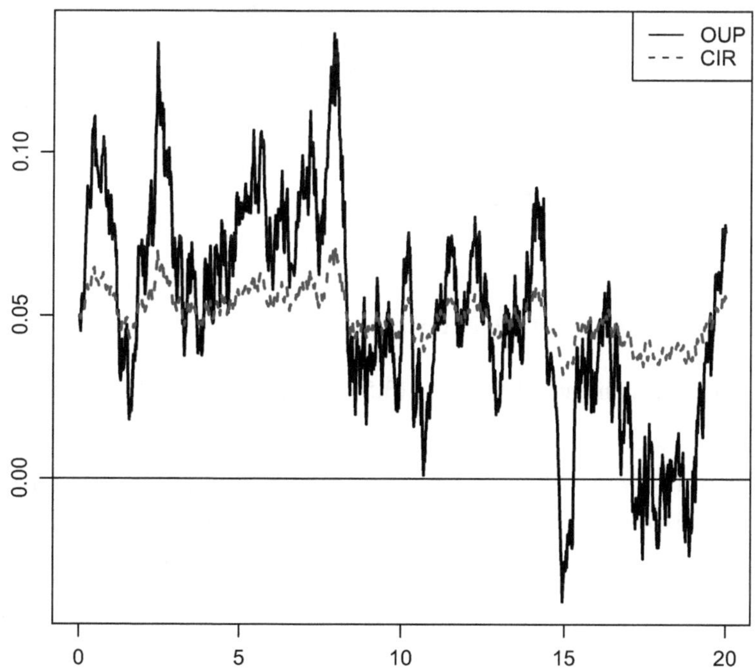

Fig. 13.6 OUP and CIR for $c_1 = -0.9$ $(X(0) = \mu = 0.05, \sigma = \sigma_2 = 0.01)$

same volatility parameter and the same drift component. The picture confirms the theoretical considerations: The process exhibits a smaller variance and does not get negative.

Further Models and Parameter Estimation

Marsh and Rosenfeld (1983) mention the variant with $\mu = 0$ as a borderline case of (13.8). Cox, Ingersoll, and Ross (1980) consider a version with $\gamma > 1$ for a special investigation:

$$dX(t) = \sigma X^{3/2}(t)\, dW(t)\,.$$

Finally, some models are applied which leave the framework of CKLS from (13.8) entirely, e.g. Constantinides and Ingersoll (1984) with

$$dX(t) = c\, X^2(t)\, dt + \sigma X^{3/2}(t)\, dW(t)\,,$$

where both drift and volatility are nonlinear.

Given the copious possibilities for specifying a diffusion process, it is not surprising that it has been tried to, first, estimate unknown parameters and second to statistically discriminate between the different model classes. Beside the work by CKLS, the papers by Broze, Scaillet, and Zakoïan (1995) and Tse (1995) should be mentioned. As a first introduction to the topic of estimation of diffusion parameters, the corresponding chapter by Gourieroux and Jasiak (2001) is recommended.

13.5 Problems and Solutions

Problems

13.1 Derive the solution (13.2) of Eq. (13.1).

13.2 Derive the moments, (13.4) and (13.5), of the OUP (a fixed starting value $X(0)$ assumed).

13.3 Discuss Eq. (13.1) for $c_1 = 0$ as a special case of the OUP (a proposal by Merton, 1973). For this purpose, consider the solution, the expected value and the variance for $c_1 \to 0$, if necessary with L'Hospital's rule. (You should be familiar with the results. By which name do you know the process as well?)

13.4 Consider now, as a combination of the interest models by Vasicek (1977) and Dothan (1978), the process from (13.7) by Brennan and Schwartz (1980),

$$dX(t) = c_1\,(X(t) - \mu)\, dt + \sigma_1 X(t)\, dW(t)\,, \quad \mu = X(0) > 0\,,$$

particularly with the starting value $X(0) = \mu$. Determine expectation and variance. How do these behave for $t \to \infty$ if it holds that $2\,c_1 < -\sigma_1^2$?

13.5 Consider the square root process (13.9) by Cox et al. (1985). Under the assumption $X(0) = \mu$ for the starting value, derive an expression for the expected value.

13.6 Again, consider the square root process (13.9) by Cox et al. (1985). Under the assumption $X(0) = \mu$ for the starting value, derive an expression for the variance.

Solutions

13.1 Equation (13.1) is a special case of (12.12) with constant coefficients. Hence, the solution results from (12.13) with

$$z(t) = \exp\left\{\int_0^t c_1 ds\right\} = e^{c_1 t}.$$

From this it follows

$$\begin{aligned}\int_0^t \frac{c_2}{z(s)}\, ds &= c_2 \int_0^t e^{-c_1 s}\, ds \\ &= -\frac{c_2}{c_1}\, e^{-c_1 s}\Big|_0^t \\ &= -\frac{c_2}{c_1}\,(e^{-c_1 t} - 1).\end{aligned}$$

Thus, from (12.13) we obtain the desired result:

$$X(t) = e^{c_1 t}\left[X(0) - \frac{c_2}{c_1}\,(e^{-c_1 t} - 1) + \int_0^t \sigma_2\, e^{-c_1 s}\, dW(s)\right].$$

13.2 The expected value function is determined from (12.14):

$$\begin{aligned}\mu_1(t) &= e^{c_1 t}\left[X(0) + \int_0^t c_2\, e^{-c_1 s}\, ds\right] \\ &= e^{c_1 t}\left[X(0) - \frac{c_2}{c_1}\,(e^{-c_1 t} - 1)\right].\end{aligned}$$

With $\mu = -\frac{c_2}{c_1}$ the formula from (13.4) results.

The variance expression follows from (12.15):

$$\begin{aligned}\mathrm{Var}(X(t)) &= \left(e^{c_1 t}\right)^2 \int_0^t \left(\frac{\sigma_2}{e^{c_1 s}}\right)^2 ds \\ &= e^{2c_1 t}\,\sigma_2^2 \int_0^t e^{-2c_1 s}\, ds \\ &= e^{2c_1 t}\,\sigma_2^2 \left(\frac{-1}{2c_1}\right) e^{-2c_1 s}\Big|_0^t \\ &= -e^{2c_1 t}\,\frac{\sigma_2^2}{2c_1}\left(e^{-2c_1 t} - 1\right).\end{aligned}$$

This expression coincides with (13.5).

13.3 L'Hospital's rule provides for $c_1 \to 0$:

$$\lim_{c_1 \to 0} \frac{e^{-c_1 t} - 1}{c_1} = \lim_{c_1 \to 0} \frac{-t\, e^{-c_1 t}}{1} = -t.$$

Hence, $X(t)$ from (13.2) merges for $c_1 \to 0$ into

$$\begin{aligned}X(t) &= \left[X(0) + c_2\, t + \int_0^t \sigma_2\, dW(s)\right] \\ &= X(0) + c_2\, t + \sigma_2\, W(t).\end{aligned}$$

Equation (13.4) is not suitable for determining the expected value as $\mu = \frac{-c_2}{c_1}$ is not defined for $c_1 \to 0$. Instead, one directly obtains (for X(0) fixed):

$$\mu_1(t) = \mathrm{E}(X(t)) = X(0) + c_2\, t + 0.$$

This linear growth is distinctive for the Brownian motion with drift, cf. Chap. 7.

In order to determine the variance from (13.5), it is again argued with L'Hospital's rule:

$$\lim_{c_1 \to 0} \frac{1 - e^{2c_1 t}}{2\, c_1} = \lim_{c_1 \to 0} \frac{-2t\, e^{2c_1 t}}{2} = -t.$$

Therefore it holds that

$$\lim_{c_1 \to 0} \mathrm{Var}(X(t)) = \sigma_2^2\, t = \mathrm{Var}(\sigma_2\, W(t)).$$

This is the familiar variance of a Brownian motion with drift. Indeed, a Brownian motion with drift is the same as the process resulting from the OUP for $c_1 = 0$ or $c_1 \to 0$.

13.4 This equation is linear and does not belong to the category "additive noise". It is rather a special case of (12.3) with

$$c_1(t) = c_1, \quad c_2(t) = -\mu c_1, \quad \sigma_1(t) = \sigma_1, \quad \sigma_2(t) = 0.$$

With

$$z(t) = e^{c_1 t}$$

due to Proposition 12.3 the expected value is given by

$$\begin{aligned}
\mathrm{E}(X(t)) &= e^{c_1 t}\left[\mathrm{E}(X(0)) - \mu c_1 \int_0^t e^{-c_1 s} ds\right] \\
&= e^{c_1 t}[\mathrm{E}(X(0)) + \mu e^{-c_1 t} - \mu] \\
&= \mu + e^{c_1 t}[\mathrm{E}(X(0)) - \mu].
\end{aligned}$$

Hence, for $c_1 < 0$ it holds:

$$\mathrm{E}(X(t)) \to \mu, \quad t \to \infty.$$

For the second moment, we determine from (12.11):

$$\mu_2(t) = \exp\{(2c_1 + \sigma_1^2)t\}\left[\mu_2(0) - \int_0^t \frac{2c_1 \mu \mu_1(s)}{\exp\{(2c_1 + \sigma_1^2)s\}} ds\right].$$

In particular for $X(0) = \mu$, this simplifies, due to $\mathrm{E}(X(t)) = \mu$, to (with $\mu_2(0) = \mu^2$):

$$\begin{aligned}
\mu_2(t) &= \exp\{(2c_1 + \sigma_1^2)t\}\left[\mu^2 - 2c_1\mu^2 \int_0^t \exp\{-(2c_1 + \sigma_1^2)s\} ds\right] \\
&= \exp\{(2c_1 + \sigma_1^2)t\}\left[\mu^2 + 2c_1\mu^2 \left[\frac{\exp\{-(2c_1 + \sigma_1^2)s\}}{2c_1 + \sigma_1^2}\right]_0^t\right] \\
&= \exp\{(2c_1 + \sigma_1^2)t\}\left[\mu^2 + \frac{2c_1\mu^2}{2c_1 + \sigma_1^2}(\exp\{-(2c_1 + \sigma_1^2)t\} - 1)\right] \\
&= \frac{2c_1\mu^2}{2c_1 + \sigma_1^2} + \frac{\sigma_1^2 \mu^2 \exp\{(2c_1 + \sigma_1^2)t\}}{2c_1 + \sigma_1^2}.
\end{aligned}$$

Thus, the variance for $X(0) = \mu$ reads:

$$\begin{aligned}\text{Var}(X(t)) &= \mu_2(t) - \mu^2 \\ &= \frac{\sigma_1^2\mu^2 \exp\{(2c_1+\sigma_1^2)t\}}{2c_1+\sigma_1^2} + \frac{2c_1\mu^2}{2c_1+\sigma_1^2} - \frac{(2c_1+\sigma_1^2)\mu^2}{2c_1+\sigma_1^2} \\ &= \frac{\sigma_1^2\mu^2 \exp\{(2c_1+\sigma_1^2)t\}}{2c_1+\sigma_1^2} - \frac{\sigma_1^2}{2c_1+\sigma_1^2}\mu^2 \\ &= \frac{\sigma_1^2\,\mu^2}{2c_1+\sigma_1^2}\left(\exp\{(2c_1+\sigma_1^2)\,t\} - 1\right) .\end{aligned}$$

If $2c_1 < -\sigma_1^2$, then it hence holds

$$\text{Var}(X(t)) \to \frac{-\sigma_1^2}{2c_1+\sigma_1^2}\mu^2 > 0\,,$$

as $t \to \infty$.

13.5 In order to determine the expected value function, we write Eq. (13.9) in integral form:

$$X(t) = X(0) + \int_0^t c_1(X(s)-\mu)ds + \sigma\int_0^t \sqrt{X(s)}dW(s).$$

By assumption, $\mu_1(0) = \text{E}(X(0)) = \text{E}(\mu) = \mu$. Thus, Propositions 8.2 and 10.3(b) yield:

$$\mu_1(t) = \mu + \int_0^t c_1(\mu_1(s)-\mu)ds + 0$$

or rather

$$d\mu_1(t) = c_1(\mu_1(t)-\mu)dt.$$

Due to (12.5), the solution of this deterministic differential equation reads:

$$\begin{aligned}\mu_1(t) &= z(t)\left[\mu_1(0) + \int_0^t \frac{(-c_1\mu)}{z(s)}ds\right] \\ &= e^{c_1 t}\left[\mu - \mu c_1\int_0^t e^{-c_1 s}ds\right] \\ &= e^{c_1 t}\left[\mu + \mu e^{-c_1 s}\big|_0^t\right] \\ &= \mu e^{c_1 t}\left[1 + e^{-c_1 t} - 1\right] \\ &= \mu.\end{aligned}$$

13.6 In order to determine the function of the second moment analogously to the expected value in Problem 13.5, we search for an integral equation for $X^2(t)$. This is provided by Ito's lemma for $g(X) = X^2$ with $g'(X) = 2X$ and $g''(X) = 2$:

$$\begin{aligned} dX^2(t) &= 2X(t)dX(t) + \frac{1}{2}2\sigma^2(t)dt \\ &= 2X(t)dX(t) + \sigma^2 X(t)dt, \end{aligned}$$

where $\sigma(t) = \sigma\sqrt{X(t)}$ from (13.9) was substituted. Plugging in the definition of $dX(t)$ further provides:

$$dX^2(t) = \left(2c_1X^2(t) - 2\mu c_1X(t) + \sigma^2X(t)\right)dt + 2\sigma X(t)\sqrt{X(t)}dW(t),$$

or rather

$$X^2(t) = X^2(0) + \int_0^t \left(2c_1X^2(s) + (\sigma^2 - 2\mu c_1)X(s)\right)ds + 2\sigma\int_0^t X(s)\sqrt{X(s)}dW(s).$$

With $X(0) = \mu$ forming expectation yields from Propositions 8.2 and 10.3(b):

$$\mu_2(t) = \mu^2 + \int_0^t \left(2c_1\mu_2(s) + (\sigma^2 - 2\mu c_1)\mu\right)ds + 0,$$

as $\mu_1(t) = \mu$ is constant. Thus, for $\mu_2(t)$ a deterministic differential equation results:

$$d\mu_2(t) = (2c_1\mu_2(t) + \sigma^2\mu - 2\mu^2c_1)dt.$$

With

$$z(t) = \exp\left\{\int_0^t 2c_1 ds\right\} = e^{2c_1t}$$

the solution reads

$$\begin{aligned} \mu_2(t) &= e^{2c_1t}\left[\mu^2 + \int_0^t \frac{\sigma^2\mu - 2\mu^2c_1}{e^{2c_1s}}ds\right] \\ &= e^{2c_1t}\left[\mu^2 - \frac{(\sigma^2\mu - 2\mu^2c_1)}{2c_1}e^{-2c_1s}\Big|_0^t\right] \\ &= \frac{e^{2c_1t}}{2c_1}\left[2c_1\mu^2 - (\sigma^2\mu - 2\mu^2c_1)(e^{-2c_1t} - 1)\right] \end{aligned}$$

$$= \frac{e^{2c_1 t}}{2c_1}\left[\sigma^2\mu - (\sigma^2\mu - 2\mu^2 c_1)e^{-2c_1 t}\right]$$
$$= \mu^2 - \frac{\sigma^2\mu}{2c_1} + \frac{\sigma^2\mu}{2c_1}e^{2c_1 t}.$$

Thus, in a last step the required variance is calculated as

$$\begin{aligned}\mathrm{Var}(X(t)) &= \mu_2(t) - \mu_1^2(t)\\ &= \frac{\sigma^2\mu}{2c_1}(e^{2c_1 t} - 1).\end{aligned}$$

References

Brennan, M. J, & Schwartz, E. S. (1980). Analyzing convertible bonds. *The Journal of Financial and Quantitative Analysis, 15*, 907–929.

Broze, L., Scaillet, O., & Zakoïan, J.-M. (1995). Testing for continuous-time models of the short-term interest rate. *Journal of Empirical Finance, 2*, 199–223.

Chan, K. C., Karolyi, G. A., Longstaff, F. A., & Sanders, A. B. (1992). An empirical comparision of alternative models of the short-term interest rate. *The Journal of Finance, XLVII*, 1209–1227.

Constantinides, G. M., & Ingersoll, J. E., Jr. (1984). Optimal bond trading with personal taxes. *Journal of Financial Economics, 13*, 299–335.

Cox, J. C., Ingersoll, J. E., Jr., & Ross S. A. (1980). An analysis of variable rate loan contracts. *The Journal of Finance, 35*, 389–403.

Cox, J. C., Ingersoll, J. E., Jr., & Ross S. A. (1985). A theory of the term structure of interest rates. *Econometrica, 53*, 385–407.

Dothan, L. U. (1978). On the term structure of interest rates. *Journal of Financial Economics, 6*, 59–69.

Gourieroux, Chr., & Jasiak, J. (2001). *Financial econometrics: Problems, models, and methods*. Princeton: Princeton University Press.

Marsh, T. A., & Rosenfeld, E. R. (1983). Stochastic processes for interest rates and equilibrium bond prices. *The Journal of Finance, XXXVIII*, 635–646.

Merton, R. C. (1973). Theory of rational option pricing. *The Bell Journal of Economics and Management Science, 4*, 141–183.

Tse, Y. K. (1995). Some international evidence on the stochastic behavior of interest rates. *Journal of International Money and Finance, 14*, 721–738.

Vasicek, O. (1977). An equilibrium characterization of the term structure. *Journal of Financial Economics, 5*, 177–188.

14 Asymptotics of Integrated Processes

14.1 Summary

This chapter aims at providing the basics in order to understand the asymptotic distributions of modern time series econometrics. In the first section, we treat the mathematical problems of a functional limit theory as solved and get to know the basic ingredients of a functional limit theory. Then, we proceed somewhat more abstractly by presenting the mathematical hurdles to be overcome in order to arrive at a functional limit theory. Finally, we consider multivariate generalizations.

14.2 Limiting Distributions of Integrated Processes

Under classical assumptions it holds that the arithmetic mean of a sample converges to the expected value of the sample variables for the growing sample size.[1] However, if the sample is generated by a random walk, then this does not hold any longer. Hence, limits and limiting distributions for so-called integrated processes will now be discussed.

Long-Run Variance

In order to technically formulate the concept of an integrated process, we need the so-called **long-run variance**. However, this involves an old acquaintance from Chap. 4.

[1]For a review on the convergence of random sequences, we recommend Pötscher and Prucha (2001).

U. Hassler, *Stochastic Processes and Calculus*, Springer Texts in Business and Economics, DOI 10.1007/978-3-319-23428-1_14

Let $\{e_t\}$ denote a stationary discrete-time process with zero expectation and the auotcovariances

$$\gamma_e(h) = \text{Cov}(e_t,\, e_{t+h}) = \text{E}(e_t\, e_{t+h})\,, \quad \text{E}(e_t) = 0\,.$$

As long-run variance we define

$$\omega_e^2 = \gamma_e(0) + 2\sum_{h=1}^{\infty} \gamma_e(h) < \infty\,. \tag{14.1}$$

Here, we rule out the case of a fractionally integrated process with long memory, $I(d)$ with $d > 0$, as introduced in Chap. 5, since we require the autocovariances to be summable: $\omega_e^2 < \infty$. In the case of a pure random process or white noise, $e_t = \varepsilon_t$, variance and long-run variance naturally coincide:

$$\omega_\varepsilon^2 = \sigma^2 = \gamma_\varepsilon(0) \quad \text{if } \gamma_\varepsilon(h) = 0\,,\ h \neq 0\,.$$

In general, it holds that the long-run variance is a multiple of the spectrum at the frequency zero, see (4.3):

$$\omega_e^2 = 2\pi f_e(0)\,.$$

Now, we further assume an MA(∞) process for $\{e_t\}$, see (3.2):

$$e_t = \sum_{j=0}^{\infty} c_j\, \varepsilon_{t-j}\,, \quad c_0 = 1\,,\ t = 1, \ldots, n\,, \tag{14.2}$$

with absolutely summable coefficients:

$$\sum_{j=0}^{\infty} |c_j| < \infty\,. \tag{14.3}$$

In order to determine the long-run variance, we establish an alternative expression for MA(∞) processes in Problem 14.1:

$$\omega_e^2 = \sigma^2 \left(\sum_{j=0}^{\infty} c_j\right)^2\,. \tag{14.4}$$

Due to $2\pi f_e(0) = \omega_e^2$, one can directly read this relation from Eq. (4.5), too.

Example 14.1 (Long-run Variance of MA(1)) We consider a moving average process of order 1,

$$e_t = \varepsilon_t + b\,\varepsilon_{t-1}\,.$$

From Example 4.3 we adopt the spectrum,

$$f(\lambda) = \left(1 + b^2 + 2b\cos(\lambda)\right)\sigma^2/2\pi\,.$$

Thus, at the origin the long-run variance results,

$$\omega^2 = 2\pi f(0) = (1+b)^2\,\sigma^2.$$

At the minimum, this expression takes on the value zero which happens for $b = -1$:

$$e_t = \varepsilon_t - \varepsilon_{t-1} = \Delta\,\varepsilon_t\,.$$

In this case $\{e_t\}$ is "overdifferenced". What is meant by this, will be explained in the following. ■

Integrated Processes

We revisit Example 14.1 and consider the differences of a stationary MA(∞) process $\{e_t\}$:

$$\begin{aligned}\Delta e_t = e_t - e_{t-1} &= \sum_{j=0}^{\infty} c_j\,\varepsilon_{t-j} - \sum_{j=0}^{\infty} c_j\,\varepsilon_{t-j-1} \\ &= c_0\,\varepsilon_t + \sum_{j=1}^{\infty}\left(c_j - c_{j-1}\right)\varepsilon_{t-j}\,.\end{aligned}$$

Therefore, $\{\Delta e_t\}$ is also a stationary process where the coefficients are now called $\{d_j\}$:

$$\Delta e_t = \sum_{j=0}^{\infty} d_j\,\varepsilon_{t-j}\,, \quad d_0 = c_0 = 1\,,\ d_j = c_j - c_{j-1}\,.$$

By definition, it hence holds

$$\sum_{j=0}^{\infty} d_j = c_0 + (c_1 - c_0) + (c_2 - c_1) + \cdots = 0\,.$$

Thus, we obtain for the long-run variance:

$$\omega_{\Delta e}^2 = \sigma^2\left(\sum_{j=0}^{\infty} d_j\right)^2 = 0\,.$$

The process $\{\Delta e_t\}$ is overdifferenced: It is differenced more often than necessary for stationarity as $\{e_t\}$ itself is already stationary.[2] Overdifferencing is reflected in the long-run variance being zero.

If the stationary, absolutely summable process $\{e_t\}$ has a positive long-run variance, then we call it integrated of order zero (in symbols: $e_t \sim I(0)$):

$$e_t \sim I(0) \quad \Longleftrightarrow \quad 0 < \omega_e^2 < \infty .$$

Verbally, this means: We have to difference zero times in order to attain stationarity, and it is not differenced once more than needed. Technically, in terms of Chap. 5, this means that the process is fractionally integrated of order $d = 0$.

Finally, the process $\{x_t\}$ with

$$x_t = \sum_{j=1}^{t} e_j , \quad e_t \sim I(0) , \quad t = 1, \ldots, n ,$$

is called integrated of order one, I(1), as it is defined as sum ("integral") of an I(0) process. The random walk from (1.8) e.g. is integrated of order one and obviously nonstationary. It holds for I(1) random walks that differencing once,

$$\Delta x_t = e_t ,$$

is required by definition to obtain stationarity. Hence, I(1) processes are sometimes called difference-stationary. Also, I(1) processes are often labelled as **unit root** processes, or are said to have an autoregressive unit root. We briefly want to elaborate on this terminology. Assume that $\{e_t\}$ is a stationary autoregressive process of order p,

$$A_e(L)e_t = \varepsilon_t \quad \text{with } A_e(L) = 1 - a_1 L - \cdots - a_p L^p .$$

Consequently, the I(1) process is autoregressive of order $p+1$ since

$$\Delta x_t = \frac{\varepsilon_t}{A_e(L)} \quad \text{or} \quad A_x(L)x_t = \varepsilon_t ,$$

where

$$\begin{aligned} A_x(L) &= A_e(L)\,(1-L) \\ &= 1 - (a_1 - 1)L - (a_2 - a_1)L^2 - \cdots - (a_p - a_{p-1})L^p + a_p L^{p+1} . \end{aligned}$$

[2]In econometrics, this overdifferencing is also described by the fact that $\{\Delta e_t\}$ is integrated of order -1, $\Delta e_t \sim$ I(-1): $\{e_t\}$ is differenced one more time although the process is already stationary.

Hence, $A_x(z)$ has a unit root, meaning that $A_x(z) = 0$ has a solution on the unit circle, namely the real root $z = 1$: $A_x(1) = 0$.

The Functional Central Limit Theorem (FCLT)

Now, we make statements on the distribution of the stochastic step function (the partial sum process)

$$X_n(s) = \frac{n^{-0.5}}{\omega_e} \sum_{j=1}^{\lfloor s\,n \rfloor} e_j\,, \quad s \in [0,1]\,. \tag{14.5}$$

Here, $\lfloor y \rfloor$ denotes the integer part of a real number y. In order to be able to divide by ω_e, the process $\{e_t\}$ has to be I(0). In (7.1), we have already considered a precursor of this step function as it holds:

$$X_n(s) = \begin{cases} \frac{1}{\omega_e \sqrt{n}} \sum_{j=1}^{i-1} e_j\,, & s \in \left[\frac{i-1}{n}, \frac{i}{n}\right)\,,\ i = 1, 2, \ldots, n \\ \frac{1}{\omega_e \sqrt{n}} \sum_{j=1}^{n} e_j\,, & s = 1\,. \end{cases}$$

For a graphical illustration, recall Fig. 7.1. The following proposition for MA(∞) processes from (14.2) holds under some additional assumptions.[3]

Proposition 14.1 (FCLT) *Let $\{e_t\}$ from (14.2) be integrated of order zero and satisfy some additional assumptions. Then it holds for $X_n(s)$ from (14.5) that,*

$$X_n(s) = \frac{n^{-0.5}}{\omega_e} \sum_{j=1}^{\lfloor s\,n \rfloor} e_j \Rightarrow W(s)\,, \quad s \in [0,1]\,, \quad n \to \infty\,,$$

where $\omega_e^2 > 0$ is from (14.1).

Note that the FCLT, so to speak, consists of infinitely many central limit theorems. For a fixed $\bar{s}$ it namely holds that

$$X_n(\bar{s}) = \frac{n^{-0.5}}{\omega_e} \sum_{j=1}^{\lfloor \bar{s}\,n \rfloor} e_j \xrightarrow{d} W(\bar{s}) \sim \mathcal{N}(0, \bar{s})\,.$$

---

[3]Phillips and Solo (1992) assume that the innovations $\{\varepsilon_t\}$ form an iid sequence and that $\sum_{j=0}^{\infty} j|c_j| < \infty$, which is more restrictive than (14.3). Phillips (1987) or Phillips and Perron (1988) do without the iid assumption, but require more technical restrictions. For a discussion of further sets of assumptions ensuring Proposition 14.1 see also Davidson (1994).

However, as this holds for each $\bar{s} \in [0, 1]$, we have quasi uncountably many central limit theorems collected in Proposition 14.1. The mathematically precise collection is the so-called weak convergence in function spaces which is symbolized by "$\Rightarrow$". For the following, an intuitive notion thereof suffices, somewhat more rigorous remarks will be given in the next section.

As the I(1) process $\{x_t\}$ was defined as the sum of the past of $\{e_t\}$, the circumstance from the proposition can also be expressed as follows: It holds for $x_t = \sum_{j=1}^{t} e_j$ that

$$\frac{n^{-0.5}}{\omega_e} x_{\lfloor s n \rfloor} \Rightarrow W(s) , \quad s \in [0, 1] .$$

The first FCLT was proved by Donsker for pure random processes (Donsker, 1951). Particularly for an iid sequence $e_t = \varepsilon_t$, one hence speaks of Donsker's theorem. Frequently, a FCTL also operates under the name "**invariance principle**" as it is invariant with respect to the distribution of $\{e_t\}$.

First Implications

The following proposition assembles some implications being of immediate relevance in application. As an exercise, we encourage the reader to come up with the proof in order to understand why which powers of the sample size n appear in the normalization of the sums; see Problems 14.3 through 14.5. Note that the weak convergence, "$\Rightarrow$", will be discussed in the next section, while "$\overset{d}{\rightarrow}$" stands for the usual convergence in distribution.

Proposition 14.2 (Some Limiting Distributions) *Let $x_t = x_{t-1} + e_t$ with $x_0 = 0$, $t = 1, \ldots, n$, i.e.*

$$x_t = \sum_{j=1}^{t} e_j ,$$

where $\{e_t\}$ is I(0) as in Proposition 14.1. Then it holds for $n \to \infty$:

$$\begin{aligned}
&(a)\ n^{-\frac{3}{2}} \sum_{t=1}^{n} x_{t-1} && \overset{d}{\rightarrow} \omega_e \int_0^1 W(s)\, ds ,\\
&(b)\ n^{-\frac{3}{2}} \sum_{t=1}^{n} t e_t && \overset{d}{\rightarrow} \omega_e \int_0^1 s\, dW(s) ,\\
&(c)\ n^{-\frac{5}{2}} \sum_{t=1}^{n} t\, x_{t-1} && \overset{d}{\rightarrow} \omega_e \int_0^1 s\, W(s)\, ds ,\\
&(d)\ n^{-0.5} \sum_{t=1}^{[s n]} (e_t - \bar{e}) \Rightarrow \omega_e \left(W(s) - s\, W(1)\right) , \ \bar{e} = \tfrac{1}{n} \sum_{t=1}^{n} e_t ,
\end{aligned}$$

$$(e)\ n^{-2} \sum_{t=1}^{n} x_{t-1}^2 \quad \xrightarrow{d} \omega_e^2 \int_0^1 W^2(s)\, ds\,,$$

$$(f)\ n^{-1} \sum_{t=1}^{n} x_{t-1}\, e_t \xrightarrow{d} \frac{\omega_e^2}{2} \left(W^2(1) - \frac{\gamma_e(0)}{\omega_e^2}\right)$$

$$= \omega_e^2 \left\{ \int_0^1 W(s)\, dW(s) + \frac{\omega_e^2 - \gamma_e(0)}{2\,\omega_e^2} \right\}\,,$$

with $\gamma_e(0) = Var(e_t)$ *and* ω_e^2 *from (14.1).*

Two remarks on the functional form of the statements shall be given.

Remark 1 Note the elegant and evocative functional analogy of the sums on the left-hand side, respectively, and the integrals on the right-hand side in (a)–(c) and (e): Here, the sums are substituted by integrals, the Wiener process corresponds to the I(1) process $\{x_t\}$, and the increments of the WP dW correspond to the I(0) increments $\Delta x_t = e_t$.

Remark 2 If $e_t = \varepsilon_t$ is white noise, then Ito's lemma in form of (10.3) also yields an accordance of the functional form of sample variables and limiting distributions in (f):

$$n^{-1} \sum_{t=1}^{n} x_{t-1}\, \varepsilon_t \xrightarrow{d} \frac{\omega_\varepsilon^2}{2} \left(W^2(1) - 1\right) = \omega_\varepsilon^2 \int_0^1 W(s)\, dW(s)\,.$$

As well, the limiting process appearing in (d) is intuitively well justified. It is a Brownian bridge with $W(1) - 1\, W(1) = 0$ which just reflects

$$\sum_{t=1}^{n} (e_t - \overline{e}) = 0$$

for $s = 1$.

Example 14.2 (Demeaned WP) From Proposition 14.2 (a) results due to $x_0 = 0$:

$$n^{-0.5}\, \overline{x} = n^{-\frac{3}{2}} \sum_{t=1}^{n} x_t = n^{-\frac{3}{2}} \left(\sum_{t=1}^{n} x_{t-1} + x_n \right) = n^{-\frac{3}{2}} \left(\sum_{t=1}^{n} x_{t-1} + \sum_{j=1}^{n} e_j \right)$$

$$\xrightarrow{d} \omega_e \int_0^1 W(s)\, ds + 0\,.$$

Hence, it holds for $\{x_t\}$ after demeaning the following FCTL:

$$\frac{x_{\lfloor sn \rfloor} - \overline{x}}{\omega_e \sqrt{n}} \Rightarrow W(s) - \int_0^1 W(r)\, dr\,,$$

where

$$\underline{W}(s) := W(s) - \int_0^1 W(r)\, dr$$

is also called a demeaned Wiener process. ■

In the example it was argued that it is negligible for the asymptotics whether we sum over x_{t-1} or x_t. This holds in Proposition 14.2 (a), (c), (e) but not in (f), where on the right-hand side the sign in front of $\gamma_e(0)$ changes. The following corollary summarizes the corresponding results, cf. as well Problem 14.2.

Corollary 14.1 (Some Limiting Distributions) *Let $x_t = x_{t-1} + e_t$ with $x_0 = 0$, $t = 1, \ldots, n$, i.e.*

$$x_t = \sum_{j=1}^{t} e_j\,,$$

where $\{e_t\}$ is I(0) as in Proposition 14.1. Then it holds for $n \to \infty$:

$$\begin{aligned}
n^{-\frac{3}{2}} \sum_{t=1}^{n} x_t &\overset{d}{\to} \omega_e \int_0^1 W(s)\, ds\,,\\
n^{-\frac{5}{2}} \sum_{t=1}^{n} t\, x_t &\overset{d}{\to} \omega_e \int_0^1 s\, W(s)\, ds\,,\\
n^{-2} \sum_{t=1}^{n} x_t^2 &\overset{d}{\to} \omega_e^2 \int_0^1 W^2(s)\, ds\,,\\
n^{-1} \sum_{t=1}^{n} x_t\, e_t &\overset{d}{\to} \tfrac{\omega_e^2}{2} \left(W^2(1) + \tfrac{\gamma_e(0)}{\omega_e^2} \right)\\
&= \omega_e^2 \left\{ \int_0^1 W(s)\, dW(s) + \frac{\omega_e^2 + \gamma_e(0)}{2\,\omega_e^2} \right\}\,,
\end{aligned}$$

with $\gamma_e(0) = Var(e_t)$ and ω_e^2 from (14.1).

14.3 Weak Convergence of Functions

In this subsection we want to briefly occupy ourselves with the mathematical concepts hiding behind Proposition 14.1. More rigorous expositions addressing an econometric audience can be found in Davidson (1994) or White (2001), see also the classical mathematical reference by Billingsley (1968).

Metric Function Spaces

Recall the stochastic step function that has led us to the WP, see (7.1),

$$X_n(t) = \begin{cases} \frac{1}{\sigma\sqrt{n}} \sum_{j=1}^{i-1} \varepsilon_j \, , \; t \in \left[\frac{i-1}{n}, \frac{i}{n}\right) \, , \; i = 1, 2, \ldots, n \\ \frac{1}{\sigma\sqrt{n}} \sum_{j=1}^{n} \varepsilon_j \, , \; t = 1 \, , \end{cases}$$

which can also be written more compactly as

$$X_n(t) = \frac{n^{-0.5}}{\sigma} \sum_{j=1}^{\lfloor tn \rfloor} \varepsilon_j \, , \quad t \in [0, 1] \, .$$

Furthermore, we define $\tilde{X}_n(t)$ as the function that coincides with $X_n(t)$ at the lower endpoint of the interval. However, it is not constant on the intervals, but varies linearly:

$$\tilde{X}_n(t) = \frac{n^{-0.5}}{\sigma} \sum_{j=1}^{\lfloor nt \rfloor} \varepsilon_j + (nt - \lfloor nt \rfloor) \frac{\varepsilon_{\lfloor nt \rfloor + 1}}{\sigma\sqrt{n}} \, , \quad t \in [0, 1] \, .$$

By construction, $\tilde{X}_n(t)$ is a continuous function on $[0, 1]$, for which we also abbreviate

$$\tilde{X}_n \in C\,[0, 1] \, .$$

In contrast, $X_n(t)$ is only right-continuous and exhibits (removable) discontinuities of the first type (i.e. jump discontinuities). It belongs to the set of so-called **cadlag**[4] functions that is denoted by $D\,[0, 1]$ due to the discontinuities:

$$X_n \in D\,[0, 1] \, .$$

Obviously, the set of continuous functions is a subset of the cadlag functions, i.e. $C\,[0, 1] \subseteq D\,[0, 1]$. Now, we want $X_n(t)$ as well as $\tilde{X}_n(t)$ to converge to a WP $W(t)$. For this purpose we need a distance measure in function spaces, a **metric** d. A precise mathematical definition follows.

Metric space: Let M be an arbitrary set and d a *metric*, i.e. a mapping,

$$d : \; M \times M \; \rightarrow \; \mathbb{R}_0^+ \, ,$$

[4]This French acronym (sometimes also "càdlàg") stands for "continue à droite, (avec une) limite à gauche": right-continuous and bounded on the left.

which assigns to x and y from M a non-negative number such that the following three conditions are satisfied:

$$d(x, y) = 0 \iff x = y,$$

$$d(x, y) = d(y, x) \quad \text{(symmetry)},$$

$$d(x, y) \leq d(x, z) + d(z, y) \quad \text{(triangle inequality)}.$$

Then, M endowed with d is called a metric space, (M, d).

Example 14.3 (Supremum Metric) Particularly $C\,[0, 1]$ or $D\,[0, 1]$ are readily endowed with the supremum metric:

$$d_s(f, g) := \sup_{0 \leq t \leq 1} |f(t) - g(t)|, \quad f, g \in D\,[0, 1].$$

In Problem 14.6 it is shown that the above-mentioned three defining properties are indeed fulfilled. ■

However, as $X_n(t)$ and $W(t)$ are stochastic functions, a convergence of $\{X_n\}$ to W cannot simply be based on $d_s(X_n, W)$. The convergence of $\{X_n\}$ to W has to be formulated rather as a statement on probabilities or expected values. In order to specify this, we need the concept of continuous functionals.

Continuous Functionals

Let the mapping h assign a real number to the function $f \in D\,[0, 1]$,

$$h : D\,[0, 1] \to \mathbb{R}.$$

As the argument of h is a function, one often speaks of a **functional**.

Now, let the set of cadlag functions be equipped with a metric d, i.e. let $(D\,[0, 1]\,, d)$ be a metric space. Then the functional h with h: $D\,[0, 1] \to \mathbb{R}$ is called continuous with respect to d if it holds for all $f, g \in D\,[0, 1]$ that

$$|h(f) - h(g)| \to 0$$

for

$$d(f, g) \to 0.$$

An alternative definition of continuity reads: h is called continuous with respect to d if there exists a $\delta > 0$ with

$$|h(f) - h(g)| < \varepsilon \ \text{ for } d(f, g) < \delta\,.$$

for each $\varepsilon > 0$.

Strictly speaking, continuity is a "pointwise" property; however, if a functional is continuous for every considered function, then one generally speaks of continuity of the functional. The integral over a function is a typical example for a continuous functional.

Example 14.4 (Three Functionals) Frequently, we encounter the following functionals in econometrics:

$$h_1(f) = \int_0^1 f(t)\,dt\,,$$

$$h_2(f) = \int_0^1 f^2(t)\,dt\,,$$

$$h_3(f) = \frac{1}{\int_0^1 f^2(t)\,dt}\,.$$

It can be shown that they are continuous on $D[0, 1]$ with respect to the supremum metric (cf. Problem 14.7). ■

Weak Convergence

We consider a set of stochastic elements, let them be random variables or stochastic functions. Let M be a set of stochastic elements and d a metric. We define somewhat loosely, see Billingsley (1968, Thm. 2.1): A sequence $S_n \in M$, $n \in \mathbb{N}$, converges weakly to $S \in M$ for $n \to \infty$ if

$$\lim_{n\to\infty} \mathrm{E}(h(S_n)) = \mathrm{E}(h(S))$$

for all real-valued mappings h that are bounded and uniformly continuous with respect to d. Symbolically, we write

$$S_n \Rightarrow S\,.$$

This definition in terms of expected values is not very illustrative as it is hard to imagine all mappings which are bounded and continuous. In order to translate **weak convergence** into a probability statement, we consider the indicator function I_a for an arbitrary real a and $x \in \mathbb{R}$:

$$I_a(x) := I_{(-\infty,a]}(x) = \begin{cases} 1, & x \leq a \\ 0, & x > a \end{cases} .$$

By linearization on $[a, a+\varepsilon]$ for an arbitrarily small $\varepsilon > 0$, the indicator function can be continuously approximated by

$$\tilde{I}_a(x) := \begin{cases} 1, & x \leq a \\ 1 - \frac{x-a}{\varepsilon}, & a \leq x \leq a+\varepsilon \\ 0, & x \geq a+\varepsilon \end{cases} .$$

The approximation can become arbitrarily close to I_a for small ε. Let us now choose $M = D\,[0,1]$. Then it holds for the stochastic cadlag processes $X_n(t)$ and $X(t)$ that

$$\mathrm{P}(X_n(t) \leq a) = \mathrm{E}\left[I_a(X_n(t))\right] \approx \mathrm{E}\left[\tilde{I}_a(X_n(t))\right],$$

$$\mathrm{P}(X(t) \leq a) = \mathrm{E}\left[I_a(X(t))\right] \approx \mathrm{E}\left[\tilde{I}_a(X(t))\right].$$

Hence, it holds for the continuous bounded functional $h = \tilde{I}_a$ for an arbitrary $a \in \mathbb{R}$ in case of weak convergence of $\{X_n(t)\}$ to $X(t)$, i.e. for $\mathrm{E}\left[\tilde{I}_a(X_n(t))\right] \to \mathrm{E}\left[\tilde{I}_a(X(t))\right]$, that

$$\mathrm{P}(X_n(t) \leq a) \approx \mathrm{P}(X(t) \leq a).$$

Hence, we have the following illustration of $\{X_n(t)\}$ converging weakly to $X(t)$: For *every* point in time t it holds that the sequence of distribution functions, $\mathrm{P}(X_n(t) \leq a)$, tends to the distribution function of $X(t)$.

If M particularly denotes the set of real random variables and if $X_n \Rightarrow X$ holds, then the same argument shows for the distribution function that:

$$F_n(a) := \mathrm{P}(X_n \leq a) \approx \mathrm{P}(X \leq a) = F(a).$$

With the definition from the end of Chap. 8, weak convergence of random variables hence implies their convergence in distribution, $X_n \xrightarrow{d} X$. The converse holds as well: For random variables $\{X_n\}$ and X, weak convergence is synonymous with convergence in distribution.

Continuous Mapping Theorem

A further ingredient of the proof of statements as in Proposition 14.2 is presented by the **continuous mapping theorem** (actually: about mappings which are discontinuous only on "infinitesimal sets"); see Billingsley (1968, Thm. 5.1), Davidson

(1994), and White (2001). We consider two versions of the proposition which are both special cases of a more general formulation.

Proposition 14.3 (Continuous Mapping Theorem (CMT)) *For continuous mappings of convergent series it holds:*

(a) Let $\{X_n\}$ be a sequence of real random variables and h, $h: \mathbb{R} \to \mathbb{R}$, a continuous function. From $X_n \xrightarrow{d} X$ for $n \to \infty$ it follows

$$h(X_n) \xrightarrow{d} h(X) .$$

(b) Let $\{X_n(s)\}$ and $X(s)$ belong to $D[0,1]$ and be h, $h: D[0,1] \to \mathbb{R}$, a continuous functional. From $X_n(s) \Rightarrow X(s)$ for $n \to \infty$ it follows

$$h(X_n(s)) \xrightarrow{d} h(X(s)) .$$

Verbally, the continuous mapping theorem means that mapping and limits can be interchanged without altering the result: It does not matter whether h is applied first and then n is let to infinity, or whether $n \to \infty$ first is followed by the mapping. At first sight, this may seem trivial which it is definitely not, see Example 14.5.

Remember that for $X = c = const$ convergence in distribution is equivalent to convergence in probability, see Sect. 8.4. Hence, the CMT holds as well for convergence in probability to a constant: From

$$X_n \xrightarrow{p} c$$

for $n \to \infty$ it follows that $h(X_n)$ tends in probability to the corresponding constant:

$$h(X_n) \xrightarrow{p} h(c) .$$

In the literature, this fact is also known as Slutsky's theorem. For this, we consider an example.

Example 14.5 (Consistency of Moment Estimators) Let $\{y_t\}$ with $y_t = \mu + \varepsilon_t$ be a white noise process with expected value μ. For the arithmetic mean of a sample of the size n we know from Example 8.4 (law of large numbers):

$$\overline{y}_n = \frac{1}{n}\sum_{t=1}^{n} y_t \xrightarrow{p} \mu ,$$

i.e. the empirical mean is a consistent estimator for the theoretical mean. Frequently, however, one is interested in a parameter which is a function of μ:

$$\theta = h(\mu) .$$

An estimator for θ constructed according to the method of moments is simply based on the substitution of the unknown expected value by its consistent estimator:

$$\hat{\theta}_n = h(\overline{y}_n)\,.$$

Slutsky's theorem as a special case of (a) from Proposition 14.3 then guarantees the consistency of the moment estimator, provided h is continuous:

$$\hat{\theta}_n \xrightarrow{p} h(\mu) = \theta\,.$$

Such an interchangeability of some operation and a mapping is by no means trivial. It does e.g. not hold for the expectation in general: For non-linear functions h one has:

$$\mathrm{E}\left(\hat{\theta}_n\right) = \mathrm{E}\left(h(\overline{y}_n)\right) \neq h\left(\mathrm{E}(\overline{y}_n)\right) = h\left(\mu\right)\,.$$

If e.g. $\{y_t\}$ is exponentially distributed with the parameter λ, i.e.

$$\mathrm{P}(y_t \leq y) = 1 - e^{-\lambda y}\,, \quad \lambda > 0\,,\ y \geq 0\,,$$

then it holds that

$$\mu = \frac{1}{\lambda}\,, \text{ and } \lambda = h(\mu) = \frac{1}{\mu}\,.$$

The function h is continuous in $\mu > 0$, which is why the moment estimator for λ is consistent:

$$\hat{\lambda}_n = \frac{1}{\overline{y}_n} \xrightarrow{p} \frac{1}{\mu} = \lambda\,.$$

However, one can show that it holds for an iid sample that

$$\mathrm{E}(\hat{\lambda}_n) = \frac{n}{n-1}\lambda \neq \frac{1}{\mathrm{E}(\overline{y}_n)} = \lambda\,,$$

which is why the estimator is not unbiased for λ in finite samples. ■

In order to justify the limit theory from the first section (i.e. in order to prove something like Proposition 14.1), mathematicians have followed two paths. First, the treatment of $\tilde{X}_n(t) \in C[0,1]$ with the ordinary supremum metric. For the proof of econometric propositions as e.g. Proposition 14.2 this has the disadvantage that the impractical "continuity appendage",

$$\tilde{X}_n(t) - X_n(t) = (nt - \lfloor nt \rfloor)\,\frac{\varepsilon_{\lfloor nt \rfloor + 1}}{\sigma\sqrt{n}}\,,$$

needs to be dragged along, cf. e.g. Tanaka (1996). Second, the treatment of the more compact cadlag functions $X_n(t)$ which, however, requires a more complicated metric (Skorohod metric) and additional considerations. These mathematical difficulties are indeed solved and do not have to bother us, see e.g. Billingsley (1968) or Davidson (1994). Hence, we always work with $\{X_n(t)\}$ in this book.

14.4 Multivariate Limit Theory

The multivariate limit theory is based, among others, on a vector variant of Proposition 14.1 and yields generalizations of Proposition 14.2 or Corollary 14.1. For the sake of simplicity, we narrow the exposition down to the bivariate case. The following elements of a functional limit theory are kept sufficiently general to cover the case of cointegration as well as the case of no cointegration in the following two chapters. Here, we deviate from the convention of Sect. 11.4 and do not use bold letters to denote vectors or matrices.

Integrated Vectors

The transposition of the vector z_t is denoted by z_t'. Let $z_t' = (z_{1,t}, z_{2,t})$ be a bivariate I(1) vector with starting value zero (we assume this for convenience), this means both components are I(1). Then it holds for the differences by definition,

$$\Delta z_t =: w_t = \begin{pmatrix} w_{1,t} \\ w_{2,t} \end{pmatrix},$$

that they are stationary with expectation zero, more precisely: Integrated of order zero. In generalization of the univariate autocovariance function we define

$$\Gamma_w(h) = \mathrm{E}\left(w_t w_{t+h}'\right) = \begin{pmatrix} \mathrm{E}(w_{1,t}w_{1,t+h}) & \mathrm{E}(w_{1,t}w_{2,t+h}) \\ \mathrm{E}\left(w_{2,t}w_{1,t+h}\right) & \mathrm{E}(w_{2,t}w_{2,t+h}) \end{pmatrix}.$$

Note that these matrices are not symmetric in h. Rather it holds that

$$\Gamma_w(-h) = \Gamma_w'(h)\,.$$

The **long-run variance matrix** is defined as a generalization of (14.1),

$$\Omega_w = \sum_{h=-\infty}^{\infty} \Gamma_w(h) = \begin{pmatrix} \omega_1^2 & \omega_{12} \\ \omega_{12} & \omega_2^2 \end{pmatrix}, \quad \omega_i^2 > 0\,,\; i = 1, 2\,. \tag{14.6}$$

This matrix is symmetric ($\Omega = \Omega'$) and positive semi-definite ($\Omega \geq 0$) by construction; sometimes we omit the subscript and write Ω instead of Ω_w. Note that

Ω cannot be equal to the zero matrix as $\{w_t\}$ is not "overdifferenced" but I(0). Nevertheless, the matrix does not have to be invertible. In the following chapter we will learn that the presence of so-called cointegration depends on the rank of the matrix.

Now, let $W(t)$ denote a vector of the length 2, namely the bivariate standard Wiener process. Its components are stochastically independent such that this vector is bivariate Gaussian with the identity matrix I_2:

$$W(t) = \begin{pmatrix} W_1(t) \\ W_2(t) \end{pmatrix} \sim \mathcal{N}_2(0, t I_2).$$

The corresponding Brownian motion is defined as a vector as follows:

$$B(t) = \begin{pmatrix} B_1(t) \\ B_2(t) \end{pmatrix} = \Omega^{0.5} W(t),$$

with

$$B(t) \sim \mathcal{N}_2(0, t\Omega),$$
$$\Omega = \Omega^{0.5}(\Omega^{0.5})'.$$

For the existence and construction of a matrix $\Omega^{0.5}$ with the given properties, which is to some extent a "square root of a matrix", we refer to the literature, e.g. Dhrymes (2000, Def. 2.35). However, concrete numerical examples are provided here.

Example 14.6 ($\Omega = \Omega^{0.5}(\Omega^{0.5})'$) First consider a matrix of rank 1,

$$\Omega_1 = \begin{pmatrix} 1 & 1 \\ 1 & 1 \end{pmatrix}.$$

Now, let us define

$$P_1 = \frac{1}{\sqrt{2}} \begin{pmatrix} 1 & 1 \\ 1 & 1 \end{pmatrix} = P_1' \quad \text{with } P_1 P_1 = \Omega_1 .$$

Multiplied by itself, P_1 just yields the starting matrix Ω_1. In the second example of a diagonal matrix with full rank,

$$\Omega_2 = \begin{pmatrix} \sigma_1^2 & 0 \\ 0 & \sigma_2^2 \end{pmatrix},$$

the construction of the square root becomes even more obvious:

$$P_2 = \begin{pmatrix} \sigma_1 & 0 \\ 0 & \sigma_2 \end{pmatrix} = P_2' \quad \text{with } P_2 P_2 = \Omega_2 .$$

As Ω_2 is a diagonal matrix, $P_2 = \Omega_2^{0.5}$ is just generated by taking the square root of the diagonal. Let us consider a third example with full rank:

$$\Omega_3 = \begin{pmatrix} 2 & 1 \\ 1 & 2 \end{pmatrix}.$$

Here it is not obvious which form $\Omega_3^{0.5}$ may have. However, one can check that

$$P_3 = \frac{1}{2}\begin{pmatrix} \sqrt{3}+1 & \sqrt{3}-1 \\ \sqrt{3}-1 & \sqrt{3}+1 \end{pmatrix} = P_3' \quad \text{with } P_3 P_3 = \Omega_3 .$$

In the case where Ω has full rank it is actually easy to come up with one specific factorization. Under full rank, Ω has a strictly positive determinant such that

$$t_{11} = \sqrt{\omega_1^2 - \frac{\omega_{12}^2}{\omega_2^2}} > 0 .$$

One may hence define the following triangular matrix factorizing Ω from (14.6),

$$T = \begin{pmatrix} t_{11} & \frac{\omega_{12}}{\omega_2} \\ 0 & \omega_2 \end{pmatrix} \quad \text{with } T\,T' = \begin{pmatrix} \omega_1^2 & \omega_{12} \\ \omega_{12} & \omega_2^2 \end{pmatrix}, \tag{14.7}$$

which is sometimes called **Cholesky decomposition** of Ω. For Ω_3 one obtains that way

$$T_3 = \frac{1}{\sqrt{2}}\begin{pmatrix} \sqrt{3} & 1 \\ 0 & 2 \end{pmatrix}.$$

We hence reinforce the postulate that $\Omega^{0.5}$ of a matrix Ω is not unique. In particular, T_3 is not symmetric, while P_3 is. Still, it holds $T_3\,T_3' = P_3 P_3 = \Omega_3$. ∎

Functional Limit Theory

Phillips (1986) and Phillips and Durlauf (1986) introduced under appropriate assumptions multivariate generalizations of Proposition 14.1 into econometrics:

$$n^{-0.5} z_{\lfloor sn \rfloor} = n^{-0.5} \sum_{t=1}^{\lfloor sn \rfloor} w_t \Rightarrow \Omega_w^{0.5}\, W(s) = \begin{pmatrix} B_1(t) \\ B_2(t) \end{pmatrix}. \tag{14.8}$$

For the individual components this means:

$$n^{-0.5} z_{1,\lfloor sn \rfloor} \Rightarrow B_1(s) \quad \text{and} \quad n^{-0.5} z_{2,\lfloor sn \rfloor} \Rightarrow B_2(s)\,,$$

where the Brownian motions are generally not independent of each other. Independence is only present if Ω_w is diagonal ($\omega_{12} = 0$) as it then holds:

$$\begin{pmatrix} B_1(t) \\ B_2(t) \end{pmatrix} = \begin{pmatrix} \omega_1 W_1(t) \\ \omega_2 W_2(t) \end{pmatrix}.$$

Under adequate technical conditions, which need not to be specified here, the following proposition holds, cf. as well Johansen (1995, Theorem B.13).

Proposition 14.4 (I(1) Asymptotics) *Let $\{z_t\}$ be a 2-dimensional integrated process and $\Delta z_t = z_t - z_{t-1} = w_t$ with $E(w_t) = (0, 0)'$ and Ω_w from (14.6). Then it holds under some additional assumptions that*

$$(a) \quad n^{-1.5} \sum_{t=1}^{n} z_t \xrightarrow{d} \Omega_w^{0.5} \int_0^1 W(s)ds,$$

$$(b) \quad n^{-2} \sum_{t=1}^{n} z_t z_t' \xrightarrow{d} \Omega_w^{0.5} \int_0^1 W(s)W'(s)ds\, (\Omega_w^{0.5})'$$

$$(c) \quad n^{-1} \sum_{t=1}^{n} z_t w_t' \xrightarrow{d} \Omega_w^{0.5} \int_0^1 W(s)dW'(s)\, (\Omega_w^{0.5})' + \sum_{h=0}^{\infty} \Gamma_w(h)$$

as $n \to \infty$.

Naturally, these results can be expressed in terms of $B = \Omega_w^{0.5} W$ as well:

$$\Omega_w^{0.5} \int_0^1 W(s)ds = \int_0^1 B(s)ds,$$

$$\Omega_w^{0.5} \int_0^1 W(s)W'(s)ds\, (\Omega_w^{0.5})' = \int_0^1 B(s)B'(s)ds\,,$$

$$\Omega_w^{0.5} \int_0^1 W(s)dW'(s)\, (\Omega_w^{0.5})' = \int_0^1 B(s)dB'(s)\,.$$

The limit from Proposition 14.4 (a) is to be read as a vector of Riemann integrals,

$$\int_0^1 W(s)\, ds = \begin{pmatrix} \int_0^1 W_1(s)ds \\ \int_0^1 W_2(s)ds \end{pmatrix}.$$

In (b), we have a square matrix:

$$\int_0^1 W(s)W'(s)\,ds = \begin{pmatrix} \int\limits_0^1 W_1^2(s)ds & \int\limits_0^1 W_1(s)W_2(s)ds \\ \int\limits_0^1 W_2(s)W_1(s)ds & \int\limits_0^1 W_2^2(s)ds \end{pmatrix}.$$

Concluding, both these outcomes are results from (14.8) and from a multivariate version of the continuous mapping theorem, cf. Proposition 14.3. The third result from Proposition 14.4, the matrix of Ito integrals, corresponds to result (f) from Proposition 14.2, also cf. Corollary 14.1:

$$\int_0^1 W(s)dW'(s) = \begin{pmatrix} \int\limits_0^1 W_1(s)dW_1(s) & \int\limits_0^1 W_1(s)dW_2(s) \\ \int\limits_0^1 W_2(s)dW_1(s) & \int\limits_0^1 W_2(s)dW_2(s) \end{pmatrix}.$$

In the multivariate setting, such a convergence cannot be elementarily derived any longer. For a proof see e.g. Phillips (1988) or Hansen (1992a).

14.5 Problems and Solutions

Problems

14.1 Derive with elementary means the expression (14.4) for the long-run variance of the MA(∞) process $\{e_t\}$ from (14.2).

14.2 Derive the limiting distribution of $n^{-1}\sum_{t=1}^{n} x_t e_t$ from Corollary 14.1.

14.3 Prove Proposition 14.2(a), (c), (d) and (e). When doing this, you may assume the functionals to be continuous with respect to an appropriate metric.

14.4 Prove Proposition 14.2(b).

14.5 Prove Proposition 14.2(f).

14.6 Check that $d_s(f,g)$ with

$$d_s(f,g) = \sup_{0\le t\le 1} |f(t)-g(t)|\ , \quad f,\ g \in D\,[0,1]\ ,$$

is a metric (supremum metric).

14.7 Show that the integral functionals h_1, h_2 and h_3 from Example 14.4 are continuous on $D\,[0,1]$ with respect to the supremum metric.

Solutions

14.1 First we adopt the autocovariance function from Proposition 3.2:

$$\gamma_e(h) = \sigma^2 \sum_{j=0}^{\infty} c_j\, c_{j+h}\,.$$

In (14.4), the long-run variance is formulated as follows:

$$\omega_e^2 = \sigma^2 \left(\sum_{j=0}^{\infty} c_j \right)^2 .$$

As the coefficients $\{c_j\}$ are absolutely summable, the infinite sums can be multiplied out:

$$\begin{aligned}
\frac{\omega_e^2}{\sigma^2} &= \left(\sum_{j=0}^{\infty} c_j \right) \left(\sum_{j=0}^{\infty} c_j \right) \\
&= c_0\, c_0 + c_0\, c_1 + c_0\, c_2 + \ldots \\
&\quad + c_1\, c_0 + c_1\, c_1 + c_1\, c_2 + \ldots \\
&\quad + c_2\, c_0 + c_2\, c_1 + c_2\, c_2 + \ldots \\
&\quad + \ldots \\
&= \sum_{j=0}^{\infty} c_j^2 + 2 \sum_{j=0}^{\infty} c_j\, c_{j+1} + 2 \sum_{j=0}^{\infty} c_j\, c_{j+2} + \ldots \\
&= \frac{1}{\sigma^2} (\gamma_e(0) + 2\gamma_e(1) + 2\gamma_e(2) + \ldots)\,.
\end{aligned}$$

Hence, the equivalence of the representations of the long-run variance from (14.1) and (14.4) is derived.

14.2 Due to $x_t = x_{t-1} + e_t$ we write

$$n^{-1} \sum_{t=1}^{n} x_t e_t = n^{-1} \sum_{t=1}^{n} x_{t-1} e_t + n^{-1} \sum_{t=1}^{n} e_t^2\,.$$

The limiting behavior of the first sum on the right-hand side is known from Proposition 14.2, and the second sum on the right-hand side tends to $Var(e_t) = \gamma_e(0)$. Thus, elementary transformations yield

$$\begin{aligned} n^{-1}\sum_{t=1}^{n} x_t e_t &\xrightarrow{d} \frac{\omega_e^2}{2}\left[W^2(1) - \frac{\gamma_e(0)}{\omega_e^2}\right] + \gamma_e(0) \\ &= \frac{\omega_e^2}{2}\left[W^2(1) + \frac{\gamma_e(0)}{\omega_e^2}\right] \\ &= \omega_e^2\left[\frac{W^2(1)-1}{2} + \frac{\omega_e^2+\gamma_e(0)}{2\omega_e^2}\right]. \end{aligned}$$

The application of Ito's lemma completes the proof.

14.3

(a) The interval $[0, 1)$ is split up into n subintervals of the same length,

$$[0,1) = \bigcup_{t=1}^{n}\left[\frac{t-1}{n}, \frac{t}{n}\right).$$

On each of these subintervals, we define the step function $X_n(s)$ as the appropriately normalized I(1) process,

$$X_n(s) = \frac{1}{\sqrt{n}\,\omega_e}\sum_{j=1}^{t-1} e_j = \frac{x_{t-1}}{\sqrt{n}\,\omega_e}, \quad s \in \left[\frac{t-1}{n}, \frac{t}{n}\right),$$

and on the endpoint, it holds for $s = 1$:

$$X_n(1) = \frac{1}{\sqrt{n}\,\omega_e}\sum_{j=1}^{n} e_j = \frac{x_n}{\sqrt{n}\,\omega_e}.$$

Due to Proposition 14.1 we have ($n \to \infty$)

$$X_n(s) \Rightarrow W(s).$$

Furthermore, we take a trick into account which allows for expressing a sum over x_{t-1} as an integral:

$$\int_{\frac{t-1}{n}}^{\frac{t}{n}} x_{t-1}\,ds = x_{t-1}\,s\Big|_{\frac{t-1}{n}}^{\frac{t}{n}} = x_{t-1}\left(\frac{t}{n} - \frac{t-1}{n}\right) = \frac{x_{t-1}}{n}.$$

Equipped with this, one explicitly obtains

$$\begin{aligned}
\frac{n^{-\frac{3}{2}}}{\omega_e}\sum_{t=1}^{n} x_{t-1} &= \frac{n^{-\frac{1}{2}}}{\omega_e}\sum_{t=1}^{n}\frac{x_{t-1}}{n} \\
&= \frac{n^{-\frac{1}{2}}}{\omega_e}\sum_{t=1}^{n}\int_{\frac{t-1}{n}}^{\frac{t}{n}} x_{t-1}\,ds \\
&= \sum_{t=1}^{n}\int_{\frac{t-1}{n}}^{\frac{t}{n}} \frac{x_{t-1}}{\sqrt{n}\,\omega_e}\,ds \\
&= \sum_{t=1}^{n}\int_{\frac{t-1}{n}}^{\frac{t}{n}} X_n(s)ds \\
&= \int_0^1 X_n(s)\,ds.
\end{aligned}$$

As we may assume that the functional

$$h(g) = \int_0^1 g(s)ds$$

is continuous with respect to an appropriate metric, Proposition 14.3 yields for $n \to \infty$:

$$\int_0^1 X_n(s)\,ds \stackrel{d}{\to} \int_0^1 W(s)\,ds\,.$$

Hence, the proof is complete.

(c) Just as for the proof of (a) we define the stochastic step function

$$X_n(s) = \frac{1}{\sqrt{n}\,\omega_e}\sum_{j=1}^{t-1} e_j = \frac{x_{t-1}}{\sqrt{n}\,\omega_e}, \quad s \in \left[\frac{t-1}{n}, \frac{t}{n}\right),$$

and in addition the analogously constructed deterministic step function

$$T_n(s) = \frac{t}{n} = \frac{\lfloor sn \rfloor + 1}{n}, \quad s \in \left[\frac{t-1}{n}, \frac{t}{n}\right),$$

$t = 1, \ldots, n$, and $T_n(1) = 1$. Then it holds that

$$\begin{aligned}
\frac{n^{-\frac{5}{2}}}{\omega_e}\sum_{t=1}^{n} t x_{t-1} &= \frac{n^{-\frac{3}{2}}}{\omega_e}\sum_{t=1}^{n} t\,\frac{x_{t-1}}{n} \\
&= \frac{n^{-\frac{3}{2}}}{\omega_e}\sum_{t=1}^{n} t \int_{\frac{t-1}{n}}^{\frac{t}{n}} x_{t-1}\,ds
\end{aligned}$$

$$= \sum_{t=1}^{n} \int_{\frac{t-1}{n}}^{\frac{t}{n}} \frac{t}{n} \frac{x_{t-1}}{\sqrt{n}\,\omega_e} \, ds$$

$$= \sum_{t=1}^{n} \int_{\frac{t-1}{n}}^{\frac{t}{n}} T_n(s)\, X_n(s)\, ds$$

$$= \int_0^1 T_n(s)\, X_n(s)\, ds.$$

For $n \to \infty$ it holds

$$T_n(s)\, X_n(s) \Rightarrow sW(s),$$

and hence, due to the continuity of the integral function, as claimed

$$\int_0^1 T_n(s)\, X_n(s)\, ds \xrightarrow{d} \int_0^1 s\, W(s)\, ds\,.$$

(d) Again, with the definition of $X_n(s)$ it is shown:

$$\frac{n^{-\frac{1}{2}}}{\omega_e} \sum_{t=1}^{\lfloor sn \rfloor} (e_t - \bar{e}) = \frac{n^{-\frac{1}{2}}}{\omega_e} \left(x_{\lfloor sn \rfloor} - \lfloor sn \rfloor \bar{e} \right)$$

$$= X_n(s) - \frac{\lfloor sn \rfloor}{n} \sum_{t=1}^{n} \frac{e_t}{\sqrt{n}\omega_e}$$

$$= X_n(s) - \frac{\lfloor sn \rfloor}{n} X_n(1)$$

$$\Rightarrow W(s) - s\, W(1).$$

(e) The proof is entirely analogous to (a),

$$\frac{n^{-2}}{\omega_e^2} \sum_{t=1}^{n} x_{t-1}^2 = \frac{1}{(\sqrt{n}\omega_e)^2} \sum_{t=1}^{n} \int_{\frac{t-1}{n}}^{\frac{t}{n}} x_{t-1}^2 \, ds$$

$$= \sum_{t=1}^{n} \int_{\frac{t-1}{n}}^{\frac{t}{n}} X_n^2(s)\, ds$$

$$= \int_0^1 X_n^2(s)\, ds$$

$$\xrightarrow{d} \int_0^1 W^2(s)\, ds.$$

14.4 The result can be shown in three steps.

(i) By definition it holds:

$$\begin{aligned} n^{-1}\sum_{t=1}^{n} x_{t-1} &= n^{-1}\{0 + e_1 + (e_1 + e_2) + \ldots + (e_1 + e_2 + \ldots + e_{n-1})\} \\ &= n^{-1}\{(n-1)e_1 + (n-2)e_2 + \ldots + e_{n-1}\} \\ &= n^{-1}\sum_{t=1}^{n}(n-t)e_t \\ &= \sum_{t=1}^{n} e_t - n^{-1}\sum_{t=1}^{n} te_t . \end{aligned}$$

(ii) Thus, the sum of interest is reduced to known quantities:

$$\begin{aligned} n^{-1}\sum_{t=1}^{n} te_t &= \sum_{t=1}^{n} e_t - n^{-1}\sum_{t=1}^{n} x_{t-1} \\ &= x_n - n^{-1}\sum_{t=1}^{n} x_{t-1} , \end{aligned}$$

or

$$\begin{aligned} n^{-\frac{3}{2}}\sum_{t=1}^{n} te_t &= \frac{x_n}{\sqrt{n}} - n^{-\frac{3}{2}}\sum_{t=1}^{n} x_{t-1} \\ &\stackrel{d}{\rightarrow} \omega_e\, W(1) - \omega_e \int_0^1 W(s)\, ds, \end{aligned}$$

where Proposition 14.1 for $s = 1$ and Proposition 14.2 (a) were applied.

(iii) From Example 9.1 (a) with $t = 1$ we know:

$$W(1) - \int_0^1 W(s)ds = \int_0^1 s dW(s).$$

Hence, the claim is proved.

14.5 The first result is based on the binomial formula applied to $x_t = x_{t-1} + e_t$:

$$x_t^2 = x_{t-1}^2 + 2x_{t-1}\, e_t + e_t^2 .$$

Solving this for the mixed term, we obtain:

$$\begin{aligned}
n^{-1}\sum_{t=1}^{n} x_{t-1}\, e_t &= \frac{n^{-1}}{2}\sum_{t=1}^{n}(x_t^2 - x_{t-1}^2 - e_t^2) \\
&= \frac{n^{-1}}{2}\left(x_n^2 - x_0^2 - \sum_{t=1}^{n} e_t^2\right) \\
&= \frac{1}{2}\left(\left(\frac{x_n}{\sqrt{n}}\right)^2 - 0 - \frac{1}{n}\sum_{t=1}^{n} e_t^2\right) \\
&\stackrel{d}{\to} \frac{1}{2}\left(\omega_e^2\, W^2(1) - \gamma_e(0)\right) \\
&= \frac{\omega_e^2}{2}\left(W^2(1) - \frac{\gamma_e(0)}{\omega_e^2}\right).
\end{aligned}$$

Thus, we come to the second claim. Obviously, it holds that

$$\frac{\omega_e^2}{2}\left(W^2(1) - \frac{\gamma_e(0)}{\omega_e^2}\right) = \frac{\omega_e^2}{2}\left(W^2(1) - 1 + \frac{\omega_e^2 - \gamma_e(0)}{\omega_e^2}\right).$$

The special case (10.3) of Ito's lemma then establishes the equality claimed.

14.6 In order to have a metric, three conditions from the text need to be fulfilled.

(i) The condition $f = g$ means

$$f(t) = g(t), \quad t \in [0, 1],$$

which is equivalent to

$$|f(t) - g(t)| = 0, \quad t \in [0, 1].$$

From this it immediately follows $d_s(f, g) = 0$. Conversely,

$$\sup_{0 \le t \le 1} |f(t) - g(t)| = 0$$

immediately implies $|f(t) - g(t)| = 0$. Therefore, two functions f and g in are indeed equal if and only if they have zero distance according to the supremum metric.

(ii) The symmetry condition is obviously given as it holds for the absolute value:

$$|f(t) - g(t)| = |g(t) - f(t)|.$$

(iii) Finally, the triangle inequality can be established by adding zero:

$$\begin{aligned} d_s(f,g) &= \sup_t |f(t) - h(t) + h(t) - g(t)| \\ &\leq \sup_t \{|f(t) - h(t)| + |h(t) - g(t)|\} \\ &\leq \sup_t |f(t) - h(t)| + \sup_t |h(t) - g(t)| \\ &= d_s(f,h) + d_s(h,g). \end{aligned}$$

The second, rather plausible inequality follows from the properties of the supremum metric, see e.g. Sydsæter, Strøm, and Berck (1999, p.77).

14.7 We consider three functionals $h_i(f),\ i = 1,2,3$, which assign a real number to the function $f(t)$.

(i) The first functional is just the integral from zero to one. Here it holds:

$$\begin{aligned} |h_1(f) - h_1(g)| &= \left| \int_0^1 f(t)dt - \int_0^1 g(t)dt \right| \\ &\leq \int_0^1 |f(t) - g(t)|dt \\ &\leq \int_0^1 \sup_{0\leq s\leq 1} |f(s) - g(s)|dt \\ &= \left(\sup_{0\leq s\leq 1} |f(s) - g(s)| \right) \int_0^1 dt \\ &= \sup_{0\leq s\leq 1} |f(s) - g(s)| \\ &= d_s(f,g). \end{aligned}$$

Hence, the smaller the deviation of f and g, the nearer are $h_1(f)$ and $h_1(g)$. This exactly corresponds to the definition of continuity.

(ii) The second functional is the integral over a quadratic function. Here we obtain with the binomial formula and the triangle inequality:

$$\begin{aligned} |h_2(f) - h_2(g)| &= \left| \int_0^1 f^2(t)dt - \int_0^1 g^2(t)dt \right| \\ &= \left| \int_0^1 (g(t) - f(t))^2 dt - 2\int_0^1 g(t)(g(t) - f(t))dt \right| \end{aligned}$$

$$\leq \int_0^1 |g(t) - f(t)|^2 dt + 2\int_0^1 |g(t)||g(t) - f(t)|dt$$

$$\leq \left(\sup_{0\leq t\leq 1} |g(t) - f(t)|\right)^2 + 2\sup_{0\leq t\leq 1} |g(t) - f(t)| \int_0^1 |g(t)|dt.$$

For the last inequality, it was approximated by the supremum as in (i) and then integrated from 0 to 1. Hence, it holds by definition

$$|h_2(f) - h_2(g)| \leq (d_s(g,f))^2 + 2d_s(g,f)\int_0^1 |g(t)|dt.$$

As $g(t)$ belongs to $D[0, 1]$ and is thus absolutely integrable, $h_2(f)$ tends to $h_2(g)$ if the distance between f and g gets smaller, which amounts to continuity.

(iii) The third functional is $\frac{1}{h_2}$. Hence, we reduce the continuity of h_3 to the one of h_2:

$$\begin{aligned}|h_3(f) - h_3(g)| &= \left|\frac{1}{\int_0^1 f^2(t)dt} - \frac{1}{\int_0^1 g^2(t)dt}\right| \\ &= \frac{|\int_0^1 g^2(t)dt - \int_0^1 f^2(t)dt|}{\int_0^1 f^2(t)dt \int_0^1 g^2(t)dt} \\ &= \frac{|h_2(g) - h_2(f)|}{\int_0^1 f^2(t)dt \int_0^1 g^2(t)dt}.\end{aligned}$$

Hence, from the (quadratic) integrability of f and g and the continuity of h_2 follows, as required, the continuity of h_3.

References

Billingsley, P. (1968). *Convergence of probability measures*. New York: Wiley.

Davidson, J. (1994). *Stochastic limit theory: An introduction for econometricians*. Oxford/New York: Oxford University Press.

Dhrymes, Ph. J. (2000). *Mathematics for econometrics* (3rd ed.). New York: Springer.

Donsker, M. D. (1951). An invariance principle for certain probability limit theorems. *Memoirs of the American Mathematical Society, 6*, 1–12.

Hansen, B. E. (1992a). Convergence to stochastic integrals for dependent heterogeneous processes. *Econometric Theory, 8*, 489–500.

Johansen, S. (1995). *Likelihood-based inference in cointegrated vector autoregressive models*. Oxford/New York: Oxford University Press.

Phillips, P. C. B. (1986). Understanding spurious regressions in econometrics. *Journal of Econometrics, 33*, 311–340.

Phillips, P. C. B. (1987). Time series regression with a unit root. *Econometrica, 55*, 277–301.

Phillips, P. C. B. (1988). Weak convergence of sample covariance matrices to stochastic integrals via martingale approximations. *Econometric Theory, 4*, 528–533.

Phillips, P. C. B., & Durlauf, S. N. (1986). Multiple time series regression with integrated processes. *Review of Economic Studies, LIII*, 473–495.

Phillips, P. C. B, & Perron, P. (1988). Testing for a unit root in time series regression. *Biometrika, 75*, 335–346.

Phillips, P. C. B., & Solo, V. (1992). Asymptotics for linear processes. *The Annals of Statistics, 20*, 971–1001.

Pötscher, B. M., & Prucha, I. R. (2001). Basic elements of asymptotic theory. In B. H. Baltagi (Ed.), *A companion to theoretical econometrics* (pp. 201–229). Malden: Blackwell.

Sydsæter, K., Strøm, A., & Berck, P. (1999). *Economists' mathematical manual* (3rd ed.). Berlin/New York: Springer.

Tanaka, K. (1996). *Time series analysis: Nonstationary and noninvertible distribution theory*. New York: Wiley.

White, H. (2001). *Asymptotic theory for econometricians* (2nd ed.). London/San Diego: Academic Press.

Trends, Integration Tests and Nonsense Regressions

15

15.1 Summary

Now we consider some applications of the propositions from the previous chapter. In particular, $\{e_t\}$ and $\{x_t\}$ are integrated of order 0 and integrated of order 1, respectively, cf. the definitions above Proposition 14.2. It turns out that the regression of a time series on a linear trend leads to asymptotically Gaussian estimators. However, test statistics constructed to distinguish between integration of order 1 and 0 are not Gaussian. Finally, we cover the problem of nonsense regressions, which occur particularly in the case of independent integrated variables.

15.2 Trend Regressions

Let $\{y_t\}$ be trending in the sense that the expectation follows a **linear time trend**, $\mathrm{E}(y_t) = \beta t$. The slope parameter is estimated following the least squares (LS) method. The residuals, $\widehat{res}_t = y_t - \hat{\beta} t$, are then the detrended series. Estimation relies on a sample of size n.

Detrending

The time series $\{y_t\}$ is regressed on a linear time trend according to the least squares method. For the sake of simplicity, we neglect a constant intercept that would have to be included in practice. The LS estimator $\hat{\beta}$ of the regression

$$y_t = \hat{\beta}\, t + \widehat{res}_t\,, \quad t = 1, \ldots, n\,, \tag{15.1}$$

U. Hassler, *Stochastic Processes and Calculus*, Springer Texts in Business and Economics, DOI 10.1007/978-3-319-23428-1_15

with the empirical residuals $\{\widehat{res}_t\}$ is

$$\hat{\beta} = \frac{\sum_{t=1}^{n} t\, y_t}{\sum_{t=1}^{n} t^2}.$$

For the denominator the following formula holds

$$\sum_{t=1}^{n} t^2 = \frac{n\,(n+1)\,(2n+1)}{6} = \frac{n^3}{3} + \frac{n^2}{2} + \frac{n}{6}, \tag{15.2}$$

such that one obtains asymptotically ($n \to \infty$)

$$\frac{1}{n^3} \sum_{t=1}^{n} t^2 \to \frac{1}{3}.$$

This limit is a special case of a more general result dealt with in Problem 15.1.

In this chapter we consider two models with a linear time trend in the mean. First, if the deviations from the linear trend are stationary, i.e. if the true model reads

$$y_t = \beta\, t + e_t\,, \quad t = 1, \ldots, n\,, \tag{15.3}$$

then we say that $\{y_t\}$ is trend stationary. More precisely, $\{e_t\}$ satisfies the assumptions of an I(0) process discussed in the previous chapter with long-run variance ω_e^2 defined in (14.1). Second, the stochastic component may be I(1). Then one says that $\{y_t\}$ is integrated with drift ($\beta \neq 0$): $\Delta y_t = \beta + e_t$. By integrating (summing up) with a starting value of zero, this translates into

$$y_t = \beta\, t + x_t\,, \quad t = 1, \ldots, n\,,\; x_t = \sum_{j=1}^{t} e_j. \tag{15.4}$$

An example will illustrate the difference between these two trend models.

Example 15.1 (Linear Time Trend) In Fig. 15.1 we see two time series, following the slope 0.1 on average, $t = 1, 2, \ldots, 250$. The upper graph shows a trend stationary series,

$$y_t^{(0)} = 0.1t + \varepsilon_t,$$

where $\varepsilon_t \sim \text{ii}\,\mathcal{N}(0,1)$. The lower diagram shows with identical $\{\varepsilon_t\}$

$$y_t^{(1)} = 0.1t + \sum_{j=1}^{t} \varepsilon_j\,,$$

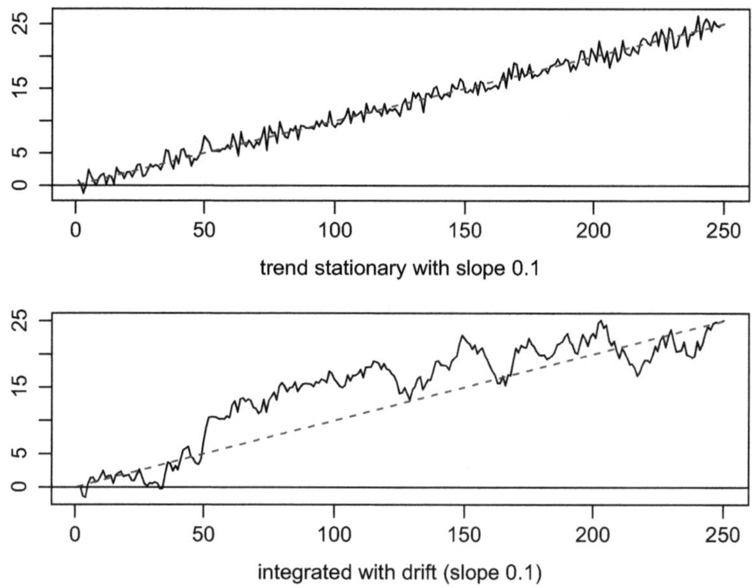

Fig. 15.1 Linear Time Trend

i.e. $y_t^{(1)}$ is $I(1)$ with drift:

$$\Delta y_t^{(1)} = 0.1 + \varepsilon_t .$$

The deviations from the linear time trend are in the lower case $I(1)$ and are hence much stronger than in the upper case. ■

Trend Stationarity

The following proposition contains the properties of LS detrending under trend stationarity.

Proposition 15.1 (Trend Stationary) *Let $\{y_t\}$ from (15.3) be trend stationary, and let $\hat{\omega}_e$ denote a consistent estimator for ω_e. It then holds for $\hat{\beta}$ from (15.1) that*

$$n^{1.5} \frac{\hat{\beta} - \beta}{\hat{\omega}_e} \xrightarrow{d} \mathcal{N}(0, 3) \tag{15.5}$$

as $n \to \infty$.

A proof is provided in Problem 15.2 relying on Proposition 14.2 (b). In practice, a consistent estimator $\hat{\omega}_e$ for ω_e will be built from LS residuals, $\widehat{res}_t = y_t - \hat{\beta}\, t$, see below for details.

Notice the fast convergence of the estimator $\hat{\beta}$ to its true value (with rate $n^{1.5}$). In real applications we would typically calculate a trend regression with intercept,

$$y_t = \bar{\alpha} + \bar{\beta}\, t + \overline{res}_t\,, \quad t = 1, \ldots, n\,.$$

Hassler (2000) showed that limiting normality and the fast convergence rate of the LS estimator with intercept, $\bar{\beta}$, pertains, although the variance is affected:

$$n^{1.5}\frac{\bar{\beta} - \beta}{\bar{\omega}_e} \xrightarrow{d} \mathcal{N}(0, 12)\,;$$

again, this requires that $\bar{\omega}_e$ constructed from the residuals is consistent.

I(1) with Drift

Now we assume $\{y_t\}$ to be integrated of order 1, possibly with drift. Note, however, that the following proposition does not require $\beta \neq 0$, as we can learn from the proof given in Problem 15.3.

Proposition 15.2 (I(1) with Drift) *Let $\{y_t\}$ from (15.4) be I(1), possibly with drift: $\Delta y_t = \beta + e_t$. Let $\hat{\omega}_e$ denote a consistent estimator for ω_e. It then holds for $\hat{\beta}$ from (15.1) that*

$$n^{0.5}\frac{\hat{\beta} - \beta}{\hat{\omega}_e} \xrightarrow{d} \mathcal{N}\left(0, \frac{6}{5}\right) \tag{15.6}$$

as $n \to \infty$.

Consistent estimation of the long-term variance will rely on the *differences* of the residuals ($\{\Delta\widehat{res}_t\}$). Here, the LS estimator obviously converges much more slowly, namely with the more usual rate $n^{0.5}$ instead of $n^{1.5}$ as in the trend stationary case. This does not come as a surprise given the impression from Fig. 15.1: Since the trend stationary series follows the straight line more closely, the estimation of the slope is more precise than in the I(1) case with drift.

A final word on Proposition 15.2: The same variance as in (15.6) is obtained in the case of a regression with intercept, see Durlauf and Phillips (1988).

Consistent Estimation of the Long-Run Variance

As we have just seen, a consistent estimation of ω_e^2 is frequently needed in practice in order to apply the functional limit theory. For this purpose, $\{e_t\}$ is to be approximated by residuals or differences thereof: In the trend stationary case we have $\widehat{res}_t = \hat{e}_t$, while for I(1) processes it holds $\widehat{res}_t = \hat{x}_t$, such that $\Delta\widehat{res}_t = \hat{e}_t$. The intuition behind an estimator $\hat{\omega}_e^2$ is readily available from (14.1), $\omega_e^2 = \gamma_e(0) + 2\sum_{h=1}^{\infty}\gamma_e(h)$, although three modifications are required. First, the theoretical autocovariances need to be replaced by the sample analogues,

$$\hat{\gamma}_e(h) = \frac{1}{n}\sum_{t=1}^{n-h}\hat{e}_t\,\hat{e}_{t+h}\,.$$

Second, the infinite sum has to be cut off as sample autocovariances can be computed only up to the lag $n-1$. Third, in order to really have a consistent (and positive) estimator, a weight function $w_B(\cdot)$ depending on a tuning parameter B is needed (whose required properties will not be discussed at this point, but see Example 15.2). In the statistics literature, $w_B(\cdot)$ is often called a kernel. Put together, we obtain as sample counterpart to (14.1):

$$\hat{\omega}_e^2 = \hat{\gamma}_e(0) + 2\sum_{h=1}^{n-1} w_B(h)\,\hat{\gamma}_e(h)\,. \tag{15.7}$$

For most kernels $w_B(\cdot)$, the parameter B (the so-called bandwidth) takes over the role of a truncation, i.e. the weights are zero for arguments greater than B:

$$\hat{\omega}_e^2 = \hat{\gamma}_e(0) + 2\sum_{h=1}^{B} w_B(h)\,\hat{\gamma}_e(h)\,.$$

In fact, the choice of B is decisive for the quality of an estimation of the long-run variance. On the one hand, B needs to tend to infinity with the sample size, on the other it needs to diverge more slowly than n. Further issues on bandwidth selection and choice of kernels have been pioneered by Andrews (1991); see also the exposition in Hamilton (1994, Sect. 10.5).

Example 15.2 (Bartlett Weights) According to Maurice S. Bartlett (English statistician, 1910–2002), a very simple weight function has a triangular form:

$$w_B(h) = \begin{cases} 1 - \frac{h}{B+1}\,, & h = 1, 2, \ldots, B+1 \\ 0\,, & \text{else} \end{cases}\,.$$

Hence, the sequence of weights reads for $h = 1, 2, \ldots B + 1$:

$$\frac{B}{B+1}, \frac{B-1}{B+1}, \ldots, \frac{B+1-h}{B+1}, \ldots, \frac{1}{B+1}, 0.$$

Plugging this in (15.7), the sum is indeed truncated at B:

$$\hat{\omega}_e^2 = \hat{\gamma}_e(0) + 2 \sum_{h=1}^{B} \frac{B+1-h}{B+1} \hat{\gamma}_e(h).$$

Although the simple Bartlett weights are by no means optimal, they are widespread in econometrics up to the present time, often also named after Newey and West (1987) who popularized them in their paper. ■

15.3 Integration Tests

We consider one test for the null hypothesis that there is integration of order 1 and one for the null hypothesis that there is integration of order 0.

Dickey-Fuller [DF] Test for Nonstationarity

The oldest and the most frequently applied test on the null hypothesis of integration of order 1 stems from Dickey and Fuller (1979).[1] In the simplest case (without deterministics) the regression model reads

$$x_t = a\, x_{t-1} + e_t\,, \quad t = 1, \ldots, n\,,$$

with the null hypothesis

$$H_0\,:\, a = 1 \quad \text{(i.e. } x_t \text{ is integrated of order 1)},$$

against the alternative of stationarity, H_1: $|a| < 1$. For the LS estimator $\hat{a}$ from

$$x_t = \hat{a}\, x_{t-1} + \hat{e}_t\,, \quad t = 1, \ldots, n\,, \tag{15.8}$$

[1]Many more procedures have been developed over the last decades, notably the test by Elliott et al. (1996) with certain optimality properties.

we obtain under H_0 the limiting distribution of the normalized LS estimator $n(\hat{a}-1)$. Often, one does not work with $n(\hat{a}-1)$, but the associated t-statistic:

$$t_a = \frac{\hat{a}-1}{s_a} \quad \text{with } s_a^2 = \frac{s^2}{\sum_{t=1}^{n} x_{t-1}^2} = \frac{n^{-1}\sum_{t=1}^{n}\hat{e}_t^2}{\sum_{t=1}^{n} x_{t-1}^2}.$$

The limiting distributions from Proposition 15.3 are established in Problem 15.4.

Proposition 15.3 (Dickey-Fuller Test) *Let the I(1) process $\{x_t\}$ ($\Delta x_t = e_t$) satisfy the assumptions from Proposition 14.2. It then holds*

(a) for $\hat{a}$ from regression (15.8) without intercept that

$$n(\hat{a}-1) \xrightarrow{d} \frac{W^2(1) - \frac{\gamma_e(0)}{\omega_e^2}}{2\int_0^1 W^2(s)\,ds},$$

(b) and for the t-statistic that

$$t_a \xrightarrow{d} \frac{\sqrt{\omega_e^2}}{\sqrt{\gamma_e(0)}} \frac{\int_0^1 W(s)\,dW(s) + \frac{\omega_e^2-\gamma_e(0)}{2\omega_e^2}}{\sqrt{\int_0^1 W^2(s)\,ds}},$$

as $n \to \infty$.

In this elegant form these limiting distributions were first given by Phillips (1987).[2] Note that the distributions depend on two parameters $\gamma_e(0)$ and ω_e^2 called "nuisance parameters" in this context. They are a nuisance because we have to somehow deal with them (remove their effect) without being economically interested in their values. Particularly, if $e_t = \varepsilon_t$ is a white noise process, then the first limit simplifies to the so-called Dickey-Fuller distribution, cf. (1.11) and (1.12):

$$n(\hat{a}-1) \xrightarrow{d} \frac{W^2(1)-1}{2\int_0^1 W^2(s)\,ds} = \frac{\int_0^1 W(s)\,dW(s)}{\int_0^1 W^2(s)\,ds}.$$

This expression does not depend on unknown nuisance parameters anymore; hence, quantiles can be simulated and approximated. One rejects for small (too strongly negative) values as the test is one-sided against the alternative of stationarity ($|a| < 1$). Similarly, for $\omega_e^2 = \gamma_e(0)$ the limit of the t-statistic simplifies to a ratio

[2]Through numerous works by Peter Phillips the functional central limit theory has found its way into econometrics. This kind of limiting distributions was then celebrated as "non-standard asymptotics"; meanwhile it has of course become standard.

free of nuisance parameters:

$$\mathcal{DF} = \frac{\int_0^1 W(s)\, dW(s)}{\sqrt{\int_0^1 W^2(s)\, ds}}\,. \tag{15.9}$$

The numerator equals that one of $\mathcal{DF}_a$ from (1.11).

In the relevant case that $\{e_t\}$ is serially correlated, one can run two paths in practice. Firstly, the test statistics can be appropriately modified by estimators for ω_e^2 and $\gamma_e(0)$. Phillips (1987) and Phillips and Perron (1988) paved this way. Secondly, one frequently calculates the regression augmented by K lags (ADF test):

$$x_t = \hat{a}\, x_{t-1} + \sum_{k=1}^{K} \hat{\alpha}_k\, \Delta x_{t-k} + \hat{\varepsilon}_t\,, \quad t = K+1, \ldots, n\,,$$

or with $\phi = a - 1$

$$\Delta x_t = \hat{\phi}\, x_{t-1} + \sum_{k=1}^{K} \hat{\alpha}_k\, \Delta x_{t-k} + \hat{\varepsilon}_t\,, \quad t = K+1, \ldots, n\,.$$

If K is so large that the error term is free of serial correlation, then the t-statistic belonging to the test on $a = 1$, i.e. $\phi = 0$, converges to the Dickey-Fuller distribution, and available tabulated percentiles serve as critical values; for further details see Said and Dickey (1984) and Chang and Park (2002). In practice, one would run a regression with intercept. This leaves the functional shape of the limiting distributions unaffected; only replace the WP W by a so-called demeaned WP, see also Problem 15.5.

KPSS Test for Stationarity

Now, the null and the alternative hypotheses are interchanged. The null hypothesis of the test suggested by Kwiatkowski, Phillips, Schmidt, and Shin (1992) claims that the time series $\{y_t\}$ is integrated of order 0 while it exhibits a random walk component under the alternative (hence, it is I(1)). Actually, this is a test for parameter constancy. The model reads

$$y_t = c_t + e_t\,, \quad t = 1, \ldots, n\,,$$

with the hypotheses

$$H_0 : \; c_t = c = \text{constant},$$

$$H_1 : \; c_t \text{ is a random walk.}$$

Under the null hypothesis, the intercept is again estimated by LS:

$$y_t = \hat{c} + \hat{e}_t , \quad \hat{c} = \overline{y} ,$$

$$\hat{e}_t = y_t - \overline{y} = e_t - \overline{e} .$$

From this, the partial sum process $\{S_t\}$ is obtained:

$$S_t := \sum_{j=1}^{t} \hat{e}_j .$$

Again, we assume that the long-run variance is consistently estimated from the residuals $\hat{e}_t$ under H_0. Then, the test statistic is formulated as

$$\eta = \frac{n^{-2}}{\hat{\omega}_e^2} \sum_{t=1}^{n} S_t^2 .$$

The limiting distribution under the null hypothesis of stationarity is provided in Problem 15.6.

Proposition 15.4 (KPSS Test) *Let the I(0) process $\{e_t\}$ satisfy the assumptions from Proposition 14.2, and $y_t = c + e_t$. It then holds that*

$$\eta \xrightarrow{d} \int_0^1 (W(s) - s\,W(1))^2 ds =: \mathcal{CM} .$$

as $n \to \infty$.

This expression does not depend on unknown nuisance parameters, and critical values are tabulated. In econometrics one often speaks about the KPSS distribution although this distribution has a long tradition in statistics where it also trades under the name of the Cramér-von-Mises ($\mathcal{CM}$) distribution. Quantiles were first tabulated by Anderson and Darling (1952). The limit $\mathcal{CM}$ is constructed from a Brownian bridge with $W(1) - 1\,W(1) = 0$, which reflects of course that $S_n = 0$ by construction.

Linear Time Trends

Many economic and financial time series are driven by a linear time trend in the mean. Consider a trend stationary process $x_t = \beta t + e_t$, which is not integrated of order 1. Still, it holds that

$$x_t = x_{t-1} + \beta + e_t - e_{t-1} .$$

Hence, it is not surprising that a regression of x_t on x_{t-1} results in a Dickey-Fuller statistic not rejecting the false null hypothesis of integration of order 1. In order to avoid the confusion of a stochastic trend (unit root process integrated of order 1) and a linear time trend, one has to include time as explanatory variable in the lag-augmented regression estimated by LS (we also add now a constant intercept)[3]:

$$x_t = \tilde{c} + \tilde{\delta}\, t + \tilde{a}\, x_{t-1} + \sum_{k=1}^{K} \tilde{\alpha}_k\, \Delta x_{t-k} + \tilde{\varepsilon}_t\,, \quad t = K+1, \ldots, n\,.$$

Under the null hypothesis that $\{x_t\}$ is integrated of order 1 (possibly with drift), the t-statistic associated with $\tilde{a} - 1$ obeys the following limiting distribution:

$$\widetilde{\mathcal{DF}} = \frac{\int_0^1 \tilde{W}(t)\, d\tilde{W}(t)}{\sqrt{\int_0^1 \tilde{W}^2(t)\, dt}}\,. \qquad (15.10)$$

The functional form is identical to that from (15.9), only that the WP is replaced by a so-called detrended WP $\tilde{W}$ defined for instance in Park and Phillips (1988, p. 474):

$$\tilde{W}(t) = W(t) - \int_0^1 W(s)ds + 12\left(\int_0^1 sW(s)ds - \frac{1}{2}\int_0^1 W(s)ds\right)\left(\frac{1}{2} - t\right).$$

We call $\widetilde{\mathcal{DF}}$ also the detrended Dickey-Fuller distribution; critical values are tabulated in the literature.

Similarly, the KPSS test may be modified to account for a linear time trend. Simply replace the demeaned series $\hat{e}_t = y_t - \overline{y}$ by the detrended one:

$$\tilde{e}_t = y_t - \tilde{c} - \tilde{\beta} t\,, \quad \tilde{S}_t = \sum_{j=1}^{t} \tilde{e}_j\,.$$

Computing the KPSS statistic $\tilde{\eta}$ from $\tilde{e}_t$ results asymptotically in a detrended Cramér-von-Mises distribution ($\widetilde{\mathcal{CM}}$ say) as long as the null hypothesis of (trend) stationarity holds true. Details on $\widetilde{\mathcal{CM}}$ and critical values thereof are given in Kwiatkowski et al. (1992): $\tilde{\eta} \xrightarrow{d} \widetilde{\mathcal{CM}}$ with

$$\widetilde{\mathcal{CM}} = \int_0^1 \left[W(s) + (2s - 3s^2)\, W(1) + (6s^2 - 6s)\int_0^1 W(r)dr\right]^2 ds\,. \qquad (15.11)$$

[3]Equivalently, one might feed detrended data into the ADF regression above.

15.4 Nonsense Regression

Nonsense or spurious regressions occur if two integrated processes are regressed on each other without being cointegrated. Hence, we need to start by defining cointegration and by briefly recapping the standard statistics from the regression model.

Cointegration

The starting point for the econometric analysis of integrated time series is the concept of **cointegration**, which is also rooted in the equilibrium paradigm of economic theory. The idea of cointegration was introduced by Granger (1981) and has firmly been embedded in econometrics by the work of Engle and Granger (1987).

Let us consider two integrated processes $\{x_t\}$ and $\{y_t\}$ integrated of order 1. Sometimes we assume that there is a linear combination with $b \neq 0$ such that

$$y_t - bx_t =: v_t \tag{15.12}$$

is integrated of order 0. Here, $y = bx$ is interpreted as a long-run equilibrium relation, as postulated by economic theory, from which, however, the empirical observations deviate at a given point in time t by v_t. If there is no linear combination of two I(1) processes that is stationary, then $\{x_t\}$ and $\{y_t\}$ are called not cointegrated.

Estimators and Statistics in the Regression Model

We consider the regression model without intercept (for the sake of simplicity) that is estimated by means of the least squares (LS) method:

$$y_t = \hat{\beta}\, x_t + \hat{u}_t\,, \; t = 1, \ldots, n\,. \tag{15.13}$$

In this section we work under the assumption that $\{x_t\}$ and $\{y_t\}$ are integrated of order 1 but not cointegrated. Hence, each linear combination, $u_t = y_t - \beta\, x_t$, is necessarily I(1) as well. Let $\{y_t\}$ and $\{x_t\}$ be both components of $\{z_t\}$:

$$z_t = \begin{pmatrix} y_t \\ x_t \end{pmatrix}.$$

Hence, it holds with (14.8):

$$n^{-0.5} y_{\lfloor sn \rfloor} \Rightarrow B_1(s) \quad \text{and} \quad n^{-0.5} x_{\lfloor sn \rfloor} \Rightarrow B_2(s)\,.$$

If the regressand y_t and the regressor x_t are stochastically independent, then this property is transferred to the limiting processes B_1 and B_2. Nevertheless, we will show that then $\hat{\beta}$ from (15.13) does not tend to the true value that is zero. Instead, a (significant) relation between the independent variables is spuriously obtained. Since Granger and Newbold (1974), this circumstance is called spurious or nonsense regression.

The LS estimator from the regression without intercept reads

$$\hat{\beta} = \frac{\sum_{t=1}^{n} x_t\, y_t}{\sum_{t=1}^{n} x_t^2}.$$

The t-statistic[4] belonging to the test on the parameter value 0 is based on the difference of estimator and hypothetical value divided by the estimated standard error of the estimator:

$$t_\beta = \frac{\hat{\beta} - 0}{s_\beta} \quad \text{with } s_\beta^2 = \frac{s^2}{\sum_{t=1}^{n} x_t^2}, \; s^2 = \frac{1}{n} \sum_{t=1}^{n} \hat{u}_t^2 .$$

As a measure of fit, the (uncentered) **coefficient of determination** is frequently calculated,[5]

$$R_{uc}^2 = 1 - \frac{\sum_{t=1}^{n} \hat{u}_t^2}{\sum_{t=1}^{n} y_t^2}.$$

Finally, the **Durbin-Watson statistic** is a well-established measure for the first order residual autocorrelation,

$$dw = \frac{\sum_{t=2}^{n} (\hat{u}_t - \hat{u}_{t-1})^2}{\sum_{t=1}^{n} \hat{u}_t^2} \approx 2 \left(1 - \frac{\frac{1}{n}\sum_{t=2}^{n} \hat{u}_t\, \hat{u}_{t-1}}{\frac{1}{n}\sum_{t=1}^{n} \hat{u}_t^2}\right).$$

Let us briefly recall the behavior of these measures if we worked with I(0) variables x and y not being correlated ($\beta = 0$). Then, $\hat{\beta}$ would tend to 0, the t-statistic would converge to a normal distribution and the coefficient of determination would tend to zero. Finally, the Durbin-Watson statistic would converge to $2\,(1 - \rho_1) > 0$, where ρ_1 denotes the first order autocorrelation coefficient of the regression errors. In the case of nonsense regressions, we obtain qualitatively entirely different asymptotic results.

[4]For the following calculation of s^2 we divide by n without correcting for degrees of freedom, which does not matter asymptotically ($n \to \infty$).

[5]"Uncentered", as the regression is calculated without intercept.

Asymptotics

Due to Proposition 14.4 (b) it holds for the denominator of the LS estimator

$$n^{-2}\sum_{t=1}^{n} x_t^2 \xrightarrow{d} \int_0^1 B_2^2(s)\,ds,$$

and for the numerator we obtain

$$n^{-2}\sum_{t=1}^{n} x_t\, y_t \xrightarrow{d} \int_0^1 B_2(s)\, B_1(s)\,ds\,.$$

Both results put together yield

$$\hat{\beta} \xrightarrow{d} \frac{\int_0^1 B_1(s)\, B_2(s)\,ds}{\int_0^1 B_2^2(s)\,ds} =: \beta_\infty\,.$$

In particular, if y_t and x_t are stochastically independent, then $\hat{\beta}$ does not tend to the true value 0 but to the random variable β_∞. And as if that was not enough, the t-statistic belonging to the test on the true parameter value $\beta = 0$ tends to infinity in absolute value! Hence, in this situation t-statistics highly significantly reject the true null hypothesis of no correlation and therefore report absurd relations as being significant. This phenomenon was experimentally discovered for small samples by Granger and Newbold (1974) and asymptotically proved by Phillips (1986). For $n \to \infty$ it namely holds that $n^{-0.5}\, t_\beta$ has a well-defined limiting distribution. In the problem section we prove in addition the further properties of the following proposition.

Proposition 15.5 (Nonsense Regression) *For I(1) processes $\{x_t\}$ and $\{y_t\}$ it holds in case of no cointegration with the notation introduced that*

$$(a)\ \hat{\beta} \xrightarrow{d} \frac{\int_0^1 B_1(s)\, B_2(s)\,ds}{\int_0^1 B_2^2(s)\,ds} =: \beta_\infty\,,$$

$$(b)\ R_{uc}^2 \xrightarrow{d} \beta_\infty^2 \frac{\int_0^1 B_2^2(s)\,ds}{\int_0^1 B_1^2(s)\,ds} =: R_\infty^2\,,$$

$$(c)\ n^{-1}\, s^2 \xrightarrow{d} \int_0^1 B_1^2(s)\,ds\,\left(1 - R_\infty^2\right) =: s_\infty^2\,,$$

$$(d)\ n^{-0.5}\, t_\beta \xrightarrow{d} \frac{\beta_\infty \sqrt{\int_0^1 B_2^2(s)\,ds}}{s_\infty}\,,$$

$$(e)\ dw \xrightarrow{p} 0\,,$$

as $n \to \infty$.

As already has been emphasized: The results (a), (b) and (d) justify to speak of a nonsense regression. A first hint at the lack of cointegration is obtained from the first order residual autocorrelation: For nonsense or spurious regressions, the Durbin-Watson statistic tends to zero.

Example 15.3 (Hendry, 1980) Hendry (1980) shows the real danger of nonsense regressions. In his polemic example, the price development (measured by the consumer price index P) is to be explained. For this purpose, first a money supply variable M is used. Then a second variable C is considered (for which we could think of consumption), and it appears that this time series explains the price development better than M does. However, there can in fact be no talk of explanation; this is a nonsense correlation as behind C the cumulated rainfalls are hidden (hence, the precipitation amount)! By the way, note that P and C are not (only) integrated of order 1 but in addition exhibit a deterministic time trend, which only aggravates the problem of nonsense regression. ∎

If integrated variables are not cointegrated, they have to be analyzed in differences, i.e. Δy_t is regressed on Δx_t, resulting in the familiar stationary regression model. Naturally, not all integrated economic variables lead to nonsense regressions. This does not happen under cointegration, which brings us to the final chapter.

15.5 Problems and Solutions

Problems

15.1 Show

$$\frac{1}{n^{k+1}} \sum_{t=1}^{n} t^k \to \frac{1}{k+1} \quad \text{for } k \in \mathbb{N}$$

as $n \to \infty$.

15.2 Prove Proposition 15.1.

15.3 Prove Proposition 15.2.

15.4 Prove Proposition 15.3.

15.5 Derive an expression for the limiting distribution of $n(\overline{a} - 1)$ under $a = 1$ if a model *with* intercept is estimated in the Dickey-Fuller test:

$$x_t = \overline{c} + \overline{a}\, x_{t-1} + \overline{\varepsilon}_t\,, \quad t = 1, \ldots, n\,.$$

Assume $\Delta x_t = \varepsilon_t$ to be white noise.

15.6 Prove Proposition 15.4.

15.7 Prove the statements (b), (c) and (d) from Proposition 15.5.

15.8 Prove statement (e) from Proposition 15.5.

Solutions

15.1 Define the continuous and hence Riemann-integrable function f on $[0, 1]$ with antiderivative F:

$$f(x) = x^k \quad \text{with} \quad F(x) = \frac{x^{k+1}}{k+1}.$$

Further, we work with the equidistant partition

$$[0, 1] = \bigcup_{i=1}^{n} [t_{i-1}, t_i], \quad t_i = \frac{i}{n}.$$

Hence,

$$\frac{1}{n^{k+1}} \sum_{t=1}^{n} t^k = \sum_{t=1}^{n} \left(\frac{t}{n}\right)^k \frac{1}{n} = \sum_{i=1}^{n} f(t_i)\,(t_i - t_{i-1})$$
$$\to \int_0^1 f(t)\,dt = F(1) - F(0),$$

which proves the claim.

15.2 According to the hints in the text, it holds

$$\hat{\beta} = \frac{\sum_{t=1}^{n} t(\beta t + e_t)}{\sum_{t=1}^{n} t^2} = \beta + \frac{\sum_{t=1}^{n} te_t}{\frac{n^3}{3} + \frac{n^2}{2} + \frac{n}{6}}$$

or

$$n^{1.5}(\hat{\beta} - \beta) = \frac{n^{-1.5} \sum_{t=1}^{n} te_t}{\frac{1}{3} + \frac{1}{2n} + \frac{1}{6n^2}}.$$

The denominator tends to $\frac{1}{3}$ whereas Proposition 14.2 (b) guarantees the following limiting distribution:

$$n^{1.5}(\hat{\beta} - \beta) \xrightarrow{d} \frac{\omega_e \int_0^1 s dW(s)}{\frac{1}{3}}.$$

From Example 9.2 it follows that

$$n^{1.5} \frac{\hat{\beta} - \beta}{3\omega_e} \xrightarrow{d} \int_0^1 s dW(s) \sim \mathcal{N}\left(0, \frac{1}{3}\right).$$

If ω_e is replaced by a consistent estimator,

$$\hat{\omega}_e \xrightarrow{p} \omega_e,$$

then the result from (15.5) is established.

15.3 The I(1) case is treated in an analogous way as the trend stationary case, see Proposition 14.2 (c):

$$\begin{aligned} n^{0.5}(\hat{\beta} - \beta) &= \frac{n^{-2.5} \sum_{t=1}^{n} t x_t}{\frac{1}{3} + \frac{1}{2n} + \frac{1}{6n^2}} \\ &\xrightarrow{d} \frac{\omega_e \int_0^1 s W(s)\, ds}{\frac{1}{3}}. \end{aligned}$$

From Corollary 8.1 (c) it follows for $c = 0$

$$\int_0^1 s\, W(s)\, ds \sim \mathcal{N}\left(0, \frac{2}{15}\right).$$

Hence, (15.6) is proved.

15.4 Under H_0, the LS estimator is given as:

$$\hat{a} = \frac{\sum_{t=1}^{n} x_{t-1} x_t}{\sum_{t=1}^{n} x_{t-1}^2} = 1 + \frac{\sum_{t=1}^{n} x_{t-1} e_t}{\sum_{t=1}^{n} x_{t-1}^2}.$$

Insofar it holds

$$n(\hat{a} - 1) = \frac{n^{-1} \sum_{t=1}^{n} x_{t-1} e_t}{n^{-2} \sum_{t=1}^{n} x_{t-1}^2}.$$

Thus, numerator as well as denominator are in a form accessible for Proposition 14.2,

$$n(\hat{a}-1) \xrightarrow{d} \frac{\frac{\omega_e^2}{2}\left(W^2(1) - \frac{\gamma_e(0)}{\omega_e^2}\right)}{\omega_e^2 \int_0^1 W^2(s)ds} = \frac{\int_0^1 W(s)dW(s) + \frac{\omega_e^2 - \gamma_e(0)}{2\omega_e^2}}{\int_0^1 W^2(s)ds},$$

which is the first distribution to be established.

To find the second limit, one has to handle the standard error s_a of $\hat{a}$. It is based on the residual variance estimation

$$\begin{aligned}
s^2 &= \frac{1}{n}\sum_{t=1}^{n} \widehat{e_t}^2 = \frac{1}{n}\sum_{t=1}^{n}(x_t - \hat{a}\,x_{t-1})^2 \\
&= \frac{1}{n}\sum_{t=1}^{n}((1-\hat{a})x_{t-1} + e_t)^2 \\
&= \frac{(1-\hat{a})^2}{n}\sum_{t=1}^{n} x_{t-1}^2 + \frac{2(1-\hat{a})}{n}\sum_{t=1}^{n} x_{t-1}e_t + \frac{1}{n}\sum_{t=1}^{n} e_t^2 \\
&= n^2(1-\hat{a})^2 \frac{\sum_{t=1}^{n} x_{t-1}^2}{n^3} + 2n(1-\hat{a})\frac{\sum_{t=1}^{n} x_{t-1}e_t}{n^2} + \frac{1}{n}\sum_{t=1}^{n} e_t^2 \\
&\xrightarrow{p} 0 + 2\cdot 0 + \gamma_e(0),
\end{aligned}$$

where Proposition 14.2 (e), (f) and $n^{-1}\sum_{t=1}^{n} e_t^2 \to \gamma_e(0)$ as well as the circumstance that $n(1-\hat{a})$ converges in distribution were used. All in all, it hence holds for the t-statistic:

$$\begin{aligned}
t_a &= \frac{\hat{a}-1}{s_a} = \frac{(\hat{a}-1)\sqrt{\sum_{t=1}^{n} x_{t-1}^2}}{s} \\
&= \frac{n(\hat{a}-1)\sqrt{n^{-2}\sum_{t=1}^{n} x_{t-1}^2}}{s} \\
&\xrightarrow{d} \frac{\left(\int_0^1 W(s)dW(s) + \frac{\omega_e^2-\gamma_e(0)}{2\omega_e^2}\right)\sqrt{\omega_e^2\int_0^1 W^2(s)ds}}{\sqrt{\gamma_e(0)}\int_0^1 W^2(s)ds} \\
&= \frac{\omega_e}{\sqrt{\gamma_e(0)}}\frac{\int_0^1 W(s)dW(s) + \frac{\omega_e^2-\gamma_e(0)}{2\omega_e^2}}{\sqrt{\int_0^1 W^2(s)ds}}.
\end{aligned}$$

This completes the proof.

15.5 With the means

$$\overline{x} = \frac{1}{n}\sum_{t=1}^{n} x_t, \quad \overline{x}_{-1} = \frac{1}{n}\sum_{t=1}^{n} x_{t-1}, \quad \overline{\varepsilon} = \frac{1}{n}\sum_{t=1}^{n} \varepsilon_t$$

it holds for the LS estimator when including an intercept under $a = 1$:

$$\begin{aligned}
\overline{a} &= \frac{\sum_{t=1}^{n}(x_{t-1} - \overline{x}_{-1})x_t}{\sum_{t=1}^{n}(x_{t-1} - \overline{x}_{-1})^2} = \frac{\sum_{t=1}^{n}(x_{t-1} - \overline{x}_{-1})(x_{t-1} + \varepsilon_t)}{\sum_{t=1}^{n}(x_{t-1} - \overline{x}_{-1})^2} \\
&= \frac{\sum_{t=1}^{n}(x_{t-1} - \overline{x}_{-1})(x_{t-1} - \overline{x}_{-1})}{\sum_{t=1}^{n}(x_{t-1} - \overline{x}_{-1})^2} + \frac{\sum_{t=1}^{n}(x_{t-1} - \overline{x}_{-1})\varepsilon_t}{\sum_{t=1}^{n}(x_{t-1} - \overline{x}_{-1})^2}.
\end{aligned}$$

Thus, we obtain

$$\begin{aligned}
n(\overline{a} - 1) &= \frac{n^{-1}\left(\sum_{t=1}^{n} x_{t-1}\varepsilon_t - \overline{x}_{-1}\sum_{t=1}^{n}\varepsilon_t\right)}{n^{-2}\left(\sum_{t=1}^{n} x_{t-1}^2 - n(\overline{x}_{-1})^2\right)} \\
&= \frac{n^{-1}\sum_{t=1}^{n} x_{t-1}\varepsilon_t - n^{-0.5}\overline{x}_{-1}n^{-0.5}x_n}{n^{-2}\sum_{t=1}^{n} x_{t-1}^2 - \left(n^{-0.5}\overline{x}_{-1}\right)^2}.
\end{aligned}$$

For $\varepsilon_t \sim \text{WN}(0, \sigma^2)$ with $\omega_\varepsilon^2 = \sigma^2 = \gamma_\varepsilon(0)$ it therefore holds due to Proposition 14.2 (a), (e) and (f) and according to Proposition 14.1 for $s = 1$:

$$\begin{aligned}
n(\overline{a} - 1) &\xrightarrow{d} \frac{\sigma^2 \int_0^1 W(s)dW(s) - \sigma \int_0^1 W(s)ds\, \sigma W(1)}{\sigma^2 \int_0^1 W^2(s)ds - (\sigma \int_0^1 W(s)ds)^2} \\
&= \frac{\int_0^1 W(s)dW(s) - W(1)\int_0^1 W(s)ds}{\int_0^1 W^2(s)ds - (\int_0^1 W(s)ds)^2}.
\end{aligned}$$

If one defines the demeaned WP,

$$\underline{W}(s) := W(s) - \int_0^1 W(r)dr,$$

then one observes two interesting identities:

$$\begin{aligned}
\int_0^1 \underline{W}(s)d\underline{W}(s) &= \int_0^1 \underline{W}(s)dW(s) \\
&= \int_0^1 W(s)dW(s) - \int_0^1 W(r)dr \int_0^1 dW(s) \\
&= \int_0^1 W(s)dW(s) - \int_0^1 W(r)drW(1),
\end{aligned}$$

and

$$\int_0^1 (\underline{W}(s))^2 ds = \int_0^1 \left(W^2(s) - 2W(s) \int_0^1 W(r)dr + \left(\int_0^1 W(r)dr \right)^2 \right) ds$$
$$= \int_0^1 W^2(s)ds - 2\int_0^1 W(s)ds \int_0^1 W(r)dr + \left(\int_0^1 W(r)dr \right)^2$$
$$= \int_0^1 W^2(s)ds - \left(\int_0^1 W(s)ds \right)^2 .$$

Hence, the limiting distribution in the case of a regression with intercept (which is equivalent to running a regression of demeaned variables!) can also be written as:

$$n(\overline{a} - 1) \xrightarrow{d} \frac{\int_0^1 \underline{W}(s)d\underline{W}(s)}{\int_0^1 (\underline{W}(s))^2 ds} .$$

Thus, one obtains the same functional form as in the case without intercept, save that $W(s)$ is replaced by $\underline{W}(s)$.

15.6 For the partial sum S_t we obtain under the null hypothesis:

$$S_t = \sum_{j=1}^{t} \hat{e}_j = \sum_{j=1}^{t} (e_j - \bar{e}).$$

The adequately normalized squared sum yields

$$n^{-2} \sum_{t=1}^{n} S_t^2 = n^{-1} \sum_{t=1}^{n} \int_{\frac{t-1}{n}}^{\frac{t}{n}} S_{t-1}^2 ds + \frac{S_n^2}{n^2}$$
$$= \sum_{t=1}^{n} \int_{\frac{t-1}{n}}^{\frac{t}{n}} \left(\frac{S_{\lfloor sn \rfloor}}{\sqrt{n}} \right)^2 ds + \frac{S_n^2}{n^2},$$

where $S_0 = 0$ was used and $S_{\lfloor sn \rfloor}$ is the step function from Proposition 14.2:

$$S_{\lfloor sn \rfloor} = \sum_{j=1}^{\lfloor sn \rfloor} (e_j - \bar{e}), \quad s \in \left[\frac{t-1}{n}, \frac{t}{n} \right).$$

Note that S_n is zero by construction:

$$S_n = \sum_{j=1}^{n} (e_j - \bar{e}) = n\bar{e} - n\bar{e} = 0.$$

Hence, we obtain

$$n^{-2} \sum_{t=1}^{n} S_t^2 = \int_0^1 \left(\frac{S_{\lfloor sn \rfloor}}{\sqrt{n}} \right)^2 ds.$$

As there is a continuous functional on the right-hand side, Proposition 14.3 with Proposition 14.2 yields:

$$\frac{n^{-2}}{\omega_e^2} \sum_{t=1}^{n} S_t^2 \xrightarrow{d} \int_0^1 (W(s) - sW(1))^2 \, ds.$$

If ω_e^2 is replaced by a consistent estimator, then we just obtain the given limiting distribution $\mathcal{CM}$ as a functional of a Brownian bridge. This completes the proof.

15.7 The limit β_∞ of $\hat{\beta}$ given in Proposition 15.5 is adopted from the text.

The s^2 is based on the LS residuals $\hat{u}_t = y_t - \hat{\beta} x_t$. Hence, the sum of squared residuals becomes

$$\begin{aligned}
\sum_{t=1}^{n} \hat{u}_t^2 &= \sum_{t=1}^{n} y_t^2 - 2\hat{\beta} \sum_{t=1}^{n} y_t x_t + \hat{\beta}^2 \sum_{t=1}^{n} x_t^2 \\
&= \sum_{t=1}^{n} y_t^2 - 2 \frac{(\sum_{t=1}^{n} y_t x_t)^2}{\sum_{t=1}^{n} x_t^2} + \left(\frac{\sum_{t=1}^{n} y_t x_t}{\sum_{t=1}^{n} x_t^2} \right)^2 \sum_{t=1}^{n} x_t^2 \\
&= \sum_{t=1}^{n} y_t^2 - \frac{(\sum_{t=1}^{n} y_t x_t)^2}{\sum_{t=1}^{n} x_t^2} \\
&= \sum_{t=1}^{n} y_t^2 - \hat{\beta}^2 \sum_{t=1}^{n} x_t^2 .
\end{aligned}$$

Thus, again with Proposition 14.4 (b), one immediately obtains for the uncentered coefficient of determination:

$$\begin{aligned}
R_{uc}^2 &= 1 - \frac{n^{-2} \sum_{t=1}^{n} \hat{u}_t^2}{n^{-2} \sum_{t=1}^{n} y_t^2} \\
&= \frac{\hat{\beta}^2 n^{-2} \sum_{t=1}^{n} x_t^2}{n^{-2} \sum_{t=1}^{n} y_t^2}
\end{aligned}$$

$$\begin{aligned}
&\xrightarrow{d} \frac{\beta_\infty^2 \int_0^1 B_2^2(s)ds}{\int_0^1 B_1^2(s)\, ds} \\
&=: R_\infty^2.
\end{aligned}$$

The relation used above between the sum of squared residuals and the coefficient of determination further yields

$$\begin{aligned}
n^{-1}s^2 &= \frac{1}{n^2}\sum_{t=1}^{n} \hat{u}_t^2 \\
&= (1 - R_{uc}^2)\, n^{-2} \sum_{t=1}^{n} y_t^2 \\
&\xrightarrow{d} (1 - R_\infty^2) \int_0^1 B_1^2(s)ds \\
&=: s_\infty^2.
\end{aligned}$$

Moreover, the asymptotics of s^2 enables us to state the behavior of the t-statistic as well. The required normalization is obvious:

$$\begin{aligned}
n^{-0.5} t_\beta &= \frac{\hat{\beta}\, \sqrt{n^{-2}\sum_{t=1}^{n} x_t^2}}{n^{-0.5} s} \\
&\xrightarrow{d} \frac{\beta_\infty}{s_\infty}\sqrt{\int_0^1 B_2^2(s)ds}.
\end{aligned}$$

15.8 Due to $\hat{u}_t = y_t - \hat{\beta}\, x_t$, the numerator of the Durbin-Watson statistic yields:

$$\begin{aligned}
n^{-1}\sum_{t=2}^{n}(\hat{u}_t - \hat{u}_{t-1})^2 &= n^{-1}\sum_{t=2}^{n}\left(\Delta y_t - \hat{\beta}\,\Delta x_t\right)^2 \\
&= n^{-1}\sum_{t=2}^{n}\left(w_{1,t} - \hat{\beta} w_{2,t}\right)^2 \\
&= n^{-1}\sum_{t=2}^{n} w_{1,t}^2 - \frac{2\,\hat{\beta}}{n}\sum_{t=2}^{n} w_{1,t}\, w_{2,t} + \frac{\hat{\beta}^2}{n}\sum_{t=2}^{n} w_{2,t}^2 \\
&\xrightarrow{d} \gamma_1(0) - 2\beta_\infty \cdot \mathrm{Cov}(w_{1,t}, w_{2,t}) + \beta_\infty^2\, \gamma_2(0),
\end{aligned}$$

where $\gamma_i(0)$ denote the variances of the processes $\{w_{i,t}\}$, $i = 1, 2$. By definition one obtains

$$n\, dw = \frac{n^{-1} \sum_{t=2}^{n} (\hat{u}_t - \hat{u}_{t-1})^2}{n^{-2} \sum_{t=1}^{n} \hat{u}_t^2}$$

$$\xrightarrow{d} \frac{\gamma_1(0) - 2\beta_\infty \cdot \mathrm{Cov}(w_{1,t}, w_{2,t}) + \beta_\infty^2 \gamma_2(0)}{s_\infty^2}.$$

Consequently, dw tends to zero in probability.

References

Anderson, T. W., & Darling, D. A. (1952). Asymptotic theory of certain "Goodness of Fit" criteria based on stochastic processes. *Annals of Mathematical Statistics, 23*, 193–212.

Andrews, D. W. K. (1991). Heteroskedasticity and autocorrelation consistent covariance matrix estimation. *Econometrica, 59*, 817–858.

Chang, Y., & Park, J. (2002). On the asymptotics of ADF tests for unit roots. *Econometric Reviews, 21*, 431–447.

Dickey, D. A., & Fuller, W. A. (1979). Distribution of the estimators for autoregressive time series with a unit root. *Journal of the American Statistical Association, 74*, 427–431.

Durlauf, S. N., & Phillips, P. C. B. (1988). Trends versus random walks in time series analysis. *Econometrica, 56*, 1333–1354.

Elliott, G., Rothenberg, T. J., & Stock, J. H. (1996). Efficient tests for an autoregressive unit root. *Econometrica, 64*, 813–836.

Engle, R. F., & Granger, C. W. J. (1987). Co-integration and error correction: Representation, estimation, and testing. *Econometrica, 55*, 251–276.

Granger, C. W. J. (1981). Some properties of time series data and their use in econometric model specification. *Journal of Econometrics, 16*, 121–130.

Granger, C. W. J., & Newbold P. (1974). Spurious regressions in econometrics. *Journal of Econometrics, 2*, 111–120.

Hamilton, J. (1994). *Time series analysis*. Princeton: Princeton University Press.

Hassler, U. (2000). Simple regressions with linear time trends. *Journal of Time Series Analysis, 21*, 27–32.

Hendry, D. F. (1980). Econometrics – alchemy or science? *Economica, 47*, 387–406.

Kwiatkowski, D., Phillips, P. C. B., Schmidt, P., & Shin Y. (1992). Testing the null hypothesis of stationarity against the alternative of a unit root. *Journal of Econometrics, 54*, 159–178.

Newey, W. K., & West, K. D. (1987). A Simple, positive semi-definite, heteroskedasticity and autocorrelation consistent covariance matrix. *Econometrica, 55*, 703–708.

Park, J. Y., & Phillips, P. C. B. (1988). Statistical inference in regressions with integrated processes: Part I. *Econometric Theory, 4*, 468–497.

Phillips, P. C. B. (1986). Understanding spurious regressions in econometrics. *Journal of Econometrics, 33*, 311–340.

Phillips, P. C. B. (1987). Time series regression with a unit root. *Econometrica, 55*, 277–301.

Phillips, P. C. B, & Perron, P. (1988). Testing for a unit root in time series regression. *Biometrika, 75*, 335–346.

Said, S. E., & Dickey, D. A. (1984). Testing for unit roots in autoregressive-moving average models of unknown order. *Biometrika, 71*, 599–607.

Cointegration Analysis 16

16.1 Summary

This chapter is addressed to the analysis of cointegrated variables. Properties like superconsistency of the LS estimator and conditions for asymptotic normality are extensively discussed. Error-correction is the reverse of cointegration, which is why we provide an introduction to the analysis of error-correction models as well. In particular, we discuss cointegration testing. In 2003, Clive W.J. Granger was awarded the Nobel prize for introducing the concept of cointegration. Finally, we stress once more the effect of linear time trends underlying the series.

16.2 Error-Correction and Cointegration

Before cointegration had been launched, so-called **error-correction models** impressed due to their empirical performance, cf. e.g. Davidson, Hendry, Srba, and Yeo (1978). Today we know that these models are just the other side of the cointegration coin. By way of example, the key statement of Granger's representation theorem from Engle and Granger (1987) is illustrated, which demonstrates the fact that error-correction and cointegration are equivalent.

Autoregressive Distributed Lag Model

Let us consider a dynamic regression model in which $\{y_t\}$ has an autoregressive structure on the one hand and which is explained by (lagged) exogenous variables x_{t-j} on the other:

$$y_t = a_1\, y_{t-1} + \cdots + a_p\, y_{t-p} + c_0\, x_t + c_1\, x_{t-1} + \cdots + c_\ell\, x_{t-\ell} + \varepsilon_t\,.$$

U. Hassler, *Stochastic Processes and Calculus*, Springer Texts in Business and Economics, DOI 10.1007/978-3-319-23428-1_16

Hence, this is an extension of the AR(p) process. Because of the additional exogenous explanatory variables, we sometimes speak of ARX(p, ℓ) models, although such processes are more often called **autoregressive distributed lag models**, ARDL(p, ℓ). We assume that $\{x_t\}$ is integrated of order one. In order to have cointegration with $\{y_t\}$, the ARDL model has to be stable. We adopt the stability condition from Proposition 3.4:

$$1 - a_1 z - \cdots - a_p z^p = 0 \quad \Rightarrow \quad |z| > 1 .$$

Example 16.1 (ARDL(2,2)) With $p = \ell = 2$ we consider

$$y_t = a_1 y_{t-1} + a_2 y_{t-2} + c_0 x_t + c_1 x_{t-1} + c_2 x_{t-2} + \varepsilon_t .$$

Due to the assumed stability, the parameter b can be defined:

$$b := \frac{c_0 + c_1 + c_2}{1 - a_1 - a_2} .$$

In fact, the denominator is not only different from zero but positive if stability is given (which we again know from Proposition 3.4): $1 - a_1 - a_2 > 0$. Thus, the following parameter γ is negative:

$$\gamma := -(1 - a_1 - a_2) < 0 .$$

Elementary manipulations lead to the reparameterization (cf. Problem 16.1)

$$\Delta y_t = \gamma \left[y_{t-1} - b\, x_{t-1} \right] - a_2 \Delta y_{t-1} + c_0 \Delta x_t - c_2 \Delta x_{t-1} + \varepsilon_t . \tag{16.1}$$

In this equation differences of y are related to their own lags and differences of x. In addition, they depend on a linear combination of lagged levels (in square brackets). This last aspect is the one constituting error-correction models. The cointegration relation $y = bx$ is understood as a long-run equilibrium relation and $v_{t-1} = y_{t-1} - b\, x_{t-1}$ as a deviation from it in $t - 1$. This deviation from the equilibrium again influences the increments of y_t. The involved linear combination of y_{t-1} and x_{t-1} needs to be stationary because $\{\Delta y_t\}$ is stationary by assumption. Hence, in the example it is obvious that such a relation between differences and levels implies cointegration. Indeed, it is the lagged deviation from the equilibrium v_{t-1} influencing the increments of Δy_t with a negative sign. If y_{t-1} is greater than the equilibrium value, $v_{t-1} > 0$, then this affects the change from y_{t-1} to y_t negatively, i.e. y is corrected towards the equilibrium, and vice versa for values below the equilibrium value. What economists know as deviation from the equilibrium is called "error" (in the sense of a deviation from a set target) in engineering, which explains the name error-correction model. ■

The example can be generalized. Cointegrated ARDL models of arbitrary order can always be formulated as error-correction models. This is not surprising against the background of Granger's representation theorem.

Granger's Representation Theorem

Error-correction adjustment is the downside to cointegration. The relation between cointegration and error-correction is explained by the following proposition where we, however, will not spell out all the technical details. The result goes back to Granger, cf. Engle and Granger (1987) or Johansen (1995, Theorem 4.2).

Proposition 16.1 (Representation Theorem) *Let $\{y_t\}$ and $\{x_t\}$ be integrated of order one. They are cointegrated if and only if they have an error-correction representation,*

$$\Delta y_t = \gamma\, v_{t-1} + \sum_{j=1}^{p} a_j\, \Delta y_{t-j} + \sum_{j=1}^{\ell} \alpha_j\, \Delta x_{t-j} + \varepsilon_t\,, \tag{16.2}$$

$$\Delta x_t = \gamma_x\, v_{t-1} + \sum_{j=1}^{p_x} a_j^{(x)}\, \Delta y_{t-j} + \sum_{j=1}^{\ell_x} \alpha_j^{(x)}\, \Delta x_{t-j} + \varepsilon_{x,t}\,, \tag{16.3}$$

$$v_t = y_t - b\, x_t \sim I(0)\,, \quad b \neq 0\,, \tag{16.4}$$

where at least one of the so-called adjustment coefficients γ or γ_x is different from zero.

Of course, not all a_j ($a_j^{(x)}$) and α_j ($\alpha_j^{(x)}$) need to be different from zero. The error sequences $\{\varepsilon_t\}$ and $\{\varepsilon_{x,t}\}$ are white noise and may be contemporaneously correlated. Frequently in practice, additional contemporaneous differences of the respective other variable are hence incorporated on the right-hand side. For Eq. (16.2) this means e.g. the inclusion of $\alpha_0\, \Delta x_t$. Then, one sometimes also speaks of the conditional or structural error-correction equation.

Cointegration and the Long-Run Variance Matrix

In (14.6) the symmetric long-run variance matrix Ω of a stationary vector has been defined. We now consider an I(1) vector $z_t' = (z_{1,t}, z_{2,t})$ such that $\Delta z_t = (w_{1,t}, w_{2,t})'$ is integrated of order zero, which is why Ω cannot be equal to the zero matrix. Nevertheless, the matrix does not have to be invertible. It rather holds: If the vector $\{z_t\}$ is cointegrated, then Ω has the reduced rank one and is not invertible. Equivalently, this means: If Ω is invertible, then $\{z_{1,t}\}$ and $\{z_{2,t}\}$ are not cointegrated. To this, we consider the following example.

Example 16.2 (Ω under Cointegration) In this example, let $\{z_{2,t}\}$ be a random walk, the cointegration parameter be one, and the deviation from the equilibrium $v_t = \varepsilon_{1,t}$ be iid and independent of $\{\Delta z_{2,t}\}$:

$$z_{1,t} = z_{2,t} + \varepsilon_{1,t} \,,$$

$$z_{2,t} = z_{2,t-1} + \varepsilon_{2,t} \,.$$

Hence, if $\{\varepsilon_{1,t}\}$ and $\{\varepsilon_{2,t}\}$ are independent with variances σ_1^2 and σ_2^2, then one shows with $w_{1,t} = \Delta z_{1,t} = \varepsilon_{2,t} + \varepsilon_{1,t} - \varepsilon_{1,t-1}$ and $w_{2,t} = \Delta z_{2,t} = \varepsilon_{2,t}$:

$$\Gamma_w(0) = \begin{pmatrix} \sigma_2^2 + 2\,\sigma_1^2 & \sigma_2^2 \\ \sigma_2^2 & \sigma_2^2 \end{pmatrix} \quad \text{and} \quad \Gamma_w(1) = \begin{pmatrix} -\sigma_1^2 & 0 \\ 0 & 0 \end{pmatrix} .$$

For $h > 1$, $\Gamma_w(h) = 0$. Thus, it holds:

$$\Omega_w = \sigma_2^2 \begin{pmatrix} 1 & 1 \\ 1 & 1 \end{pmatrix} ,$$

i.e. the matrix is of rank one and not invertible. ■

Conversely, it holds as well that full rank of Ω follows from the absence of cointegration. This is to be illustrated by the following example.

Example 16.3 (Ω without Cointegration) Now, let $\{z_{1,t}\}$ and $\{z_{2,t}\}$ be two random walks independent of each other:

$$z_{1,t} = z_{1,t-1} + \varepsilon_{1,t} \,,$$

$$z_{2,t} = z_{2,t-1} + \varepsilon_{2,t} \,.$$

Since $\{\varepsilon_{1,t}\}$ and $\{\varepsilon_{2,t}\}$ are independent with variances σ_1^2 and σ_2^2, one shows with $w_{1,t} = \varepsilon_{1,t}$ and $w_{2,t} = \varepsilon_{2,t}$:

$$\Gamma_w(0) = \begin{pmatrix} \sigma_1^2 & 0 \\ 0 & \sigma_2^2 \end{pmatrix} .$$

For $h > 0$, $\Gamma_w(h) = 0$. Thus, it holds:

$$\Omega_w = \Gamma_w(0) \,,$$

where this matrix has full rank 2 in the case of positive variances and is thus invertible. ■

Hence, the presence of cointegration of the I(1) vector $\{z_t\}$ depends on the matrix Ω. The examples show that no cointegration of $\{z_t\}$ is equivalent to the full rank of Ω, cf. Phillips (1986).

Linearly Independent Cointegration Vectors

For more than two I(1) variables linearly independent cointegration vectors can exist. In such a situation there can be no talk of "the true" cointegration vector. Each and every linear combination of independent cointegration vectors is itself again a cointegration vector. Although this cannot occur under our assumption of a bivariate vector, we want to become aware of this problem by means of a three-dimensional example.

Example 16.4 (Three Interest Rates) Assume z_1, z_2 and z_3 to be interest rates for one-month, two-month and three-month loans integrated of order one. Then one expects (due to the expectations hypothesis of the term structure) the interest rate differentials $z_1 - z_2$ and $z_2 - z_3$ to provide stable relations:

$$z_{1,t} - z_{2,t} = v_{1,t} \sim I(0),$$
$$z_{2,t} - z_{3,t} = v_{2,t} \sim I(0).$$

Here, $b_1' = (1, -1, 0)$ and $b_2' = (0, 1, -1)$ are linearly independent, and both are cointegrating vectors for $z_t' = (z_{1,t}, z_{2,t}, z_{3,t})$ as $v_{1,t}$ and $v_{2,t}$ are both assumed to be I(0):

$$\begin{pmatrix} b_1' \\ b_2' \end{pmatrix} \begin{pmatrix} z_{1,t} \\ z_{2,t} \\ z_{3,t} \end{pmatrix} = \begin{pmatrix} v_{1,t} \\ v_{2,t} \end{pmatrix}.$$

Hence, the uniqueness of the cointegration vector is lost. It is rather that b_1 and b_2 form a basis for the cointegration space. Each vector contained in the plane they span, i.e. each linear combination of b_1 and b_2, is itself again a cointegration vector. For example for

$$\beta_\alpha = b_1 + (1 - \alpha) b_2 = \begin{pmatrix} 1 \\ -\alpha \\ \alpha - 1 \end{pmatrix}$$

one obtains a stationary relation comprising all three variables from z_t,

$$z_{1,t} = \alpha z_{2,t} + (1 - \alpha) z_{3,t} + v_{1,t} + (1 - \alpha) v_{2,t}\,,$$

where $v_{1,t} + (1 - \alpha) v_{2,t}$ is I(0). In particular for $\alpha = 0$ one sees that $z_{1,t}$ and $z_{3,t}$ alone are also cointegrated with the cointegration vector $\beta_0' = b_1' + b_2' = (1, 0, -1)$. The cointegration vectors b_1, b_2 as well as β_0 provide economically reasonable and theoretically secured statements on the interest rate differentials. However, β_α for an arbitrary α is just as good a cointegrating vector. Hence,

$$z_1 = \alpha z_2 + (1 - \alpha) z_3$$

for $\alpha \neq 1$ or $\alpha \neq 0$ is as well a "true" long-run equilibrium relation, even if, in contrast to the interest rate differentials, it is not readily amenable to an economic interpretation. These problems with the interpretation of more than one linearly independent cointegration vectors are of a fundamental nature and cannot be solved unless one makes a priori (economically plausible) assumptions on the form of the cointegrating vectors. The cointegration analysis is not the life belt saving us from the lack of identification: With purely statistical methods one generally cannot make economic statements. ■

For only two I(1) variables $\{y_t\}$ and $\{x_t\}$, linearly independent cointegration vectors cannot exist. We show this by contradiction. Hence, let us assume with $b_1 \neq b_2$ from $\mathbb{R}$ that two linearly independent relations exist. We collect them row-wise in the matrix B:

$$B = \begin{pmatrix} 1 & -b_1 \\ 1 & -b_2 \end{pmatrix}.$$

Then, it holds by assumption that

$$B \begin{pmatrix} y_t \\ x_t \end{pmatrix} = \begin{pmatrix} y_t - b_1 x_t \\ y_t - b_2 x_t \end{pmatrix} = \begin{pmatrix} v_{1,t} \\ v_{2,t} \end{pmatrix}$$

is a vector of I(0) variables $\{v_{1,t}\}$ and $\{v_{2,t}\}$. Due to the independence of the cointegration vectors, B is invertible:

$$\begin{pmatrix} y_t \\ x_t \end{pmatrix} = B^{-1} \begin{pmatrix} v_{1,t} \\ v_{2,t} \end{pmatrix}.$$

This yields $\{y_t\}$ and $\{x_t\}$ as linear combinations of $\{v_{i,t}\}$, $i = 1, 2$, from which it follows that $\{y_t\}$ and $\{x_t\}$ themselves have to be I(0), which contradicts the assumption.

16.3 Cointegration Regressions

In the case of bivariate cointegration, the LS estimator of a static regression of y_t on x_t tends to the true value with the sample size (i.e. with rate n). For this fast rate of convergence the term **superconsistency** has been coined in the literature. At the same time limiting normality only arises under additional assumptions.

Superconsistent Estimation

We consider the LS estimator regressing y_t on x_t under the assumption (16.4) that cointegration is present. For the sake of simplicity, we again do not allow for an intercept (which is why we assume that the I(1) processes have the starting value zero):

$$y_t = \hat{b}\,x_t + \hat{v}_t\,, \quad t = 1, \ldots, n\,. \tag{16.5}$$

Then we write for the LS estimator:

$$\hat{b} - b = \frac{\sum_{t=1}^{n} x_t v_t}{\sum_{t=1}^{n} x_t^2}$$

or

$$n\left(\hat{b} - b\right) = \frac{n^{-1} \sum_{t=1}^{n} x_t v_t}{n^{-2} \sum_{t=1}^{n} x_t^2}\,.$$

To be able to apply the functional limit theory from Chap. 14, we now define

$$w_t = \begin{pmatrix} v_t \\ \Delta x_t \end{pmatrix}, \quad \text{i.e.} \quad z_t = \begin{pmatrix} \sum_{i=1}^{t} v_i \\ x_t \end{pmatrix} \tag{16.6}$$

instead of $z_t' = (y_t, x_t)$ as in Sect. 15.4 without cointegration. Then it holds with the results from Proposition 14.4(b) and (c):

$$n(\hat{b} - b) \xrightarrow{d} \frac{\int_0^1 B_2(s)dB_1(s) + \sum_{h=0}^{\infty} \mathrm{E}(\Delta x_t v_{t+h})}{\int_0^1 B_2^2(s)ds}\,.$$

As the LS estimator tends to the true value with the sample size n instead of with only $n^{0.5}$ as for the stationary regression model, since Stock (1987) and Engle and Granger (1987) it has become common usage to speak of superconsistency of the static cointegration estimator from (16.5); however, this result has been known from Phillips and Durlauf (1986) already.

Note that the estimation of b is consistent despite possible correlation between error term v_t and regressor x_t (or Δx_t). Insofar the cointegration regression knocks out the simultaneity bias (or “Haavelmo bias”): Superconsistency is a strong asymptotic argument for single equation regressions despite possibly existing dependencies through simultaneous relations between the individual equations, i.e. despite correlation between regressors and error term. According to this, the cointegration approach can be understood as a reaction to the simultaneous equation methodology of former decades. At the same time, it is not clear anymore which variable constitutes the endogenous left-hand side and which quantity identifies the exogenous regressor. Beside (16.4), it also holds as a "true relation" that

$$x_t = \frac{y_t}{b} - \frac{v_t}{b}.$$

Hence, if x_t is regressed on y_t, then one would obtain analogously a superconsistent estimator for b^{-1}. This vehemently contrasts the results of the stationary standard econometrics where the (asymptotic) validity of LS crucially depends on the correct specification of the single equation and exogeneity assumptions.

Further Asymptotic Properties

In Problems 16.2 and 16.3 we prove the further properties of the following proposition. Here, the standard errors of the t-statistic, the uncentered coefficient of determination and the Durbin-Watson statistic are defined as in Sect. 15.4.

Proposition 16.2 (Cointegration Regression) *For cointegrated I(1) processes $\{x_t\}$ and $\{y_t\}$ it holds with (16.4) and the notation introduced that*

$$\begin{aligned}
(a)\ n(\hat{b} - b) &\xrightarrow{d} \frac{\int_0^1 B_2(s)dB_1(s) + \Delta_{xv}}{\int_0^1 B_2^2(s)ds}, \\
(b)\ n(1 - R_{uc}^2) &\xrightarrow{d} \frac{\gamma_1(0)}{b^2 \int_0^1 B_2^2(s)ds}, \\
(c)\ s^2 &\xrightarrow{p} \gamma_1(0), \\
(d)\ t_b = \frac{\hat{b}-b}{s_b} &\xrightarrow{d} \frac{\int_0^1 B_2(s)dB_1(s) + \Delta_{xv}}{\sqrt{\gamma_1(0) \int_0^1 B_2^2(s)ds}}, \\
(e)\ dw &\xrightarrow{p} 2\,(1 - \rho_v(1)),
\end{aligned}$$

where

$$\Delta_{xv} := \sum_{h=0}^{\infty} E(\Delta x_t v_{t+h}) \quad and \quad \gamma_1(0) := Var(v_t)$$

as $n \to \infty$.

Three remarks are to help with the interpretation. (i) Note again that superconsistency holds even if the regressors x_t correlate with the error terms v_t. This nice property, however, comes at a price: Unfortunately, the limiting distributions from Proposition 16.2(a) and (d) are (without further assumptions) generally not Gaussian. (ii) As a rule, for regressions involving trending (integrated) variables one empirically observes values of the coefficient of determination near one, which is explained by Proposition 16.2(b). Due to that, for trending (integrated) time series the coefficient of determination cannot be interpreted as usual: Since $\{y_t\}$ does not have a constant variance, the coefficient of determination does not give the percentage of the variance explained by the regression. (iii) At no point it was assumed that the error terms, $\{v_t\}$, are white noise. For the first order residual autocorrelation, it holds

$$\hat{\rho}_{\hat{v}}(1) = \frac{\sum_{t=1}^{n-1} \hat{v}_t \hat{v}_{t+1}}{\sum_{t=1}^{n} \hat{v}_t^2} \xrightarrow{p} \rho_v(1) = \frac{\mathrm{E}(v_t v_{t+1})}{\mathrm{Var}(v_t)},$$

which is just reflected by the behavior of the Durbin-Watson statistic.

Asymptotic Normality

The price for the superconsistency without exogeneity assumption is that the limiting distribution of the t-statistic is generally not Gaussian anymore. However, if v_t and Δx_s are stochastically independent for all t and s, then Krämer (1986) shows that asymptotic normality arises. This assumption, however, is stronger than necessary. It suffices to require, first, $\Delta_{xv} = 0$ and, second, that the Brownian motions B_1 and B_2 are independent. For independence of B_1 and B_2 we only need

$$\omega_{12} = \sum_{h=-\infty}^{\infty} \mathrm{E}(\Delta x_t v_{t+h}) = 0 \tag{16.7}$$

as under this condition it holds that $B_i = \omega_i W_i$. Due to Proposition 10.4 the following corollary is obtained (also cf. Problem 16.4).

Corollary 16.1 (Asymptotic Normality) *If Δ_{xv} from Proposition 16.2 is zero, then it holds under (16.7) that*

$$t_b \xrightarrow{d} \mathcal{N}\left(0, \frac{\omega_1^2}{\gamma_1(0)}\right),$$

as $n \to \infty$.

If one has consistent estimators for the variance and the long-run variance, then the t-statistic can be modified as follows and can be applied with standard normal distribution asymptotics under Corollary 16.1:

$$\tau_b := \sqrt{\frac{\hat{\gamma}_1(0)}{\hat{\omega}_1^2}}\, t_b \xrightarrow{d} \mathcal{N}(0,1).$$

For the estimation of the long-run variance, we refer to the remarks in Sect. 15.2, in particular Example 15.2. As $\{v_t\}$ itself is not observable, $\hat{\gamma}_1(0)$ and $\hat{\omega}_1^2$ have to be calculated from $\hat{v}_t = y_t - \hat{b}x_t$.

Efficient Estimation

The assumptions $\Delta_{xv} = 0$ and $\omega_{12} = 0$ required for Corollary 16.1 are often not met in practice. We now consider modifications of LS resulting in limiting normality that get around such restrictions. To that end we have a closer look at the LS limit from Proposition 16.2 called now $\mathcal{L}(\hat{b})$:

$$n\,(\hat{b} - b) \xrightarrow{d} \mathcal{L}(\hat{b})\,.$$

We define the process $B_{1\cdot 2}$,

$$B_{1\cdot 2}(s) := B_1(s) - \omega_{12}\omega_2^{-2} B_2(s), \tag{16.8}$$

in such a way that it does not correlate with B_2:

$$\mathrm{E}(B_{1\cdot 2}(r)B_2(s)) = \min(r,s)\omega_{12} - \min(r,s)\omega_{12}\omega_2^{-2}\omega_2^2 = 0.$$

Because of normality the two processes are thus independent. The variance of the new process is

$$\mathrm{Var}(B_{1\cdot 2}(s)) = \mathrm{E}(B_{1\cdot 2}^2(s)) = s\omega_{1\cdot 2}^2\,, \quad \omega_{1\cdot 2}^2 := \omega_1^2 - \omega_{12}^2\omega_2^{-2}.$$

With

$$B_1(s) = B_{1\cdot 2}(s) + B_2(s)\omega_2^{-2}\omega_{12}$$

we may now decompose the LS limit as follows:

$$\begin{aligned}\mathcal{L}(\hat{b}) &= \left(\int_0^1 B_2^2(s)ds\right)^{-1}\left[\int_0^1 B_2(s)dB_{1\cdot 2}(s) + \int_0^1 B_2(s)dB_2(s)\omega_2^{-2}\omega_{12} + \Delta_{xv}\right]\\ &= \mathcal{L}_1 + \mathcal{L}_2 + \mathcal{L}_3\,.\end{aligned}$$

The first component, $\mathcal{L}_1$, is conditionally normal with mean zero, i.e.

$$\mathcal{L}_1 | B_2 \sim \mathcal{N}\left(0, \omega_{1\cdot 2}^2 \left(\int_0^1 B_2^2(s)ds\right)^{-1}\right),$$

which is true by Proposition 10.4. The second component defined as a multiple of $\int_0^1 B_2(s)dB_2(s) = (B_2^2(1) - 1)/2$ is stochastic and introduces skewness into $\mathcal{L}(\hat{b})$, while finally this distribution is shifted deterministically by Δ_{xv}. Consequently, for arbitrary $\varepsilon > 0$ one has

$$\mathrm{P}(|\mathcal{L}_1| < \varepsilon) > \mathrm{P}(|\mathcal{L}(\hat{b})| < \varepsilon),$$

see Saikkonen (1991, Theorem 3.1) Saikkonen. In that sense, $\mathcal{L}_1$ is more concentrated around zero than $\mathcal{L}(\hat{b})$ in general, such that intuitively the LS estimator is closest to zero for $\omega_{12} = \Delta_{xv} = 0$. This is the intuition for the following definition: Let $\hat{b}^+$ denote a cointegration estimator with

$$n(\hat{b}^+ - b) \xrightarrow{d} \mathcal{L}_1 \quad \text{where } \mathcal{L}_1 | B_2 \sim \mathcal{N}\left(0, \omega^2 \left(\int_0^1 B_2^2(s)ds\right)^{-1}\right), \qquad (16.9)$$

for some positive constant ω; then $\hat{b}^+$ is said to be efficient (see Saikkonen 1991 for a more general discussion). To further justify this notion of efficiency we note that full information maximum likelihood estimation of a cointegrated system results in exactly this distribution, see Phillips (1991).

Efficient cointegration regressions are not only interesting because they achieve the lower bound for the standard error; more importantly, related t-type statistics are asymptotically normal, which allows for standard inference. Suppose we have an efficient estimator satisfying (16.9), and that $\hat{\omega}$ (typically computed from cointegration residuals) is consistent for ω. Then we define the t-type statistic

$$t^+ = \frac{\hat{b}^+ - b}{\hat{\omega}} \sqrt{\sum_{t=1}^n x_t^2}.$$

From the previous discussion it follows, see Phillips and Park (1988):

$$t^+ \xrightarrow{d} \mathcal{N}(0, 1).$$

Consequently, the caveat of superconsistent LS cointegration regression, lacking normality in general, is overcome by efficient estimators.

Let us repeat once more: LS cointegration estimation is efficient under the not very realistic assumption that $\omega_{12} = \Delta_{xv} = 0$. Several modifications of LS achieving efficiency without this assumption have been proposed. First, the so-called dynamic LS estimator suggested independently by Saikkonen (1991) and Stock and Watson (1993) is settled in the time domain, see also Phillips and Loretan (1991); second, so-called frequency domain based modifications of LS have been suggested by Phillips and Hansen (1990) ("fully modified LS") or Park (1992) ("canonical cointegrating regression"); they all meet (16.9) and $t^+ \sim \mathcal{N}(0,1)$, asymptotically.

Linear Time Trends

In Sect. 15.2 we considered linear time trends and I(1) processes, i.e. so-called integrated processes with drift:

$$x_t = \mu + x_{t-1} + e_t\,, \quad \mu \neq 0. \tag{16.10}$$

By repeated substitution one obtains

$$x_t = x_0 + \mu\, t + \sum_{j=1}^{t} e_j,$$

i.e. $\{x_t\}$ consists of a linear trend of the slope μ and an I(1) component; and of a starting value whose influence can be neglected such that we set $x_0 = 0$ w.l.o.g. In addition, let the cointegration relation (16.4) hold true. Consequently, $\{y_t\}$ as well exhibits a linear trend of the slope $b\,\mu$. The cointegration relation (16.4) simultaneously eliminates the deterministic linear time trend and the stochastic I(1) trend from both series. In this case the static LS regression from (16.5) yields an even faster rate of convergence ($n^{1.5}$ instead of n) and simultaneously, the limiting distribution is Gaussian. The following proposition is a special case of the more general results from West (1988). We prove it in Problem 16.5.

Proposition 16.3 (West) *We assume cointegrated I(1) processes $\{x_t\}$ and $\{y_t\}$ with drift (i.e. (16.10) with (16.4)). Then it holds for the regression (16.5) without intercept that*

$$n^{1.5}(\hat{b} - b) \stackrel{d}{\to} \mathcal{N}\left(0, \frac{3\omega_1^2}{\mu^2}\right)$$

as $n \to \infty$, where ω_1^2 is the long-run variance of $\{v_t\}$ from (16.4).

For the sake of completeness, note that the asymptotics from Proposition 16.3 is qualitatively retained if the regression is run *with* intercept. However, the variance of the Gaussian distribution is changed. It becomes $12\omega_1^2/\mu^2$. For practical inference, μ^2 has to be estimated from $\{x_t\}$ or $\{\Delta x_t\}$, while ω_1^2 can be estimated consistently from the cointegration residuals $\hat{v}_t$. For $\mu = 1$ the limiting distribution from Proposition 16.3 equals that from (15.5), which is not coincidental, see also the proof in Problem 16.5: The scalar I(1) regressor with drift is dominated by the linear time trend; hence, the cointegration regression amounts to a trend stationary regression.

Note, however, that Proposition 16.3 holds only in our special case that x_t is a scalar I(1) variable. If one has a vector of I(1) regressors with drift instead of a scalar variable, then the asymptotic normality in general does not hold anymore, and the very fast convergence with rate $n^{1.5}$ is lost as well. Instead, Hansen (1992) proved results in line with Proposition 16.2 if there are several I(1) regressors of which at least one has a drift.

16.4 Cointegration Testing

If one regresses nonstationary (integrated) time series on each other, the interpretation of the regression outcome largely depends on whether the series are cointegrated or not. Hence, one has to test for the absence or presence of cointegration.

Residual-Based Dickey-Fuller Test

The idea of the following test for the null hypothesis of no cointegration dates back to Engle and Granger (1987), although a rigorous asymptotic treatment was provided later by Phillips and Ouliaris (1990). The idea is very simple. Without cointegration any linear combination of I(1) variables results in a series that too has a unit root. Hence, the Dickey-Fuller test is applied to LS residuals, which are computed from a regression with intercept:

$$\hat{u}_t = y_t - \hat{\alpha} - \hat{\beta} x_t \,.$$

Due to the included intercept, $\{\hat{u}_t\}$ are zero mean by construction. Hence, the DF regression in the second step may be run w.l.o.g. without intercept:

$$\hat{u}_t = \overline{a}\,\hat{u}_{t-1} + \overline{e}_t \,, \quad t = 1, \ldots, n \,.$$

The LS estimator $\overline{a}$ converges to 1 under the null hypothesis of a residual unit root with the rate known from Sect. 15.3. However, $\hat{\beta}$ does not converge to 0, but a limit characterized in Proposition 15.5. Consequently, the asymptotic distribution of $n(\overline{a} - 1)$ does not only depend on one WP but rather on two. Let $\overline{t}_a(2)$, involving 2 I(1) processes, denote the t-statistic related to $\overline{a} - 1$. Then the limit depends on two standard Wiener processes W_1 and W_2:

$$\bar{t}_a(2) \xrightarrow{d} \overline{\mathcal{DF}}(W_1, W_2) \, . \tag{16.11}$$

Interestingly, this limit is free of nuisance parameters as long as $\{\Delta y_t\}$ and $\{\Delta x_t\}$ are white noise; in particular, it does not depend on the eventual correlation between $\{\Delta y_t\}$ and $\{\Delta x_t\}$; see also Problem 16.7. If $\{\Delta y_t\}$ and $\{\Delta x_t\}$ are not white noise processes, then $\overline{a}$ may be computed from a lag-augmented regression, or a modification to the test statistic in line with Phillips (1987) has to be applied. Critical values or p-values are most often taken from MacKinnon (1991, 1996). Here, we do not present a definition of the functional shape of the limit $\overline{\mathcal{DF}}(W_1, W_2)$; rather, to give at least an idea thereof, we consider now explicitly the less complicated case without constant. So, $\hat{u}_t$ and $\hat{\beta}$ are now from a regression without intercept,

$$\hat{u}_t = y_t - \hat{\beta} x_t \, .$$

We use a new notation to denote the subsequent DF regression

$$\hat{u}_t = \breve{a}\, \hat{u}_{t-1} + \breve{\varepsilon}_t \, , \quad t = 1, \ldots, n \, .$$

In Problem 16.7 the following result is given.

Proposition 16.4 (Phillips & Ouliaris) *Let $\{z_t\}$ with $z_t' = (y_t, x_t)$ be a random walk such that $\{\Delta y_t\}$ and $\{\Delta x_s\}$ are white noise processes not correlated for $t \neq s$, although we do allow for contemporaneous correlation. In case of no cointegration it holds with the notation introduced above that*

$$n(\breve{a} - 1) \xrightarrow{d} \frac{\int_0^1 U(t)\, dU(t)}{\int_0^1 U^2(t)\, dt} \, , \qquad n \to \infty \, ,$$

with

$$U(t) := W_1(t) - \frac{\int_0^1 W_1(s) W_2(s)\, ds}{\int_0^1 W_2^2(s)\, ds} W_2(t) \, ,$$

where W_1 and W_2 are two independent Wiener processes.

The functional shape of this limit corresponds exactly to the one in (1.11); only that the WP W is replaced by U, which is, however, no longer a WP. Not surprisingly, the new process U is defined as residual from a projection of W_1 (corresponding to y) on W_2 (corresponding to x). Once more this shows the power and elegance of the functional limit theory approach introduced in Chap. 14.

Residual-Based KPSS Test

It comes in natural to apply also the KPSS test to regression residuals. As in Sect. 15.3 the hypotheses are now exchanged: We test for the null hypothesis of cointegration against the alternative of no cointegration. Working with LS cointegration residuals, $\hat{v}_t = y_t - \hat{b}x_t$, we have under the null hypothesis

$$\hat{v}_t = v_t - (\hat{b} - b)\, x_t \,,$$

where

$$n\,(\hat{b} - b) \xrightarrow{d} b_\infty$$

with the limiting distribution b_∞ characterized in Proposition 16.2. Interestingly, we observe by a FCLT (Proposition 14.1) that

$$n^{0.5}\,(\hat{b} - b)\, x_{\lfloor rn \rfloor} \;\Rightarrow\; b_\infty B_2(r)\,,$$

with B_2 being the Brownian motion behind $\{x_t\}$ from $z_t' = \left(\sum_{j=1}^{t} v_j,\, x_t\right)$. Therefore, it holds that $(\hat{b} - b)\, x_t$ converges to zero with growing sample size, and the empirical residuals are proxies of the unobserved cointegration deviation: $\hat{v}_t \approx v_t$. Hence, it is tempting to believe that a KPSS test applied to the sequence $\{\hat{v}_t\}$ behaves as if applied to $\{v_t\}$. This, however, is not correct, as we will demonstrate next, since the limit characterized in Proposition 15.4 is not recovered when working with cointegration residuals.

The KPSS test builds on the partial sum process $S_t = \sum_{j=1}^{t} \hat{v}_j$. Mimicking the proof of Proposition 14.2(a) in Problem 14.3, we obtain the following FCLT for the partial sum process:

$$\begin{aligned} n^{-0.5} S_{\lfloor rn \rfloor} &= n^{-0.5} \sum_{j=1}^{\lfloor rn \rfloor} v_j - n\,(\hat{b} - b)\, n^{-1.5} \sum_{j=1}^{\lfloor rn \rfloor} x_j \\ &\Rightarrow B_1(r) - b_\infty \int_0^r B_2(s)\, ds \,. \end{aligned}$$

Notwithstanding that $\hat{v}_t \approx v_t$ we must thus not jump at the conclusion that the residual effect is negligible: The more careful analysis showed that the limit of the partial sum process depends on the distribution b_∞ arising from the cointegration regression.

What is more, we know that the LS limit b_∞ from Proposition 16.2 is plagued by the nuisance parameters Δ_{xv} and ω_{12}, except for the special case of Corollary 16.1. Therefore, Shin (1994) suggested to apply the KPSS test not with LS residuals but with residuals from an efficient regression (now with intercept),

$$S_t^+ = \sum_{j=1}^{t} \hat{v}_j^+ , \quad \hat{v}_j^+ = y_t - \hat{a}^+ - \hat{b}^+ x_t ,$$

where $\hat{b}^+$ is efficient in the sense of (16.9). Efficient cointegration regressions rely on removing Δ_{xv} and ω_{12} consistently. Hence, Shin (1994) showed that the limiting distribution of the KPSS test applied to efficient residuals is free of nuisance parameters, and he provided critical values. Let $\overline{\eta}^+(2)$ denote the residual-based KPSS statistic building on $v_j^+ = y_t - \hat{a}^+ - \hat{b}^+ x_t$, thus involving two I(1) variables. Under the null hypothesis of cointegration it holds asymptotically that (Shin, 1994, Thm. 2)

$$\overline{\eta}^+(2) \xrightarrow{d} \overline{\mathcal{CM}}(W_1, W_2) ,$$

where

$$\overline{\mathcal{CM}}(W_1, W_2) = \int_0^1 \left[W_1(s) - sW_1(1) - \frac{\int_0^s \underline{W}_2(r)dr \int_0^1 \underline{W}_2(r)dW_1(r)}{\int_0^1 \underline{W}_2^2(r)dr} \right]^2 ds , \tag{16.12}$$

and $\underline{W}$ is again short for a demeaned Wiener process. For related work see Harris and Inder (1994) or Leybourne and McCabe (1994), although the latter paper considered only the case of LS residuals.

Error-Correction Test

The third test we look into is not residual-based. The analysis rather relies on the error-correction equation (16.2):

$$\Delta y_t = \gamma \, v_{t-1} + \text{differences} + \varepsilon_t .$$

The fact that we restrict the analysis to the error-correction equation of $\{y_t\}$ and that we ignore Eq. (16.3) has to be justified by the assumption

$$\gamma_x = 0 . \tag{16.13}$$

This assumptions implies that, when cointegration is present, only Δy_t reacts to the deviation from the equilibrium of the previous period. Because of the assumption (16.13), absence of cointegration means $\gamma = 0$, which is the null hypothesis. In order that there is an adjustment to the equilibrium in the case of cointegration, $\gamma < 0$ under the alternative:

$$H_0 : \gamma = 0 \qquad \text{vs.} \qquad H_1 : \gamma < 0 .$$

To provide further intuition for the test statistic, we rewrite the error-correction equation (16.2) by inserting the definition of v_{t-1}:

$$\begin{aligned} \Delta y_t &= \gamma (y_{t-1} - b x_{t-1}) + \sum_{j=1}^{p} a_j \, \Delta y_{t-j} + \sum_{j=1}^{\ell} \alpha_j \, \Delta x_{t-j} + \varepsilon_t \\ &= \gamma y_{t-1} + \theta x_{t-1} + \sum_{j=1}^{p} a_j \, \Delta y_{t-j} + \sum_{j=1}^{\ell} \alpha_j \, \Delta x_{t-j} + \varepsilon_t \, . \end{aligned} \qquad (16.14)$$

Here, we defined

$$\theta = -\gamma b \, ,$$

where the null hypothesis of course implies $\theta = 0$. Hence, one may test the null hypothesis by means of an F-type test statistic for $\gamma = \theta = 0$, which has been investigated by Boswijk (1994). Alternatively, one may employ a t-type test specifically for $\gamma = 0$ only as proposed by Banerjee, Dolado, and Mestre (1998). The following proposition characterizes the asymptotic behavior of the LS estimator $\hat{\gamma}$, cf. Banerjee, Dolado, and Mestre (1998, Proposition 1). The Wiener processes W_1 and W_2 are adopted from Proposition 14.4 with $z_t' = (y_t, x_t)$. In order to get a limiting distribution free of nuisance parameters, we assume that Δx_t and ε_s are uncorrelated at arbitrary points in time:

$$\mathrm{E}(\Delta x_t \varepsilon_s) = 0 \, ; \qquad (16.15)$$

see also the proof in Problem 16.6.

Proposition 16.5 (BDM) *Let the I(1) processes $\{x_t\}$ and $\{y_t\}$ be not cointegrated, and let the exogeneity assumption (16.15) be fulfilled. Then, it holds for the LS estimator from the regression (16.14) that*

$$n\,\hat{\gamma} \xrightarrow{d} \frac{\int_0^1 W_2^2(s)\,ds \int_0^1 W_1(s)\,dW_1(s) - \int_0^1 W_1(s)\,W_2(s)\,ds \int_0^1 W_2(s)\,dW_1(s)}{\int_0^1 W_2^2(s)ds \int_0^1 W_1^2(s)ds - \left(\int_0^1 W_1(s) W_2(s) ds \right)^2}$$

as $n \to \infty$.

Obviously, this limiting distribution can be reshaped into the following form (which lends itself for a multivariate generalization with vectors $\{x_t\}$):

$$n\hat{\gamma} \xrightarrow{d} \frac{\int_0^1 W_1(s)dW_1(s) - \int_0^1 W_1(s)W_2(s)ds \left(\int_0^1 W_2^2(s)ds\right)^{-1} \int_0^1 W_2(s)dW_1(s)}{\int_0^1 W_1^2(s)ds - \int_0^1 W_1(s)W_2(s)ds \left(\int_0^1 W_2^2(s)ds\right)^{-1} \int_0^1 W_1(s)W_2(s)ds} .$$

In fact, Banerjee et al. (1998) suggest the computation of the t-statistic relating to $\hat{\gamma}$: $t_\gamma(2)$. As limiting distribution under H_0 one obtains:

$$t_\gamma(2) \xrightarrow{d} \mathcal{BDM}(W_1, W_2) \tag{16.16}$$

$$= \frac{\int_0^1 W_1(s)dW_1(s) - \int_0^1 W_1(s)W_2(s)ds \left(\int_0^1 W_2^2(s)ds\right)^{-1} \int_0^1 W_2(s)dW_1(s)}{\sqrt{\int_0^1 W_1^2(s)ds - \int_0^1 W_1(s)W_2(s)ds \left(\int_0^1 W_2^2(s)ds\right)^{-1} \int_0^1 W_1(s)W_2(s)ds}} .$$

Again, in most practical situations an intercept will be included in (16.14). If the corresponding test statistic is called $\bar{t}_\gamma(2)$, then it holds under the null hypothesis that

$$\bar{t}_\gamma(2) \xrightarrow{d} \overline{\mathcal{BDM}}(W_1, W_2) . \tag{16.17}$$

This limit has the same functional shape as $\mathcal{BDM}(W_1, W_2)$, only that W_i have to be replaced by the demeaned analogs $\underline{W}_i$, $i = 1, 2$.

Simulated critical values for conducting the tests can be found in Banerjee et al. (1998). One rejects for small values. From Ericsson and MacKinnon (2002) p-values are available, too.

Linear Time Trends

It has been mentioned in this and the previous chapter that in practice one would run regressions with an intercept to account for non-zero means of the series. But how should one proceed if the mean function follows a linear time trend, i.e. if the series are I(1) with drift? This is a quite realistic assumption for many economic and financial time series where positive growth rates are plausible. One might consider the analysis of detrended data, see Sect. 15.2. Note that the regression of detrended series is equivalent to including a linear time trend in the regression (see Frisch & Waugh, 1933):

$$y_t = \hat{\alpha} + \hat{\delta}\, t + \hat{\beta}\, x_t + \hat{u}_t .$$

This is why we call such regressions also **detrended regressions**. Similarly, one may augment the error-correction regression (16.14) by a linear time trend (and a constant). Many economists, however, do not run detrended regression or detrend the series even if the data display a linear time trend by eyeball inspection. Economically, it is often more meaningful to "explain" one trend by another instead of regressing deviations from linear trends on each other. Also statistically the regression of detrended data may not seem advisable since power losses are to be expected (see Hamilton, 1994, Sect. 19.2).

Running regressions with intercept only, i.e. without detrending, in the presence of linear time trends in the regressors has some subtle implications, however. Generally, the presence of a linear time trend in the data not accounted for in the regression will affect the limiting distributions. Just remember that a simple (or bivariate) cointegration regression in the presence of a linear time trend (Proposition 16.3) resembles more the detrending of a trend stationary process (Proposition 15.1) than a cointegration regression without linear trend (Proposition 16.2). More precisely, a linear time trend in $\{x_t\}$ will dominate the stochastic unit root in the following sense: If $\{x_t\}$ is I(1) with drift, $\mathrm{E}(\Delta x_t) = \mu + e_t$, $\mu \neq 0$, or

$$x_t = x_0 + \mu\, t + \sum_{j=1}^{t} e_j\,, \quad t = 1, \ldots n\,,$$

then this process grows with rate n (and not $n^{0.5}$, see Proposition 14.1):

$$\frac{x_{\lfloor rn \rfloor}}{n} \Rightarrow 0 + \mu\, r + 0\,, \quad \mu \neq 0\,.$$

This provides an intuition for the following finding by Hansen (1992, Theorem 7): If $\{x_t\}$ is I(1) with drift and $\{y_t\}$ and $\{x_t\}$ are not cointegrated, and if a regression-based DF test for no cointegration is computed from a regression with intercept but without detrending, then the limiting distribution of the t-type DF statistic is not given by $\overline{\mathcal{DF}}(W_1, W_2)$ from (16.11), but rather by the detrended univariate distribution $\widetilde{\mathcal{DF}}$ given in (15.10). A corresponding result was established for the error-correction test by Hassler (2000a): If $\{x_t\}$ and $\{y_t\}$ are I(1) but not cointegrated, and if $\{x_t\}$ is integrated with drift and the error-correction test for no cointegration is computed from a regression with intercept but without detrending, then the limiting distribution of the t-statistic is not given by $\overline{\mathcal{BDM}}(W_1, W_2)$ from (16.17), but by the detrended Dickey-Fuller distribution $\widetilde{\mathcal{DF}}$. And similarly: If $\{x_t\}$ is I(1) with drift and cointegrated with $\{y_t\}$, and if a regression-based KPSS test for cointegration is computed from a regression with intercept only, then the limiting distribution of the KPSS statistic is not given by $\overline{\mathcal{CM}}(W_1, W_2)$ from (16.12), but rather by the detrended univariate distribution $\widetilde{\mathcal{CM}}$ given in (15.11); see Hassler (2000b). Hence, we have the following proposition.

Proposition 16.6 (Hansen & Hassler) *Consider the test statistics $\bar{t}_a(2)$, $\bar{t}_\gamma(2)$ or $\bar{\eta}^+(2)$ computed without detrending to test for the null hypothesis of (no)* *cointegration of the* $I(1)$ *processes* $\{x_t\}$ *and* $\{y_t\}$*. With* $\widetilde{\mathcal{DF}}$ *from (15.10) and* $\widetilde{\mathcal{CM}}$ *from (15.11) it holds under the respective null hypotheses that*

$$(a)\ \bar{t}_a(2) \xrightarrow{d} \begin{cases} \overline{\mathcal{DF}}(W_1, W_2), & \text{if } E(\Delta x_t) = 0 \\ \widetilde{\mathcal{DF}}, & \text{if } E(\Delta x_t) \neq 0 \end{cases},$$

$$(b)\ \bar{t}_\gamma(2) \xrightarrow{d} \begin{cases} \overline{\mathcal{BDM}}(W_1, W_2), & \text{if } E(\Delta x_t) = 0 \\ \widetilde{\mathcal{DF}}, & \text{if } E(\Delta x_t) \neq 0 \end{cases},$$

$$(c)\ \bar{\eta}^+(2) \xrightarrow{d} \begin{cases} \overline{\mathcal{CM}}(W_1, W_2), & \text{if } E(\Delta x_t) = 0 \\ \widetilde{\mathcal{CM}}, & \text{if } E(\Delta x_t) \neq 0 \end{cases},$$

as $n \to \infty$.

Proposition 16.6 is not restricted to bivariate regressions, but carries over to the general multiple regression case as follows:

Consider single-equation regressions estimated by LS (or efficient variants thereof); regressions with intercept only on k I(1) regressors, of which at least one has a drift, result under the null hypothesis (of cointegration or no cointegration, respectively) in a limit as if one runs a detrended regression on $k-1$ *I(1) regressors.*

For $k = 1$ this reproduces Proposition 16.6. For a proof for the residual-based DF test with $k > 1$ see again Hansen (1992), and also the lucid discussion by Hamilton (1994, p. 596, 597); for a proof for the error-correction test see Hassler (2000a), and for the residual-based KPSS test see Hassler (2001) for $k > 1$.

In view of Proposition 16.6 one may identify two strategies when testing cointegration from regressions with intercept only; we restrict the discussion to the case of a scalar regressor x_t. First, one might ignore the possibility of linear trends and always work with critical values from $\overline{\mathcal{DF}}(W_1, W_2)$, $\overline{\mathcal{BDM}}(W_1, W_2)$ or $\overline{\mathcal{CM}}(W_1, W_2)$ provided for the case of regressions with intercept only under no drift; we call this strategy S_I (I for "ignoring"), and of course it is not correct if $\{x_t\}$ displays a linear trend in mean. Second, one may always account for the possibility of linear time trends and work with critical values from $\widetilde{\mathcal{DF}}$ or $\widetilde{\mathcal{CM}}$; let us call this strategy S_A (A for "account"), and note that it is only appropriate if $\{x_t\}$ is indeed dominated by a linear time trend. In a numerical example we discuss the consequences of S_I and S_A.

Example 16.5 (Testing under the Suspicion of Time Trends) We consider tests at a nominal significance level of 5 %. Let $\tilde{c}_1$ and $\bar{c}_2$ denote critical values from $\widetilde{\mathcal{DF}}$ or $\widetilde{\mathcal{CM}}$ and $\overline{\mathcal{DF}}(W_1, W_2)$, $\overline{\mathcal{BDM}}(W_1, W_2)$ or $\overline{\mathcal{CM}}(W_1, W_2)$, respectively. For the residual-based DF test by Phillips and Ouliaris (1990) we take asymptotic critical values from MacKinnon (1991): $\tilde{c}_1 = -3.41$ and $\bar{c}_2 = -3.34$. Coincidently, these critical values are not very distant. Strategy S_I results in a slightly too liberal test (rejecting more often than with 5 % probability) in the presence of a drift, while S_A is mildly conservative (rejecting less often than in 5 % of all cases) in the absence of a linear trend in the regressor. So, for the residual-based DF test, Proposition 16.6 is not so relevant, since the distributions happen to differ not that much, and $\tilde{c}_1 \approx \overline{c}_2$. For the error-correction test by Banerjee et al. (1998), however, matters are not quite so harmless, since stronger size distortions are caused by a larger difference of the asymptotic critical values: $\tilde{c}_1 = -3.41$ and $\overline{c}_2 = -3.19$. With the residual-based KPSS test, things change qualitatively and quantitatively. Critical values from Kwiatkowski, Phillips, Schmidt, and Shin (1992) and Shin (1994) are $\tilde{c}_1 = 0.146$ and $\overline{c}_2 = 0.314$, and thus differ dramatically. Since one rejects for too large values, strategy S_I implies in the presence of a linear time trend a very conservative test; it will hardly reject the null hypothesis, which comes at a price of power of course. The other way round, without linear time trends strategy S_A will reject the true null hypothesis much too often resulting in an intolerably liberal test. ■

Clearly, none of the strategies S_I or S_A is generally acceptable when testing under the possibility of linear time trends. Fortunately, one often has strong a priori beliefs regarding the absence or presence of a linear time trend in the regressor. If one is convinced that a linear time trend is present in $\{x_t\}$, then one would apply e.g. $\bar{t}_a(2)$ or $\bar{t}_\gamma(2)$ with critical values from $\widetilde{\mathcal{DF}}$; if one believes that there is no linear time trend behind $\{x_t\}$, then critical values from $\overline{\mathcal{DF}}(W_1, W_2)$ or $\overline{\mathcal{BDM}}(W_1, W_2)$ must be recommended.

If one is not sure about the absence or presence of a linear time trend in the data, then there are (at least) two more strategies beyond S_I or S_A one may employ. As a third strategy, one may always test from detrended data. This clearly circumvents size distortions, but comes at a price of power losses as has been acknowledged for instance by Hansen (1992) or Hamilton (1994, p. 598). Fourth, one may rely on a pretest whether the regressor follows a linear trend or not. If a linear time trend in $\{x_t\}$ is significant, then one would apply e.g. $\overline{\eta}^+(2)$ with critical values from $\widetilde{\mathcal{CM}}$; if not, then critical values from $\overline{\mathcal{CM}}(W_1, W_2)$ should be applied. Such a pretesting strategy, however, will be troubled in small samples by the problem of controlling the significance level when carrying out a sequence of conditional tests (multiple testing). A recommendation whether the strategy of generally detrending or the strategy of pretesting is to be preferred, when the presence or absence of a linear time trend is debatable, will require future research.

16.5 Problems and Solutions

Problems

16.1 Show the equivalence of (16.1) and the ARDL(2,2) parameterization from Example 16.1.

16.2 Prove statements (b), (c) and (d) from Proposition 16.2.

16.3 Prove statement (e) from Proposition 16.2.

16.4 Prove Corollary 16.1.

16.5 Prove Proposition 16.3.

16.6 Prove Proposition 16.5 for the special case that no lagged differences are required to obtain white noise errors $\{\varepsilon_t\}$:

$$\Delta y_t = \gamma y_{t-1} + \theta x_{t-1} + \varepsilon_t .$$

16.7 Prove Proposition 16.4.

Solutions

16.1 Using γ and b, one obtains with $\Delta = 1 - L$ from (16.1):

$$\begin{aligned} y_t &= y_{t-1} - (1 - a_1 - a_2)y_{t-1} + (c_0 + c_1 + c_2)x_{t-1} \\ &\quad - a_2(y_{t-1} - y_{t-2}) + c_0(x_t - x_{t-1}) - c_2(x_{t-1} - x_{t-2}) + \varepsilon_t \\ &= a_1 y_{t-1} + a_2 y_{t-2} + c_0 x_t + c_1 x_{t-1} + c_2 x_{t-2} + \varepsilon_t . \end{aligned}$$

Hence, the claim is already proved.

16.2 We proceed in the same way as in the proof of Proposition 15.5, only that we work under (16.6) when appealing to Proposition 14.4. We start with

$$n\left(1 - R_{uc}^2\right) = \frac{n^{-1} \sum\limits_{t=1}^{n} \hat{v}_t^2}{n^{-2} \sum\limits_{t=1}^{n} y_t^2} = \frac{s^2}{n^{-2} \sum\limits_{t=1}^{n} y_t^2} .$$

The numerator on the right-hand side is just

$$\begin{aligned} s^2 &= n^{-1} \sum_{t=1}^{n} \left(y_t - \hat{b}\, x_t\right)^2 \\ &= n^{-1} \sum_{t=1}^{n} \left(b\, x_t - \hat{b}\, x_t + v_t\right)^2 \\ &= n^{-1} \sum_{t=1}^{n} \left[(b - \hat{b})^2 x_t^2 + 2\,(b - \hat{b}) x_t\, v_t + v_t^2\right] \\ &= n\,(b - \hat{b})^2 \frac{\sum\limits_{t=1}^{n} x_t^2}{n^2} + 2\,(b - \hat{b}) \frac{\sum\limits_{t=1}^{n} x_t\, v_t}{n} + \frac{\sum\limits_{t=1}^{n} v_t^2}{n} . \end{aligned}$$

The first of the three remaining terms tends to zero as $\sum x_t^2$ is of order n^2 and $(b - \hat{b})$ is of order n^{-1}; correspondingly, the second expression tends to zero as $\sum x_t\, v_t$ grows with n; finally, the third term converges to $\mathrm{Var}(v_t)$ as a law of large numbers holds for $\{v_t^2\}$. This proves Proposition 16.2(c). By the same arguments, one establishes:

$$n^{-2} \sum_{t=1}^{n} y_t^2 = n^{-2} \sum_{t=1}^{n} (b^2\, x_t^2 + 2\, b\, x_t\, v_t + v_t^2) \xrightarrow{d} b^2 \int_0^1 B_2^2(s)\, ds + 0 + 0\,.$$

Hence, Proposition 16.2(b) is proved as well.

Finally, the behavior of the t-statistic with $s_{\hat{b}}^2 = s^2 / \sum x_t^2$ is clear again by Proposition 14.4:

$$t_b = \frac{\hat{b} - b}{s_b} = \frac{\frac{1}{n} \sum\limits_{t=1}^{n} x_t\, v_t}{s \sqrt{\frac{1}{n^2} \sum\limits_{t=1}^{n} x_t^2}} \xrightarrow{d} \frac{\int\limits_0^1 B_2(s) dB_1(s) + \sum\limits_{h=0}^{\infty} \mathrm{E}\,(\Delta x_t\, v_{t+h})}{\sqrt{\gamma_1(0) \int\limits_0^1 B_2^2(s) ds}}\,.$$

16.3 In order to analyze the behavior of the Durbin-Watson statistic, we only need to study the numerator,

$$n^{-1} \sum_{t=2}^{n} (\hat{v}_t - \hat{v}_{t-1})^2 = n^{-1} \sum_{t=2}^{n} \left((b - \hat{b}) \Delta\, x_t + \Delta\, v_t \right)^2$$

$$= (b - \hat{b})^2 \frac{\sum\limits_{t=2}^{n} (\Delta\, x_t)^2}{n} + 2(b - \hat{b}) \frac{\sum\limits_{t=2}^{n} \Delta\, x_t\, \Delta\, v_t}{n} + \frac{\sum\limits_{t=2}^{n} (\Delta\, v_t)^2}{n}\,.$$

As $(b - \hat{b})$ tends to zero, there remains asymptotically

$$n^{-1} \sum_{t=2}^{n} (\Delta \hat{v}_t)^2 \xrightarrow{p} \mathrm{Var}(\Delta\, v_t) = 2\, \mathrm{Var}(v_t) - 2\, \mathrm{Cov}(v_t, v_{t-1})\,.$$

Hence, it holds as claimed:

$$dw = \frac{n^{-1} \sum\limits_{t=2}^{n} (\Delta \hat{v}_t^2)}{s^2} \xrightarrow{p} 2\,(1 - \rho_v(1))\,,$$

as s^2 approaches $\mathrm{Var}(v_t)$ with n growing.

16.4 The result will follow from Proposition 16.2. By (16.7), we have $\omega_{12} = 0$. Due to the resulting diagonality of Ω, $\Omega^{0.5}$ is diagonal as well, cf. Ω_2 from

Example 14.6. Hence it holds:

$$\begin{pmatrix} B_1(s) \\ B_2(s) \end{pmatrix} = \begin{pmatrix} \omega_1 W_1(s) \\ \omega_2 W_2(s) \end{pmatrix},$$

and ω_2 cancels from the limiting distribution of Proposition 16.2(d) (using the second assumption $\Delta_{xv} = 0$):

$$\frac{\omega_2 \int_0^1 W_2(s)dW_1(s)\omega_1}{\sqrt{\gamma_1(0)\omega_2^2 \int_0^1 W_2^2(s)ds}} = \frac{\omega_1}{\sqrt{\gamma_1(0)}} \frac{\int_0^1 W_2(s)dW_1(s)}{\sqrt{\int_0^1 W_2^2(s)ds}}.$$

According to Proposition 10.4, the stochastic quotient on the right-hand side follows a standard normal distribution. Hence, it holds that

$$\frac{\sqrt{\gamma_1(0)}}{\omega_1} t_b \xrightarrow{d} \mathcal{N}(0,1),$$

which proves the corollary.

16.5 By assumption, it holds:

$$n^{-3} \sum_{t=1}^{n} x_t^2 = n^{-3} \left[\mu^2 \sum_{t=1}^{n} t^2 + 2\mu \sum_{t=1}^{n} \left(t \sum_{j=1}^{t} e_j \right) + \sum_{t=1}^{n} \left(\sum_{j=1}^{t} e_j \right)^2 \right].$$

We know from Proposition 14.2(c) and (e) that the second and the third expression in square brackets have to be divided by $n^{2.5}$ and n^2, respectively, such that they converge. However, in front of the square bracket there is n^{-3}, such that it holds with (15.2):

$$n^{-3} \sum_{t=1}^{n} x_t^2 \xrightarrow{d} \frac{\mu^2}{3}.$$

Hence we have the denominator of the LS estimator under control:

$$\hat{b} = b + \frac{\sum_{t=1}^{n} x_t v_t}{\sum_{t=1}^{n} x_t^2}.$$

In order to crack the numerator, we consider

$$\frac{x_{\lfloor sn \rfloor}}{n} = \frac{x_0}{n} + \frac{\mu \lfloor sn \rfloor}{n} + \frac{\sum_{j=1}^{\lfloor sn \rfloor} e_j}{n},$$

and due to Proposition 14.1 it holds

$$\frac{x_{\lfloor sn \rfloor}}{n} \approx \frac{\mu \lfloor sn \rfloor}{n} \xrightarrow{d} \mu\, s\,,$$

and x_t is dominated by a linear trend, i.e. x_t behaves just as μt. Thus, as for the detrending in the trend stationary case, we obtain with a standard Wiener process V that (see Sect. 15.2):

$$\begin{aligned} n^{1.5}(\hat{b} - b) &= \frac{n^{-1.5} \sum_{t=1}^{n} x_t v_t}{n^{-3} \sum_{t=1}^{n} x_t^2} \\ &\xrightarrow{d} \frac{\mu \omega_1 \int_0^1 s dV(s)}{\mu^2/3} \\ &= \frac{3\,\omega_1}{\mu} \int_0^1 s dV(s) \\ &\sim \mathcal{N}\left(0, \frac{3\,\omega_1^2}{\mu^2}\right), \end{aligned}$$

where $\omega_1 V(s)$ is the Brownian motion corresponding to $\sum_{j=1}^{t} v_j$, and ω_1^2 is the long-run variance of $\{v_t\}$. The normality of the integral follows from Example 9.2. Hence, the claim is proved.

16.6 Because of the simplifying assumption we consider the regression of (16.14) without differences. As LS estimator for the vector

$$\psi := \begin{pmatrix} \gamma \\ \theta \end{pmatrix}$$

one hence obtains for a sample $t = 1, \ldots, n$:

$$\hat{\psi} = D^{-1} \sum_{t=1}^{n} \begin{pmatrix} y_{t-1} \\ x_{t-1} \end{pmatrix} \Delta y_t\,, \tag{16.18}$$

where D has the form

$$D = \begin{pmatrix} \sum_{t=1}^{n} y_{t-1}^2 & \sum_{t=1}^{n} y_{t-1} x_{t-1} \\ \sum_{t=1}^{n} x_{t-1} y_{t-1} & \sum_{t=1}^{n} x_{t-1}^2 \end{pmatrix}.$$

Plugging in $\Delta y_t = \varepsilon_t$ under H_0, we obtain

$$\hat{\psi} = D^{-1} \sum_{t=1}^{n} \begin{pmatrix} y_{t-1} \\ x_{t-1} \end{pmatrix} \varepsilon_t \, .$$

In the case of no cointegration, for using Proposition 14.4 we choose

$$z_t = \begin{pmatrix} y_t \\ x_t \end{pmatrix} .$$

Then it holds for the matrix D:

$$n^{-2} D \overset{d}{\to} \begin{pmatrix} \int_0^1 B_1^2(s)\, ds & \int_0^1 B_1(s)\, B_2(s)\, ds \\ \int_0^1 B_1(s)\, B_2(s)\, ds & \int_0^1 B_2^2(s)\, ds \end{pmatrix} . \tag{16.19}$$

For the inverse, this implies

$$n^2 D^{-1} \overset{d}{\to} \frac{1}{\det} \begin{pmatrix} \int_0^1 B_2^2(s)\, ds & -\int_0^1 B_1(s)\, B_2(s)\, ds \\ -\int_0^1 B_1(s)\, B_2(s)\, ds & \int_0^1 B_1^2(s)\, ds \end{pmatrix} ,$$

where "det" stands for the determinant of the limiting matrix from (16.19). Here, the known inversion formula for (2×2)-matrices was applied:

$$\begin{pmatrix} a\, b \\ c\, d \end{pmatrix}^{-1} = \frac{1}{\det} \begin{pmatrix} d & -b \\ -c & a \end{pmatrix}, \qquad \det = a\, d - b\, c \, . \tag{16.20}$$

Note that Proposition 14.4(b) was applied with z_{t-1} instead of z_t. This is unproblematic and can be justified by similar arguments like those leading to Corollary 14.1 in the univariate case.

Next, we analyze with Proposition 14.4(c)

$$\begin{aligned} n^{-1} \sum_{t=1}^{n} \begin{pmatrix} y_{t-1} \\ x_{t-1} \end{pmatrix} \Delta y_t &= n^{-1} \sum_{t=1}^{n} (z_t - w_t)\, w_{1,t} \\ &\overset{d}{\to} \int_0^1 B(s)\, dB_1(s) + \sum_{h=0}^{\infty} \mathrm{E}(w_t w_{1,t+h}) - \mathrm{E}(w_t w_{1,t}) \end{aligned}$$

$$= \int_0^1 B(s)\, dB_1(s) + \sum_{h=1}^{\infty} \mathrm{E}(w_t w_{1,t+h})$$

$$= \int_0^1 B(s)\, dB_1(s) + 0\,,$$

where we used that $w_{1,t} = \Delta y_t = \varepsilon_t$ is free from serial correlation and is uncorrelated with Δx_s at each point in time, see (16.15):

$$\mathrm{E}(w_t w_{1,t+h}) = 0\,, \quad h > 0\,.$$

Thus, under the null hypothesis of no cointegration, we obtain that $\hat{\gamma}$ tends to zero. For this purpose we consider the first row of the limit of $n^2 D^{-1}$ multiplied by $\int B(s)\, dB_1(s)$:

$$n\,\hat{\gamma} \xrightarrow{d} \frac{\int_0^1 B_2^2(s)\, ds \int_0^1 B_1(s)\, dB_1(s) - \int_0^1 B_1(s) B_2(s)\, ds \int_0^1 B_2(s)\, dB_1(s)}{\det}\,.$$

This is almost the claim as "det" is defined as the determinant of the limit of $n^{-2}D$. Finally, note that $B_i = \omega_i W_i$ holds as $\Delta x_t = w_{2,t}$ and $\Delta y_s = \varepsilon_s = w_{1,s}$ are uncorrelated. Thus, the long-run variances cancel from the limiting distribution and one obtains the required result.

16.7 Under the null hypothesis of no cointegration we define $z_t' = (y_t, x_t)$ with long-run variance matrix of full rank. The corresponding vector Brownian motion $B' = (B_1, B_2)$ can be written in terms of independent WPs $W' = (W_1, W_2)$ as

$$B(t) = T\,W(t) = \begin{pmatrix} t_{11} W_1(t) + \frac{\omega_{12}}{\omega_2} W_2(t) \\ \omega_2 W_2(t) \end{pmatrix}.$$

Here, T is the triangular decomposition given in (14.7) with $TT' = \Omega$. The limit of $\hat{\beta}$ from Proposition 15.5(a) hence becomes

$$\omega_2 \beta_\infty = \frac{\int_0^1 B_1(s) W_2(s) ds}{\int_0^1 W_2^2(s) ds} = \frac{t_{11} \int_0^1 W_1(s) W_2(s) ds + \frac{\omega_{12}}{\omega_2} \int_0^1 W_2^2(s) ds}{\int_0^1 W_2^2(s) ds}\,. \quad (16.21)$$

With this result one obtains a FCLT for the residuals $\hat{u}_t = y_t - \hat{\beta} x_t$:

$$n^{-0.5} \hat{u}_{\lfloor rn \rfloor} \Rightarrow B_1(r) - \beta_\infty B_2(r)$$
$$= (1, -\beta_\infty)\, T\, W(r)$$

$$= t_{11}\left(W_1(r) - \frac{\int_0^1 W_1(s)W_2(s)\,ds}{\int_0^1 W_2^2(s)\,ds} W_2(r)\right)$$

$$= t_{11}U(r)\,.$$

Unfortunately, however, Proposition 14.2 does not apply directly since $U(r)$ is not a WP. Still, with the techniques we used to prove Proposition 14.2(e) and (f) in Problems 14.3 and 14.5, we can establish (omitting details)

$$n^{-2}\sum_{t=1}^{n}\hat{u}_{t-1}^2 \xrightarrow{d} t_{11}^2 \int_0^1 U^2(t)dt\,,$$

$$n^{-1}\sum_{t=1}^{n}\hat{u}_{t-1}\Delta\hat{u}_t \xrightarrow{d} t_{11}^2\frac{U^2(1)}{2} - \frac{1}{2}\left(\omega_1^2 - 2\beta_\infty\omega_{12} + \beta_\infty^2\omega_2^2\right),$$

where the last limit arises because $\{\Delta z_t\}$ is white noise such that ω_i^2 and ω_{12} coincide with the (co)variances. Further, note by $B = TW$ that

$$\omega_{12}\beta_\infty = \frac{t_{11}\omega_{12}}{\omega_2}\frac{\int_0^1 W_1(s)W_2(s)ds}{\int_0^1 W_2^2(s)ds} + \frac{\omega_{12}^2}{\omega_2^2}\,. \tag{16.22}$$

Remember from (14.7) that

$$t_{11}^2 = \omega_1^2 - \frac{\omega_{12}^2}{\omega_2^2}\,.$$

Consequently, for

$$\breve{a} = 1 + \frac{\sum_{t=1}^n \hat{u}_{t-1}\Delta\hat{u}_t}{\sum_{t=1}^n \hat{u}_{t-1}^2}$$

we get by (16.21) and (16.22) that

$$n\left(\breve{a}-1\right) \xrightarrow{d} \frac{\frac{U^2(1)}{2} - \frac{1}{2}\left(1 + \left(\frac{\int_0^1 W_1(s)W_2(s)ds}{\int_0^1 W_2^2(s)ds}\right)^2\right)}{\int_0^1 U^2(t)dt}\,.$$

The numerator of this limit may be condensed. Use the product rule from Example 11.5 to obtain

$$W_1(1)W_2(1) = \int_0^1 W_1(s)dW_2(s) + \int_0^1 W_2(s)dW_1(s)\,.$$

It follows that

$$\frac{U^2(1)}{2} = \frac{W_1^2(1)}{2} - \frac{\int_0^1 W_1(s)W_2(s)ds}{\int_0^1 W_2^2(s)ds}\left(\int_0^1 W_1(s)dW_2(s) + \int_0^1 W_2(s)dW_1(s)\right)$$
$$+ \left(\frac{\int_0^1 W_1(s)W_2(s)ds}{\int_0^1 W_2^2(s)ds}\right)^2 \frac{W_2^2(1)}{2}.$$

Once more by Ito's lemma $\frac{W_i^2(1)}{2} = \int_0^1 W_i(s)dW_i(s) + \frac{1}{2}$, such that

$$\frac{U^2(1)}{2} = \int_0^1 W_1(s)dW_1(s) + \frac{1}{2}$$
$$-\frac{\int_0^1 W_1(s)W_2(s)ds}{\int_0^1 W_2^2(s)ds}\left(\int_0^1 W_1(s)dW_2(s) + \int_0^1 W_2(s)dW_1(s)\right)$$
$$+ \left(\frac{\int_0^1 W_1(s)W_2(s)ds}{\int_0^1 W_2^2(s)ds}\right)^2 \left(\int_0^1 W_2(s)dW_2(s) + \frac{1}{2}\right)$$
$$= \int_0^1 U(t)\,dU(t) + \frac{1}{2}\left(1 + \left(\frac{\int_0^1 W_1(s)W_2(s)ds}{\int_0^1 W_2^2(s)ds}\right)^2\right).$$

This provides the expression for the limiting distribution given in Proposition 16.4 as required.

References

Banerjee, A., Dolado, J. J., & Mestre R. (1998). Error-correction mechanism tests for cointegration in a single-equation framework. *Journal of Time Series Analysis, 19*, 267–283.

Boswijk, H. P. (1994). Testing for an unstable root in conditional and structural error correction models. *Journal of Econometrics, 63*, 37–60.

Davidson, J., Hendry, D. F., Srba, F., & Yeo S. (1978). Econometric modelling of the aggregate time-series relationship between consumers' expenditure and income in the United Kingdom. *Economic Journal, 88*, 661–692.

Engle, R. F., & Granger, C. W. J. (1987). Co-integration and error correction: Representation, estimation, and testing. *Econometrica, 55*, 251–276.

Ericsson, N. R., & MacKinnon, J. G. (2002). Distributions of error correction tests for cointegration. *Econometrics Journal, 5*, 285–318.

Frisch, R., & Waugh, F. V. (1933). Partial time regressions as compared with individual trends. *Econometrica, 1*, 387–401.

Hamilton, J. (1994). *Time series analysis*. Princeton: Princeton University Press.

Hansen, B. E. (1992). Efficient estimation and testing of cointegrating vectors in the presence of deterministic trends. *Journal of Econometrics, 53*, 87–121.

Harris, D., & Inder, B. (1994). A test of the null hypothesis of cointegration. In C. P. Hargreaves (Ed.), *Nonstationary time series analysis and cointegration* (pp. 133–152). Oxford/New York: Oxford University Press.

Hassler, U. (2000a). Cointegration testing in single error-correction equations in the presence of linear time trends. *Oxford Bulletin of Economics and Statistics, 62*, 621–632.

Hassler, U. (2000b). The KPSS test for cointegration in case of bivariate regressions with linear trends. *Econometric Theory, 16*, 451–453.

Hassler, U. (2001). The effect of linear time trends on the KPSS test for cointegration. *Journal of Time Series Analysis, 22*, 283–292.

Johansen, S. (1995). *Likelihood-based inference in cointegrated vector autoregressive models.* Oxford/New York: Oxford University Press.

Krämer, W. (1986). Least squares regression when the independent variable follows an ARIMA process. *Journal of the American Statistical Association, 81*, 150–154.

Kwiatkowski, D., Phillips, P. C. B., Schmidt, P., & Shin Y. (1992). Testing the null hypothesis of stationarity against the alternative of a unit root. *Journal of Econometrics, 54*, 159–178.

Leybourne, S. J., & McCabe, B. P. M. (1994). A simple test for cointegration. *Oxford Bulletin of Economics and Statistics, 56*, 97–103.

MacKinnon, J. G. (1991). Critical values for co-integration tests. In R. F. Engle, & C. W. J. Granger (Eds.), *Long-run economic relationships* (pp. 267–276). Oxford/New York: Oxford University Press.

MacKinnon, J. G. (1996). Numerical distribution functions for unit root and cointegration tests. *Journal of Applied Econometrics, 11*, 601–618.

Park, J. Y. (1992). Canonical cointegrating regressions. *Econometrica, 60*, 119–143.

Phillips, P. C. B. (1986). Understanding spurious regressions in econometrics. *Journal of Econometrics, 33*, 311–340.

Phillips, P. C. B. (1987). Time series regression with a unit root. *Econometrica, 55*, 277–301.

Phillips, P. C. B. (1991). Optimal inference in cointegrated systems. *Econometrica, 59*, 283–306.

Phillips, P. C. B., & Durlauf, S. N. (1986). Multiple time series regression with integrated processes. *Review of Economic Studies, LIII*, 473–495.

Phillips, P. C. B., & Hansen, B. E. (1990). Statistical inference in instrumental variables regression with I(1) processes. *Review of Economic Studies, 57*, 99–125.

Phillips, P. C. B., & Loretan, M. (1991). Estimating long-run economic equilibria. *Review of Economic Studies, 58*, 407–436.

Phillips, P. C. B., & Ouliaris, S. (1990). Asymptotic properties of residual based tests for cointegration. *Econometrica, 58*, 165–193.

Phillips, P. C. B., & Park, J. Y. (1988). Asymptotic equivalence of ordinary least squares and generalized least squares in regressions with integrated regressors. *Journal of the American Statistical Association, 83*, 111–115.

Saikkonen, P. (1991). Asymptotically efficient estimation of cointegration regressions. *Econometric Theory, 7*, 1–21.

Shin, Y. (1994). A residual-based test of the Null of cointegration against the alternative of no cointegration. *Econometric Theory, 10*, 91–115.

Stock, J. H. (1987). Asymptotic properties of least squares estimators of cointegrating vectors. *Econometrica, 55*, 1035–1056.

Stock, J. H., & Watson, M. W. (1993). A simple estimator of cointegrating vectors in higher order integrated systems. *Econometrica, 61*, 783–820.

West, K. D. (1988). Asymptotic normality, when regressors have a unit root. *Econometrica, 56*, 1397–1418.

References

Anderson, T. W., & Darling, D. A. (1952). Asymptotic theory of certain "Goodness of Fit" criteria based on stochastic processes. *Annals of Mathematical Statistics, 23*, 193–212.
Andrews, D. W. K. (1991). Heteroskedasticity and autocorrelation consistent covariance matrix estimation. *Econometrica, 59*, 817–858.
Andrews, D. W. K., & Chen, H.-Y. (1994). Approximately median-unbiased estimation of autoregressive models. *Journal of Business & Economic Statistics, 12*, 187–204.
Baillie, R. T. (1996). Long memory processes and fractional integration in econometrics. *Journal of Econometrics, 73*, 5–59.
Banerjee, A., Dolado, J. J., Galbraith, J. W., & Hendry, D. F. (1993). *Co-integration, error correction, and the econometric analysis of non-stationary data*. Oxford/New York: Oxford University Press.
Banerjee, A., Dolado, J. J., & Mestre R. (1998). Error-correction mechanism tests for cointegration in a single-equation framework. *Journal of Time Series Analysis, 19*, 267–283.
Bickel, P. J., & Doksum, K. A. (2001). *Mathematical statistics: Basic ideas and selected topics, volume 1* (2nd ed.). Upper Saddle River: Prentice-Hall.
Billingsley, P. (1968). *Convergence of probability measures*. New York: Wiley.
Billingsley, P. (1986). *Probability and measure* (2nd ed.). New York: Wiley.
Black, F., & Scholes, M. (1973). The pricing of options and corporate liabilities. *The Journal of Political Economy, 81*, 637–654.
Bollerslev, T. (1986). Generalized autoregressive conditional heteroskedasticity. *Journal of Econometrics, 31*, 307–327.
Bondon, P., & Palma, W. (2007). A class of antipersistent processes. *Journal of Time Series Analysis, 28*, 261–273.
Boswijk, H. P. (1994). Testing for an unstable root in conditional and structural error correction models. *Journal of Econometrics, 63*, 37–60.
Breiman, L. (1992). *Probability* (2nd ed.). Philadelphia: Society for Industrial and Applied Mathematics.
Brennan, M. J, & Schwartz, E. S. (1980). Analyzing convertible bonds. *The Journal of Financial and Quantitative Analysis, 15*, 907–929.
Brockwell, P. J., & Davis, R. A. (1991). *Time series: Theory and methods* (2nd ed.). New York: Springer.
Broze, L., Scaillet, O., & Zakoïan, J.-M. (1995). Testing for continuous-time models of the short-term interest rate. *Journal of Empirical Finance, 2*, 199–223.
Campbell, J. Y., & Mankiw, N. G. (1987). Are output fluctuations transitory? *Quarterly Journal of Economics, 102*, 857–880.
Chan, K. C., Karolyi, G. A., Longstaff, F. A., & Sanders, A. B. (1992). An empirical comparision of alternative models of the short-term interest rate. *The Journal of Finance, XLVII*, 1209–1227.
Chang, Y., & Park, J. (2002). On the asymptotics of ADF tests for unit roots. *Econometric Reviews, 21*, 431–447.

U. Hassler, *Stochastic Processes and Calculus*, Springer Texts in Business and Economics, DOI 10.1007/978-3-319-23428-1

Cochrane, J. H. (1988). How big is the random walk in GNP? *Journal of Political Economy, 96*, 893–920.

Cogley, T., & Sargent T. S. (2005). Drifts and volatilities: Monetary policies and outcomes in the post WWII US. *Review of Economic Dynamics, 8*, 262–302.

Constantinides, G. M., & Ingersoll, J. E., Jr. (1984). Optimal bond trading with personal taxes. *Journal of Financial Economics, 13*, 299–335.

Cox, J. C., Ingersoll, J. E., Jr., & Ross S. A. (1980). An analysis of variable rate loan contracts. *The Journal of Finance, 35*, 389–403.

Cox, J. C., Ingersoll, J. E., Jr., & Ross S. A. (1985). A theory of the term structure of interest rates. *Econometrica, 53*, 385–407.

Davidson, J. (1994). *Stochastic limit theory: An introduction for econometricians*. Oxford/New York: Oxford University Press.

Davidson, J., Hendry, D. F., Srba, F., & Yeo S. (1978). Econometric modelling of the aggregate time-series relationship between consumers' expenditure and income in the United Kingdom. *Economic Journal, 88*, 661–692.

Demetrescu, M., Kuzin, V., & Hassler, U. (2008). Long memory testing in the time domain. *Econometric Theory, 24*, 176–215.

Dhrymes, Ph. J. (2000). *Mathematics for econometrics* (3rd ed.). New York: Springer.

Dickey, D. A., & Fuller, W. A. (1979). Distribution of the estimators for autoregressive time series with a unit root. *Journal of the American Statistical Association, 74*, 427–431.

Donsker, M. D. (1951). An invariance principle for certain probability limit theorems. *Memoirs of the American Mathematical Society, 6*, 1–12.

Dothan, L. U. (1978). On the term structure of interest rates. *Journal of Financial Economics, 6*, 59–69.

Durlauf, S. N., & Phillips, P. C. B. (1988). Trends versus random walks in time series analysis. *Econometrica, 56*, 1333–1354.

Elliott, G., Rothenberg, T. J., & Stock, J. H. (1996). Efficient tests for an autoregressive unit root. *Econometrica, 64*, 813–836.

Engle, R. F. (1982). Autoregressive conditional heteroskedasticity with estimates of the variance of U.K. inflation. *Econometrica, 50*, 987–1008.

Engle, R. F. (2002). New frontiers for ARCH models. *Journal of Applied Econometrics, 17*, 425–446.

Engle, R. F., & Bollerslev T. (1986). Modelling the persistence of conditional variances. *Econometric Reviews, 5*, 1–50.

Engle, R. F., & Granger, C. W. J. (1987). Co-integration and error correction: Representation, estimation, and testing. *Econometrica, 55*, 251–276.

Engle, R. F., Lilien, D. M., & Robins, R. P. (1987). Estimating time-varying risk premia in the term structure: the ARCH-M model. *Econometrica, 55*, 391–407.

Ericsson, N. R., & MacKinnon, J. G. (2002). Distributions of error correction tests for cointegration. *Econometrics Journal, 5*, 285–318.

Frisch, R., & Waugh, F. V. (1933). Partial time regressions as compared with individual trends. *Econometrica, 1*, 387–401.

Fuller, W. A. (1996). *Introduction to statistical time series* (2nd ed.). New York: Wiley.

Giraitis, L., Koul, H. L., & Surgailis, D. (2012). *Large sample inference for long memory processes*. London: Imperial College Press.

Gourieroux, Chr., & Jasiak, J. (2001). *Financial econometrics: Problems, models, and methods*. Princeton: Princeton University Press.

Gradshteyn, I. S., & Ryzhik, I. M. (2000). *Table of integrals, series, and products* (6th ed.). London/San Diego: Academic Press.

Granger, C. W. J. (1981). Some properties of time series data and their use in econometric model specification. *Journal of Econometrics, 16*, 121–130.

Granger, C. W. J., & Joyeux, R. (1980). An Introduction to long-memory time series models and fractional differencing. *Journal of Time Series Analysis, 1*, 15–29.

Granger, C. W. J., & Newbold P. (1974). Spurious regressions in econometrics. *Journal of Econometrics, 2*, 111–120.

Grimmett, G. R., & Stirzaker, D. R. (2001). *Probability and random processes* (3rd ed.). Oxford: Oxford University Press.

Hamilton, J. (1994). *Time series analysis*. Princeton: Princeton University Press.

Hansen, B. E. (1992). Efficient estimation and testing of cointegrating vectors in the presence of deterministic trends. *Journal of Econometrics, 53*, 87–121.

Hansen, B. E. (1992a). Convergence to stochastic integrals for dependent heterogeneous processes. *Econometric Theory, 8*, 489–500.

Harris, D., & Inder, B. (1994). A test of the null hypothesis of cointegration. In C. P. Hargreaves (Ed.), *Nonstationary time series analysis and cointegration* (pp. 133–152). Oxford/New York: Oxford University Press.

Hassler, U. (2000). Simple regressions with linear time trends. *Journal of Time Series Analysis, 21*, 27–32.

Hassler, U. (2000a). Cointegration testing in single error-correction equations in the presence of linear time trends. *Oxford Bulletin of Economics and Statistics, 62*, 621–632.

Hassler, U. (2000b). The KPSS test for cointegration in case of bivariate regressions with linear trends. *Econometric Theory, 16*, 451–453.

Hassler, U. (2001). The effect of linear time trends on the KPSS test for cointegration. *Journal of Time Series Analysis, 22*, 283–292.

Hassler, U. (2012). Impulse responses of antipersistent processes. *Economics Letters, 116*, 454–456.

Hassler, U. (2014). Persistence under temporal aggregation and differencing. *Economics Letters, 124*, 318–322.

Hassler, U., & Hosseinkouchack, M. (2014). Effect of the order of fractional integration on impulse responses. *Economics Letters, 125*, 311–314.

Hassler, U., & Kokoszka (2010). Impulse responses of fractionally integrated processes with long memory. *Econometric Theory, 26*, 1855–1861.

Hendry, D. F. (1980). Econometrics – alchemy or science? *Economica, 47*, 387–406.

Hosking, J. R. M. (1981). Fractional differencing. *Biometrika, 68*, 165–176.

Johansen, S. (1995). *Likelihood-based inference in cointegrated vector autoregressive models*. Oxford/New York: Oxford University Press.

Johnson, N. L., Kotz, S., & Balakrishnan, N. (1994). *Continuous univariate distributions, Volume 1* (2nd ed.). New York: Wiley.

Kirchgässner, G., Wolters, J., & Hassler, U. (2013). *Introduction to modern time series analysis* (2nd ed.). Berlin/New York: Springer.

Klebaner, F. C. (2005). *Introduction to stochastic calculus with applications* (2nd ed.). London: Imperical College Press.

Krämer, W. (1986). Least squares regression when the independent variable follows an ARIMA process. *Journal of the American Statistical Association, 81*, 150–154.

Kwiatkowski, D., Phillips, P. C. B., Schmidt, P., & Shin Y. (1992). Testing the null hypothesis of stationarity against the alternative of a unit root. *Journal of Econometrics, 54*, 159–178.

Leybourne, S. J., & McCabe, B. P. M. (1994). A simple test for cointegration. *Oxford Bulletin of Economics and Statistics, 56*, 97–103.

Maasoumi, E., & McAleer, M. (2008). Realized volatility and long memory: An overview. *Econometric Reviews, 27*, 1–9.

MacKinnon, J. G. (1991). Critical values for co-integration tests. In R. F. Engle, & C. W. J. Granger (Eds.), *Long-run economic relationships* (pp. 267–276). Oxford/New York: Oxford University Press.

MacKinnon, J. G. (1996). Numerical distribution functions for unit root and cointegration tests. *Journal of Applied Econometrics, 11*, 601–618.

Marsh, T. A., & Rosenfeld, E. R. (1983). Stochastic processes for interest rates and equilibrium bond prices. *The Journal of Finance, XXXVIII*, 635–646.

Merton, R. C. (1973). Theory of rational option pricing. *The Bell Journal of Economics and Management Science, 4*, 141–183.

Mikosch, Th. (1998). *Elementary stochastic calculus with finance in view*. Singapore: World Scientific Publishing.

Mills, T. C., & Markellos, R. N. (2008). *The econometric modelling of financial time series* (3rd ed.). Cambridge/New York: Cambridge University Press.

Nelson, D. B. (1991). Conditional heteroskedasticity in asset returns: A new approach. *Econometrica, 59*, 347–370.

Nelson, D. B., & Cao, Ch. Q. (1992). Inequality constraints in the univariate GARCH model. *Journal of Business & Economic Statistics, 10*, 229–235.

Newey, W. K., & West, K. D. (1987). A Simple, positive semi-definite, heteroskedasticity and autocorrelation consistent covariance matrix. *Econometrica, 55*, 703–708.

Øksendal, B. (2003). *Stochastic differential equations: An introduction with applications* (6th ed.). Berlin/New York: Springer.

Park, J. Y. (1992). Canonical cointegrating regressions. *Econometrica, 60*, 119–143.

Park, J. Y., & Phillips, P. C. B. (1988). Statistical inference in regressions with integrated processes: Part I. *Econometric Theory, 4*, 468–497.

Phillips, P. C. B. (1986). Understanding spurious regressions in econometrics. *Journal of Econometrics, 33*, 311–340.

Phillips, P. C. B. (1987). Time series regression with a unit root. *Econometrica, 55*, 277–301.

Phillips, P. C. B. (1988). Weak convergence of sample covariance matrices to stochastic integrals via martingale approximations. *Econometric Theory, 4*, 528–533.

Phillips, P. C. B. (1991). Optimal inference in cointegrated systems. *Econometrica, 59*, 283–306.

Phillips, P. C. B., & Durlauf, S. N. (1986). Multiple time series regression with integrated processes. *Review of Economic Studies, LIII*, 473–495.

Phillips, P. C. B., & Hansen, B. E. (1990). Statistical inference in instrumental variables regression with I(1) processes. *Review of Economic Studies, 57*, 99–125.

Phillips, P. C. B., & Loretan, M. (1991). Estimating long-run economic equilibria. *Review of Economic Studies, 58*, 407–436.

Phillips, P. C. B., & Ouliaris, S. (1990). Asymptotic properties of residual based tests for cointegration. *Econometrica, 58*, 165–193.

Phillips, P. C. B., & Park, J. Y. (1988). Asymptotic equivalence of ordinary least squares and generalized least squares in regressions with integrated regressors. *Journal of the American Statistical Association, 83*, 111–115.

Phillips, P. C. B, & Perron, P. (1988). Testing for a unit root in time series regression. *Biometrika, 75*, 335–346.

Phillips, P. C. B., & Solo, V. (1992). Asymptotics for linear processes. *The Annals of Statistics, 20*, 971–1001.

Pötscher, B. M., & Prucha, I. R. (2001). Basic elements of asymptotic theory. In B. H. Baltagi (Ed.), *A companion to theoretical econometrics* (pp. 201–229). Malden: Blackwell.

Ross, S. (2010). *A first course in probability* (8th ed.). Upper Saddle River: Prentice-Hall.

Rudin, W. (1976). *Principles of mathematical analyis* (3rd ed.). New York: McGraw-Hill.

Said, S. E., & Dickey, D. A. (1984). Testing for unit roots in autoregressive-moving average models of unknown order. *Biometrika, 71*, 599–607.

Saikkonen, P. (1991). Asymptotically efficient estimation of cointegration regressions. *Econometric Theory, 7*, 1–21.

Shin, Y. (1994). A residual-based test of the Null of cointegration against the alternative of no cointegration. *Econometric Theory, 10*, 91–115.

Soong, T. T. (1973). *Random differential equations in science and engineering*. New York: Academic Press.

Stock, J. H. (1987). Asymptotic properties of least squares estimators of cointegrating vectors. *Econometrica, 55*, 1035–1056.

Stock, J. H., & Watson, M. W. (1993). A simple estimator of cointegrating vectors in higher order integrated systems. *Econometrica, 61*, 783–820.

Sydsæter, K., Strøm, A., & Berck, P. (1999). *Economists' mathematical manual* (3rd ed.). Berlin/New York: Springer.

Tanaka, K. (1996). *Time series analysis: Nonstationary and noninvertible distribution theory*. New York: Wiley.

Taylor, S.J. (1994). Modeling stochastic volatlity: A review and comparative study. *Mathematical Finance, 4*, 183–204.

Trench, W. F. (2013). *Introduction to real analysis*. Free Hyperlinked Edition 2.04 December 2013. Downloaded on 10th May 2014 from http://digitalcommons.trinity.edu/mono/7.

Tsay, R. S. (2005). *Analysis of financial time series* (2nd ed.). New York: Wiley.

Tse, Y. K. (1995). Some international evidence on the stochastic behavior of interest rates. *Journal of International Money and Finance, 14*, 721–738.

Vasicek, O. (1977). An equilibrium characterization of the term structure. *Journal of Financial Economics, 5*, 177–188.

West, K. D. (1988). Asymptotic normality, when regressors have a unit root. *Econometrica, 56*, 1397–1418.

White, H. (2001). *Asymptotic theory for econometricians* (2nd ed.). London/San Diego: Academic Press.

Wold, H. O. A. (1938). *A study in the analysis of stationary time series*. Stockholm: Almquist & Wiksell.

Index

U. Hassler, *Stochastic Processes and Calculus*, Springer Texts in Business and Economics, DOI 10.1007/978-3-319-23428-1

MIX
Papier aus verantwortungsvollen Quellen
Paper from responsible sources
FSC® C105338
FSC
www.fsc.org

If you have any concerns about our products,
you can contact us on
ProductSafety@springernature.com

In case Publisher is established outside the EU,
the EU authorized representative is:
Springer Nature Customer Service Center GmbH
Europaplatz 3, 69115 Heidelberg, Germany

Printed by Libri Plureos GmbH
in Hamburg, Germany